Microbiology and Biogeochemistry of HYPERSALINE ENVIRONMENTS

The Microbiology of
EXTREME AND UNUSUAL ENVIRONMENTS

SERIES EDITOR
RUSSELL H. VREELAND

Titles in the Series

The Biology of Halophilic Bacteria
Russell H. Vreeland and Lawrence Hochstein

The Microbiology of Deep-Sea Hydrothermal Vents
David M. Karl

The Microbiology of Solid Waste
Anna C. Palmisano

The Microbiology of the Terrestrial Subsurface
Penny S. Amy and Dana L. Haldeman

The Microbiology and Biogeochemistry of Hypersaline Environment
Aharon Oren

Microbiology and Biogeochemistry of HYPERSALINE ENVIRONMENTS

Edited by
AHARON OREN

Division of Microbial and Molecular Ecology
The Institute of Life Sciences
and The Moshe Shilo Minerva Center
for Marine Biogeochemistry
The Hebrew Institute of Jerusalem

CRC Press
Boca Raton   London   New York   Washington, D.C.

Acquiring Editor: Harvey Kane
Project Editor: Sylvia Wood
Marketing Manager: Becky McEldowney
Cover design: Dawn Boyd

Library of Congress Cataloging-in-Publication Data

Microbiology and biogeochemistry of hypersaline environments / edited by Aharon Oren
    p. cm.
  Includes bibliographical references and index.
  ISBN 0-8493-8363-3 (alk. paper)
  1. Halophilic microorganisms. 2. Salinity. 3. Microbial ecology.
4. Biogeochemistry. I. Oren, Aharon, 1952–
QR97.S3M53 1998
579′.1714—dc21
                                98-18052
                                      CIP

This book contains information obtained from authentic and highly regarded sources. Reprinted material is quoted with permission, and sources are indicated. A wide variety of references are listed. Reasonable efforts have been made to publish reliable data and information, but the author and the publisher cannot assume responsibility for the validity of all materials or for the consequences of their use.

Neither this book nor any part may be reproduced or transmitted in any form or by any means, electronic or mechanical, including photocopying, microfilming, and recording, or by any information storage or retrieval system, without prior permission in writing from the publisher.

All rights reserved. Authorization to photocopy items for internal or personal use, or the personal or internal use of specific clients, may be granted by CRC Press LLC, provided that $.50 per page photocopied is paid directly to Copyright Clearance Center, 27 Congress Street, Salem, MA 01970 USA. The fee code for users of the Transactional Reporting Service is ISBN 0-8493-8363-3/99/$0.00+$.50. The fee is subject to change without notice. For organizations that have been granted a photocopy license by the CCC, a separate system of payment has been arranged.

The consent of CRC Press LLC does not extend to copying for general distribution, for promotion, for creating new works, or for resale. Specific permission must be obtained in writing from CRC Press LLC for such copying.

Direct all inquiries to CRC Press LLC, 2000 Corporate Blvd., N.W., Boca Raton, Florida 33431.

Cover photograph: Mixed community of halophilic microorganisms from a saltern crystallizer pond near Alicante, Spain, collected by filtration and viewed by scanning electron microscopy. (Photograph courtesy of F. Rodríguez-Valera, Alicante.)

© 1999 by CRC Press LLC

No claim to original U.S. Government works
International Standard Book Number 0-8493-8363-3
Library of Congress Card Number 98-18052
Printed in the United States of America 1 2 3 4 5 6 7 8 9 0
Printed on acid-free paper

Preface

A workshop on Microbiology and Biogeochemistry of Halophilic Microorganisms was held from June 22 to 26, 1997 in Jerusalem, Israel. The organizing committee consisted of Aharon Oren (The Hebrew University of Jerusalem), Moshe Mevarech (Tel Aviv University) and Aaron Kaplan (The Hebrew University of Jerusalem).

The meeting brought together close to 70 scientists involved in halophilic microbiology. The lectures and posters presented an up-to-date picture of all aspects of halophile science: taxonomy, ecology, biogeochemistry, biochemistry, physiology, molecular biology and genetics. Five years had passed since the previous international meeting entirely devoted to halophilic microorganisms (the ASM conference on Biology of Halophilic Bacteria: Research Priorities and Biotechnological Potential for the 1990s, Williamsburg, VA, USA in 1992). Thus, the Jerusalem workshop presented a unique opportunity to obtain an updated overview of our present understanding of microbial life at high salt concentrations. This volume is based on lectures presented at the workshop in Jerusalem, and offers a comprehensive picture of the state of research on halophilic Archaea, Bacteria, and Eukarya, and the place of these intriguing microorganisms in nature.

Editor

Aharon Oren, associate professor of microbial ecology at the Hebrew University of Jerusalem, Israel, was born in Zwolle, The Netherlands, in 1952. He received B.S. (1972) and M.S. degrees (1974) in microbiology and biochemistry from the University of Groningen (The Netherlands). In 1974 he immigrated to Israel, and he received his Ph.D. degree in microbiology from the Hebrew University of Jerusalem (1979), following research on anoxygenic photosynthesis in cyanobacteria. He spent two years of post-doctoral research (1982–1984) at the University of Illinois at Urbana-Champaign with Prof. Ralph S. Wolfe (research on the biochemistry of methanogenesis) and Prof. Carl R. Woese (bacterial phylogeny), and from 1983–1984, he served as a visiting assistant professor at the University of Illinois at Urbana-Champaign.

Since 1980, Dr. Oren has studied the biology of halophilic microorganisms. The microbial ecology of the Dead Sea has been his main subject of interest, but his studies have also involved the communities of halophilic Archaea and Bacteria in solar saltern ponds, regulation of intracellular salt concentrations in halophilic microorganisms, osmotic solutes in halophilic photosynthetic prokaryotes, and taxonomy and comparative enzymology of halophilic Archaea.

Dr. Oren is recipient of the Moshe Shilo prize of the Israel Society for Microbiology (1993). He is an editor of *FEMS Microbiology Letters*, and a member of the editorial board of the *International Journal of Salt Lake Research*. He serves as secretary of the International Committee of Systematic Bacteriology subcommittee on the taxonomy of Halobacteriaceae, and is a member of the subcommittee on the taxonomy of photosynthetic prokaryotes. In addition, he is a member of the International Committee of the International Symposia on Environmental Biogeochemistry. He has participated in the organization of international symposia and workshops on halophilic microorganisms in Jerusalem (1986) and Williamsburg, VA (1992), and was chairman of the organizing committee of the workshop on Microbiology and Biogeochemistry of Hypersaline Environments, Jerusalem (1997).

Dr. Oren has published approximately 130 research articles and other publications, including microbiology textbooks in Hebrew for high school and university students, and a number of review articles on halophilic microorganisms.

Since 1995, Dr. Oren has been director of the Moshe Shilo Minerva Center for Marine Biogeochemistry at the Hebrew University of Jerusalem, and in 1998 he was elected chairman of the Division of Microbial and Molecular Ecology at the Institute of Life Sciences, the Hebrew University of Jerusalem.

Acknowledgments

I would like to thank the editors of the CRC Press series on *Microbiology of Extreme and Unusual Environments* for enabling the publication of the 27 chapters in this volume.

Further, I acknowledge those organizations that financially supported the Jerusalem workshop:

- The Israel Science Foundation, founded by the Israel Academy of Sciences and Humanities.
- The Moshe Shilo Minerva Center for Marine Biogeochemistry — Minerva Stiftung für die Forschung m.b.H., München, Germany, funded by the German Federal Ministry for Science, Education, Research and Technology.
- The Hebrew University of Jerusalem.

Last but not least, I wish to express my sincere thanks to all chapter authors and reviewers for their timely submission and rapid handling of the manuscripts, which resulted in the publication of the present volume so soon after the workshop.

Aharon Oren

Contributors

Aharon Oren, Yehuda Cohen, and Etana Padan
Division of Microbial and Molecular Ecology, and the Moshe Shilo Minerva Center for Marine Biogeochemistry, The Hebrew University of Jerusalem, Jerusalem, Israel

Masahiro Kamekura
Noda Institute for Scientific Research, Noda-shi, Chiba-ken, Japan

Francisco Rodríguez-Valera, Silvia G. Acinas, and Josefa Antón
Departmento de Genética y Microbiología, Universidad de Alicante, Spain

Carol D. Litchfield and Amy Irby
Department of Biology, George Mason University, Fairfax, VA

Russell H. Vreeland
Department of Biology, West Chester University, PA

Dennis W. Powers
Consulting Geologist, Anthony, TX

Anna A. Gorbushina and Wolfgang E. Krumbein
Geomicrobiology, ICBM, Carl von Ossietzky Universität, Oldenburg, Germany

Zeev Aizenshtat, Irena Miloslavski and Dorit Ashengrau
The Casali Institute, Department of Organic Chemistry, and The Moshe Shilo Minerva Center for Marine Biogeochemistry, The Hebrew University of Jerusalem, Jerusalem, Israel

Naama Gazit-Yaari, Boaz Lazar, and Jonathan Erez
The Institute of Earth Sciences, and The Moshe Shilo Minerva Center for Marine Biogeochemistry, The Hebrew University of Jerusalem, Jerusalem, Israel

Ittai Gavrieli, Michael Beyth, and Yoseph Yechieli
Geological Survey of Israel, Jerusalem, Jerusalem, Israel

Antonio Ventosa and David R. Arahal
Department of Microbiology and Parasitology, Faculty of Pharmacy, University of Sevilla, Spain

Benjamin E. Volcani
Marine Biology Research Division, Scripps Institution of Oceanography, University of California, San Diego, CA

Mark Marvin DiPasquale and Ronald S. Oremland
U.S. Geological Survey, Menlo Park, CA

Martin Hageman, Arne Schoor, Stefan Mikkat, Uta Effmert, Ellen Zuther, Kay Marin, Sabine Fulda, Josef Vinnemeier, Anjan Kunert, Carsten Milkowski, Christian Probst, and Norbert Erdmann
University of Rostock, Department of Biology, Rostock, Germany

Erwin A. Galinski
Institut für Biochemie, Westfälische Wilhelms-Universität, Münster, Germany

Petra Louis
Institut für Mikrobiologie & Biotechnologie, Rheinische Friedrich-Wilhelms- Universität, Bonn, Germany

Irena Gokhman, Uri Pick, and Ada Zamir
Department of Biological Chemistry, The Weizmann Institute of Science, Rehovot, Israel

Morly Fisher
Israel Institute for Biological Research, Ness Ziona, Israel

Donn J. Kushner
Department of Botany and Microbiology, University of Toronto, Canada

Christine Ebel, Pierre Faou, Bruno Franzetti, Blandine Kernel, Dominque Madern, Mihaela Pascu, Claude Pfister, Stéphane Richard, and Giuseppe Zaccai
Institut de Biologie Structurale Jean Pierre Ebel, Grenoble, France

Michael J. Danson, Keith A. Jolley, Deborah G. Maddocks, and David W. Hough
Centre for Extremophile Research, Department of Biology and Biochemistry, University of Bath, UK

Melissa L. Holmes and Michael L. Dyall-Smith
Department of Microbiology and Immunology, University of Melbourne, Parkville, Australia

Jorge Söppa
Universität Frankfurt, Biozentrum Niederursel, Institut für Mikrobiologie, Frankfurt, Germany; Max-Planck-Institut für Biochemie, Martinsried, Germany

Petra Vatter and Alexandra zur Mühlen
Universität Frankfurt, Biozentrum Niederursel, Institut für Mikrobiologie, Frankfurt, Germany

Thomas A. Link
Universitätsklinikum Frankfurt, ZBC, Institut für Biochemie, Frankfurt, Germany

Andreas Ruepp
Max-Planck-Institut für Biochemie, Martinsried, Germany

Lawrence I. Hochstein and Roberto Bogomolni
Department of Chemistry and Biochemistry, University of California at Santa Cruz, CA

Felicitas Pfeifer, Andrea Mayr, Sonja Offner, and Richard Röder
Institut für Mikrobiologie und Genetik, Technische Universität Darmstadt, Germany

Richard F. Shand, Lance B. Price, and Elizabeth M. O'Connor
Department of Biological Sciences, Northern Arizona University, Flagstaff, AZ

Robert L. Charlebois
Evolutionary Biology Program, Canadian Institute for Advanced Research and Department of Biology, University of Ottawa, Canada

Patrick P. Dennis
Department of Biochemistry and Molecular Biology, University of British Columbia, Vancouver, Canada

Ron Ortenberg, Ronen Tchelet, and Moshe Mevarech
Department of Molecular Microbiology and Biotechnology, George S. Wise Faculty of Life Sciences, Tel Aviv University, Israel

Contents

Chapter 1
Microbiology and Biogeochemistry of Halophilic Microorganisms –
An Overview 1
A. Oren

Section I   Diversity and Ecology

Chapter 2
Diversity of Members of the Family *Halobacteriaceae* 13
M. Kamekura

Chapter 3
Contribution of Molecular Techniques to the Study of Microbial
Diversity in Hypersaline Environments 27
F. Rodríguez-Valera, S.G. Acinas, and J. Anton

Chapter 4
The Microbial Ecology of Solar Salt Plants 39
C.D. Litchfield, A. Irby, and R. H. Vreeland

Chapter 5
Considerations for Microbiological Sampling of Crystals from
Ancient Salt Formations 53
R.H. Vreeland and D.W. Powers

Chapter 6
Poikilotrophic Response of Microorganisms to Shifting Alkalinity,
Salinity, Temperature and Water Potential 75
A.A. Gorbushina and W.E. Krumbein

Section II Biogeochemistry

Chapter 7
Hypersaline Depositional Environments and their Relation to
Oil Generation 89
Z. Aizenshtat, I. Miloslavski, D. Aschengrau, and A. Oren

Chapter 8
Field Evidence for ^{13}C Depletion Due to Atmospheric CO_2 Invasion in
Hypersaline Microbial Mats 109
N. Gazit-Yaari, B. Lazar, and J. Erez

Section III The Dead Sea

Chapter 9
The Dead Sea — A Terminal Lake in the Dead Sea Rift:
A Short Overview 121
I. Gavrieli, M. Beyth, and Y. Yechieli

Chapter 10
The Rise and Decline of a Bloom of Halophilic Algae and
Archaea in the Dead Sea: 1992–1995 129
A. Oren

Chapter 11
Studies on the Microbiota of the Dead Sea — 50 Years Later 139
A. Ventosa, D.R. Arahal, and B.E. Volcani

Chapter 12
Radiotracer Studies of Bacterial Methanogenesis in Sediments from
the Dead Sea and Solar Lake (Sinai) 149
M.M. DiPasquale, A. Oren, Y. Cohen, and R.S. Oremland

Section IV Ion Metabolism and Osmotic Regulation

Chapter 13
The Molecular Mechanism of Regulation of the Na^+/H^+ Antiporter of
Escherichia Coli, a Paradigm for an Adaptation to Na^+ and N^+ 163
E. Padan

Chapter 14
The Biochemistry and Genetics of the Synthesis of Osmoprotective
Compounds in Cyanobacteria 177
*M. Hagemann, A. Schoor, S. Mikkat, U. Effmert, E. Zuther, K. Marin,
S. Fulda, J. Vinnemeier, A. Kunert, C. Milkowski, C. Probst, and
N. Erdmann*

Chapter 15
Compatible Solutes: Ectoine Production and Gene Expression 187
E.A. Galinski and P. Louis

Chapter 16
New Insights into the Extreme Salt Tolerance of the Unicellular
Green Alga *Dunaliella*.. 203
I. Gokhman, M. Fisher, U. Pick, and A. Zamir

Section V   Biochemistry and Molecular Biology

Chapter 17
What is Halophilic and What is Archaeal?.. 217
D.J. Kushner

Chapter 18
Molecular Interactions in Extreme Halophiles — The Solvation-
Stabilization Hypothesis for Halophilic Proteins.. 227
*C. Ebel, P. Faou, B. Franzetti, B. Kernel, D. Madern, M. Pascu, C. Pfister,
S. Richard, and G. Zaccai*

Chapter 19
New Insights into the Molecular Enzymology of Pyruvate
Metabolism in the Halophilic Archaea 239
*M.J. Danson, K.A. Jolley, D.G. Maddocks, M.L. Dyall-Smith, and
D.W. Hough*

Chapter 20
Regulation of Gene Expression in *Halobacterium Salinarum*:
The *arcRACB* Gene Cluster and the TATA Box-Binding Protein 249
J. Soppa, P. Vatter, A. zur Mühlen, T. Link, and A. Ruepp

Chapter 21
Cloning, Sequence and Heterologous Expression of *bgaH*, a
Beta-Galactosidase Gene of *"Haloferax alicantei"* 265
M. Holmes and M.L. Dyall-Smith

Chapter 22
What Do Extreme Halophiles Tell Us About the Evolution of the
Proton-Translocating ATPases? 273
L.I. Hochstein and R. Bogomolni

Chapter 23
Comparative Analysis of the Halobacterial Gas Vesicle Gene Clusters 281
F. Pfeifer, A. Mayr, S. Offner, and R. Röder

Chapter 24
Halocins: Protein Antibiotics from Hypersaline Environments 295
R.F. Shand, L.B. Price, and E.M. O'Connor

Section VI   Genetics and Genomics

Chapter 25
Evolutionary Origins of the Haloarchaeal Genome... 309
R.L. Charlebois

Chapter 26
Expression of Ribosomal RNA Operons in Halophilic Archea 319
P.P. Dennis

Chapter 27
A Model for the Genetic Exchange System of the Extremely
Halophilic Archaeon *Haloferax volcanii* 331
R. Ortenberg, R. Rechlet, and M. Mevarech

Index 339

1 Microbiology and Biogeochemistry of Halophilic Microorganisms — An Overview

Aharon Oren

CONTENTS

1.1 Introduction 1
1.2 Biodiversity of Halophilic Archaea and Bacteria 2
1.3 Ecology of Halophilic Microorganisms 3
1.4 Biogeochemistry 4
1.5 The Dead Sea 4
1.6 Adaptation and Adaptability to High and Changing Salt Concentrations 5
1.7 Retinal Proteins of Halophilic Archaea 6
1.8 Biochemistry and Molecular Biology of Halophilic Prokaryotes 6
1.9 Genetics and Genomics 7
1.10 Biotechnological Aspects 8
1.11 Epilogue 8
References 9

1.1 INTRODUCTION

It is not necessary to be trained as a microbiologist to recognize the occurrence of salt-tolerant and salt-requiring microorganisms in such environments as hypersaline lakes and salterns. The display of colors by the dense communities of microorganisms inhabiting these ecosystems is often very dramatic. The higher the salt concentration of the environment, the denser these communities can become, mainly as a result of the lack of predators able to withstand the extreme salinities. Especially at the highest salt concentrations, at or close to the solubility limit of NaCl, dense communities of halophilic Archaea (family *Halobacteriaceae*) can impart a bright red color to environments such as the Great Salt Lake, UT and saltern crystallizer ponds all over the world (Madigan, 1997). Even the upper water layers of the Dead Sea, with its extremely high magnesium concentration (presently exceeding 1.8 M) turns red from time to time as a result of mass development of Archaea rich in carotenoid pigments. Development of benthic microbial mats composed of halophilic representatives of the domain Bacteria (cyanobacteria, purple sulfur bacteria, sulfate-reducing bacteria, and many others) is often prominent in salt lakes of lower salinities (up to 200–250 g l^{-1}). Halophilic or halotolerant Eukarya, such as different species of the green alga *Dunaliella*, can be found over the entire range of salinities, from seawater to brines saturated with NaCl.

The existence of profuse microbial life at high salt concentrations raises a number of fundamental questions relating to the mode of adaptation of the cells to osmotic stress, in many cases requiring enzymatic activities to occur in the presence of high ionic strengths. No less intriguing is the adaptability of many halophiles, especially those belonging to the domains Bacteria and Eukarya, which display a high degree of versatility in their ability to rapidly adjust to changes in the osmotic pressure of their surrounding medium. With the increase in our understanding of the diversity of halophilic microorganisms, it has become clear that nature has devised an extensive variety of strategies to cope with life in hypersaline environments.

The basic knowledge accumulated on the properties of halophilic Archaea, Bacteria and Eukarya has led to the development of some interesting biotechnological applications, including the use of their organic osmotic solutes as enzyme protectants, the production of salt-resistant enzymes, and the use of certain *Dunaliella* strains for mass production of β-carotene.

Much progress has been made in the study of all aspects of halophile microbial life since the appearance of the previous general monographs on the subject (Rodriguez-Valera, 1988, 1991; Vreeland and Hochstein, 1993) and the book by Javor (1989) on the ecology and biogeochemistry of hypersaline environments.

Our volume contains a compilation of lectures by halophile experts, including taxonomists, ecologists, biogeochemists, physiologists, biochemists, and molecular biologists, who gathered at a meeting in Jerusalem in June 1997. While recognizing that it is impossible to cover here all aspects of hypersaline microbiology, this is an attempt to present a representative overview of the rapidly expanding knowledge in this interesting field of biology.

1.2 BIODIVERSITY OF HALOPHILIC ARCHAEA AND BACTERIA

Little more than 20 years have passed since the 8th edition of *Bergey's Manual of Determinative Bacteriology* was published. That edition of the authoritative handbook of prokaryote taxonomy recognized just three species of what we now know as halophilic Archaea (Gibbons, 1974): *Halobacterium salinarium, Halobacterium halobium* (in the meantime, these two were united into one species: *Halobacterium salinarum*; Ventosa and Oren, 1996), and the coccoid *Halococcus morrhuae*. The 1989 edition of *Bergey's Manual* recognizes six genera and 15 species (Grant and Larsen, 1989). The recent explosive increase in the number of recognized genera and species (at the time or writing, the family *Halobacteriaceae* consisted of 10 genera with 29 species) is due to several new developments. First, a wider variety of enrichment and isolation media enabled the isolation of hitherto unknown physiological types. Moreover, analysis of polar lipids in the cell membrane demonstrated great diversity (Tindall, 1993). However, it was the comparison of small subunit ribosomal RNA sequences that revolutionized prokaryote taxonomy and led to the recognition of the Archaea as a third form of life (Woese and Fox, 1977; Woese et al., 1990), and that gave us a more profound insight into the complexity of halophilic archaeal taxonomy and the possible phylogenetic relationships between the different taxa. An updated view of the taxonomy of halophilic Archaea is given by Kamekura in Chapter 2.

The progress in our understanding of the diversity and taxonomy of halophilic Bacteria is equally dramatic. Only 18 years have passed since the first description of the genus *Halomonas* (Vreeland et al., 1980), a genus of extremely versatile Gram-negative bacteria, now known to be widely distributed in nature, that contains some of the most exhaustively studied organisms among the halophilic Bacteria. Systematic attempts to isolate and characterize salt-tolerant and salt-requiring Bacteria from different habitats, and the use of 16S rRNA sequencing to elucidate their phylogenetic position, has led to the recognition of many new genera and species (Ventosa et al., 1998) whose properties have just begun to be investigated. Similarly, anaerobic life in hypersaline environments also has started to be explored in depth (Ollivier et al., 1994; Rainey et al., 1995).

1.3 ECOLOGY OF HALOPHILIC MICROORGANISMS

In comparison with our rapidly growing knowledge on the physiology and biochemistry of halophilic bacteria, our understanding of the distribution of the different types of halophiles in nature and the factors that influence their development and decline in their natural habitats are still relatively little understood (Oren, 1994). Therefore, an effort toward "rediscovering the ecology of halobacteria" (Norton, 1992) is long overdue.

The application of newly developed techniques of molecular biology has shed new light on the structure of natural communities of halophiles. When 16S rRNA genes amplified directly from Spanish saltern crystallizer ponds were analyzed, it appeared that the gene sequences recovered were different from all known halophilic archaeal taxa (Rodríguez-Valera et al., Chapter 3). Thus, the dominant types in nature possibly still await isolation in pure culture, once more encouraging new attempts toward isolation and characterization of as many different strains as possible. Such a study, involving comparison of salterns from different geographical sites, and employing a variety of methods of isolation and characterization, is in progress (Litchfield et al., Chapter 4). With the possible exception of the Dead Sea, very few ecological studies of salt lakes such as the Great Salt Lake, UT and the interesting Antarctic hypersaline lakes have been performed in recent years.

Adaptation of microorganisms to high — and often varying — salinities in hypersaline environments in nature represents only one aspect of the complex array of adaptations to such variables. Other variables include shifting alkalinity, salinity, temperature, and water potential, and are collectively termed "poikilotrophic adaptation" by Gorbushina and Krumbein (Chapter 6). Bacteria can adapt to life, or at least long-term survival, even inside salt crystals, and the techniques for studying biodiversity and survival of bacteria within salt crystals in salt mines are rapidly improving (Vreeland and Powers, Chapter 5).

1.4 BIOGEOCHEMISTRY

The dense microbial communities occurring in the water body and in the benthic microbial mats of hypersaline lakes often exhibit high activities of photosynthesis, dissimilatory sulfate reduction, and other microbial processes, thereby exerting a profound influence on the biogeochemical cycles of carbon, nitrogen, sulfur, and other elements (Javor, 1989; Oren, 1998). In the case of the carbon cycle, stable isotope analysis of CO_2 often shows an unusual depletion of ^{13}C in hypersaline environments (Gazit-Yaari et al., Chapter 8). Also unusual is the fact that most methane formed in anaerobic hypersaline environments is not derived from the conventional methanogenic processes based on reduction of CO_2 by hydrogen or by splitting of acetate (energy sources that are more efficiently used by sulfate-reducing bacteria), but rather originates from breakdown of "non-competitive" substrates such as methylated amines and dimethylsulfide. These in their turn can be derived from microbial degradation of glycine betaine, dimethylsulfoniopropionate, and other methylated compounds that serve as organic osmotic solutes in many halophilic microorganisms (Oremland and King, 1989; Marvin DiPasquale et al., Chapter 12).

In-depth studies of the sulfur cycle in the hypersaline Solar Lake (Sinai peninsula, Egypt) have shown extremely active sulfate reduction by a varied community of sulfate reducing bacteria, with a spatial, temporal, and functional separation of activities among the species present, the properties of which became established recently using microscale analysis and molecular biological techniques.

On a geological scale, all these processes have a profound influence on the properties of geological formations derived from hypersaline depositional environments (Aizenshtat et al., Chapter 7).

1.5 THE DEAD SEA

The Dead Sea, a terminal lake at the border between Israel and Jordan, has been subjected to intensive interdisciplinary studies during the last 20 years (Niemi et al., 1997). The lake has witnessed drastic changes in its water balance during this century. The net drop of the water level had a profound influence on the lake's chemical and physical properties (Gavrieli et al., Chapter 9).

Microbiological studies of the Dead Sea were initiated in the late 1930s by Benjamin Elazari-Volcani. Some of the enrichments set up during that period have been preserved, and were recently revived and studied in further detail (Ventosa et al., Chapter 11). Investigations of the microbiology of the Dead Sea have yielded a considerable number of novel microorganisms, including Archaea (*Haloarcula marismortui, Haloferax volcanii, Halorubrum sodomense, Halobaculum gomorrense*), aerobic Bacteria (*Halomonas halmophila, Halomonas israelensis, Chromohalobacter marismortui*), and anaerobic fermentative halophilic Bacteria of the family *Haloanaerobiaceae* (*Halobacteroides halobius, Sporohalobacter lortetii, Orenia marismortui*) (Rainey et al., 1995). An understanding of the population dynamics of the Archaea and the eukaryotic algae (*Dunaliella*) in the lake is emerging. Seasonal and long-term hydrological changes greatly influence the interactions between these two groups of organisms (Oren, 1988; Oren, Chapter 10). Evidence for microbiological processes in the sediments of the Dead Sea came from radiotracer studies, showing that bacterial methanogenesis can occur in the lake (Marvin DiPasquale et al., Chapter 12). The bacteria responsible for the process are still awaiting isolation.

1.6 ADAPTATION AND ADAPTABILITY TO HIGH AND CHANGING SALT CONCENTRATIONS

To be able to live at high salt concentrations, halophilic and halotolerant microorganisms must maintain a cytoplasm that is osmotically isotonic with the outside medium. To achieve osmotic equilibrium, two fundamentally different strategies exist. The first option, used by the aerobic halophilic Archaea of the family *Halobacteriaceae* and the anaerobic halophilic Bacteria of the order Haloanaerobiales involves the maintenance of high intracellular ionic concentrations, with K^+ rather than Na^+ being the dominant cation, and adaptation of the entire intracellular enzymatic machinery to function in the presence of high salt. This strategy permits little flexibility and adaptability to changing conditions, as many enzymes and other proteins will require the continuous presence of high salt for activity and stability. The adaptive evolution of proteins and salinity-mediated selection of their properties has recently been reviewed (Dennis and Shimmin, 1997).

The second option, realized in most halophilic and halotolerant representatives of the Bacteria, Eukarya, and also in halophilic methanogenic Archaea, involves the maintenance of a cytoplasm much lower in salt concentration, and the accumulation of "compatible" osmotic solutes that serve to achieve osmotic equilibrium while not being too inhibitory to enzymatic activity (Galinski, 1995; Ventosa et al., 1998). The concentrations of these osmotic solutes are regulated according to the salt concentration in which the cells are found, and can be rapidly adjusted as required when the outside salinity is changed. The list of compounds known to be used as compatible solutes by halophilic microorganisms (either by *de novo* synthesis or by accumulation from the medium, when available) is steadily growing. Many prokaryotic cells contain cocktails of different compatible solutes, rather than relying on a single compound (Galinski, 1995).

In either case, intracellular sodium concentrations are kept as low as possible, and outward-directed sodium pumps in the cytoplasmic membrane are of the utmost importance, both in maintaining the proper intracellular ionic environment, and in pH regulation. The mechanism of sodium/proton antiporters is being studied in depth in *Escherichia coli* as a model organism (Padan, Chapter 13), and the conclusions obtained can easily be extrapolated to an understanding of the metabolism of more halophilic and halotolerant microorganisms.

Adaptation and adaptability of halophilic Bacteria depend on the regulation of the synthesis of such organic osmolytes as glycine betaine, ectoine (1,4,5,6-tetrahydro-2-methyl-4-pyrimidine carboxylic acid), glucosylglycerol, and others. In recent years, much progress has been made in our understanding of the processes of osmotic regulation, thanks mostly to the application of molecular biological techniques. These studies have led to an insight into the biosynthesis of ectoine in *Marinococcus halophilus* (Galinski and Louis, Chapter 15), and the production of glucosylglycerol in cyanobacteria of the genus *Synechocystis* (Hagemann et al., Chapter 16). Much progress has also been made in the study of the effects of changing salt concentrations on the alga *Dunaliella* (Gokhman et al., Chapter 16), a eukaryotic organism that accumulates glycerol as the osmotic solute.

1.7 RETINAL PROTEINS OF HALOPHILIC ARCHAEA

The discovery of the retinal proteins bacteriorhodopsin, halorhodopsin and the sensory rhodopsins in *Halobacterium salinarum* did much to increase the interest of the scientific community in halophilic microorganisms. Bacteriorhodopsin as a light-driven proton pump and halorhodopsin as an inward chloride pump became perfect models for energy conversion leading to the conservation of light energy in a form directly usable by the cell. In-site-directed mutagenesis studies of amino acid residues within these molecules have elucidated the mechanism of the transport of these ions across the membrane. Retinal proteins have been found in other representatives of the *Halobacteriaceae* as well, and comparative studies of the retinal-based ion pumps found within the family are rapidly progressing. An updated view of our present knowledge of the retinal proteins can be found in papers by Lanyi and Váró (1995), Oesterhelt (1995), and Spudich et al. (1995). Bacteriorhodopsin and halorhodopsin are not "real" halophilic proteins. In contrast to other proteins of halophilic Archaea that require the presence of high salt concentrations for activity and stability, these light-driven ion pumps are perfectly stable and active in the absence of salt. The properties of bacteriorhodopsin — being highly stable, not requiring salt, easily amenable to manipulation, and available in large amounts — opened interesting biotechnological perspectives for the use of this molecule in different applications, including holographic techniques and information storage (Birge, 1995; Oesterhelt et al., 1991).

1.8 BIOCHEMISTRY AND MOLECULAR BIOLOGY OF HALOPHILIC PROKARYOTES

Comparative studies of halophilic microorganisms in all their diversity have led to the recognition that certain properties are directly related to adaptation to high salt concentrations, while other unique traits found in different halophiles may be primarily related to their phylogenetic position within the domains of Archaea, Bacteria, and Eukarya. The question, "What is halophilic and what is archaeal?" is discussed in an essay by Kushner (Chapter 17).

The biochemistry and molecular biology of the halophilic Archaea, with their special adaptations of the intracellular enzymatic machinery to the presence of high salt concentrations, have been most extensively studied. The malate dehydrogenase and the ferredoxin of *Haloarcula marismortui* have become models for the study of the three-dimensional structure at high resolution of halophilic proteins (for a review see Eisenberg, 1995), and techniques of physical chemistry are used to understand the complex interactions between salt and proteins of extreme halophiles (Ebel et al., Chapter 18).

The bacteriorhodopsin system is only one of the mechanisms some halophilic Archaea may use to obtain their energy, and their bioenergetics are based on a careful regulation of different modes of energy generation, including aerobic and anaerobic respiration, arginine fermentation, and use of light energy. Much information has recently become available on the signals involved in the regulation of these processes. The biochemistry and molecular biology of the use of arginine as energy source by certain members of the *Halobacteriaceae* has also been studied in depth. In the course of a study of the arginine deiminase gene cluster as a model system to investigate gene-specific regulation, elements of the general transcription apparatus have been characterized (Soppa et al., Chapter 20).

In addition, the study of the enzymatic properties and regulation of the synthesis of enzymes involved in pyruvate metabolism (Danson et al., Chapter 19), proton-translocating ATPases (Hochstein and Bogomolni, Chapter 22), and β-galactosidases from halophilic Archaea (Holmes and Dyall-Smith, Chapter 21) have enhanced our insight into the nature of enzymes functioning at high salt concentrations.

Several of the halophilic archaeal species possess gas vesicles built of protein subunits, which provide the cells with buoyancy. The presence of gas vesicles possibly aids the cells in obtaining oxygen in an environment where, due to the low solubility of gases in concentrated brines, oxygen could easily become a limiting factor. The genes involved in gas vesicle production in *Haloferax mediterranei* and their regulation have been characterized, and much information is now available on the factors determining the formation of gas vesicles in halophilic Archaea (Pfeifer et al., Chapter 23).

Another intriguing group of proteins produced by many of the halophilic Archaea is protein antibiotics, termed halocins. These bacteriocins inhibit the growth of other members of the group that may compete for the same resources. A number of halocins have been characterized, and the regulation of their synthesis is presently an active field of study (Shand et al., Chapter 24).

1.9 GENETICS AND GENOMICS

The latest developments in methodology, enabling the sequencing of complete genomes, have also had their impact on our understanding of the organization of the genomes of halophilic Archaea. No full genomic sequence of any halophilic Archaeon is known as yet, but the amount of information obtained on genome organization is rapidly increasing (Charlebois, Chapter 25). It can be expected that, within a relatively short time, one of the halophiles will also be included in the growing list of prokaryotes whose complete genome has been sequenced. No doubt the information thus obtained will enable extensive comparisons with nonhalophilic Archaea and shed more light on the special adaptations related to life at high salt concentrations. One specific field in which these comparative studies have made rapid progress is the elucidation of the structure of the ribosomal RNA operons (Dennis, Chapter 26).

Progress on the understanding of the genetics of halophilic Archaea is to a large extent dependent on the development of genetic systems for the transfer of genes between strains. *Haloferax volcanii* has become an organism of choice in genetic studies, as its genetic exchange system has now been well characterized (Ortenberg et al., Chapter 27).

1.10 BIOTECHNOLOGICAL ASPECTS

Many of the early studies on halophilic microorganisms were initiated in an attempt to achieve an understanding of bacteria that cause deterioration of salted fish, other salted foods, and salted hides. However, halophilic microorganisms have many positive aspects as well. Pigmented halophilic Archaea and microalgae absorb light energy in saltern ponds, thereby raising the water temperature, increasing the rate of evaporation and hastening the deposition of salt. In addition, some valuable products can be obtained from halophilic microorganisms (Galinski and Tindall, 1992): the use of carotenoid-rich *Dunaliella* strains for the commercial production of β-carotene, for use as health food and food additive (Ben-Amotz and Avron, 1989); and the extraction of ectoine from moderately halophilic Bacteria, to be used as enzyme protectant and as a moisturizer in the cosmetics industry (Galinski and Louis, Chapter 15).

It can safely be assumed that the halophilic microorganisms will have a much greater, yet-to-be-exploited, potential for biotechnological applications.

1.11 EPILOGUE

Thanks to the enormous diversity of halophilic microorganisms dispersed over the three domains of life, each with its own interesting and unique properties, there is hardly a hypersaline niche in nature that is not occupied by some halophile. A few exceptions include halophilic methanogenic bacteria able to grow on hydrogen + CO_2 or on acetate as an energy source, and halophilic nitrifying bacteria. The first halophilic aerobic methylotrophic bacteria have been described only very recently (Doronina and Trotsenko, 1997). In any case, the existing types of Archaea, Bacteria, and Eukarya that are able to withstand the stress exerted by salt concentrations up to halite saturation exhibit a large metabolic diversity that empowers hypersaline ecosystems to function.

The chapters in this book attempt to summarize our present understanding of a large number of aspects related to halophilic life. Hopefully, the information presented will not only serve as an up-to-date picture of hypersaline microbiology, but will also stimulate further work and exploration of the hypersaline environments and the microorganisms inhabiting them.

REFERENCES

Ben-Amotz, A. and M. Avron. 1989. The biotechnology of mass culturing *Dunaliella* for products of commercial interest, in *Algal and Cyanobacterial Biotechnology*, R.C. Cresswell, T.A.V. Rees, and N. Shah, Eds. Longman, Harlow, UK / John Wiley & Sons, New York. 90–114.

Birge, R.R. 1995. Protein-based computers. *Sci. Am.*, March 1995:66–71.

Dennis, P.P. and L.C. Shimmin. 1997. Evolutionary divergence and salinity-mediated selection in halophilic archaea. *Microbiol. Mol. Biol. Rev.* 61:90–104.

Doronina, N.V. and Yu. A. Trotsenko. 1997. Aerobic methylotrophic bacterial communities of hypersaline ecosystems. *Microbiology* (Russia) 66: 130–136.

Eisenberg, H. 1995. Life in unusual environments: progress in understanding the structure and function of enzymes from extreme halophilic bacteria. *Arch. Biochem. Biophys.* 318:1–5.

Galinski, E.A. 1995. Osmoadaptation in bacteria. *Adv. Microb. Physiol.* 37:273–328.

Galinski, E.A. and B.J. Tindall. 1992. Biotechnological prospects for halophiles and halotolerant micro-organisms, in *Molecular Biology and Biotechnology of Extremophiles*, R.A. Herbert and R.J. Sharp, Eds. Blackie, Glasgow. 76–114.

Gibbons, N.E. 1974. Family V. *Halobacteriaceae* fam. nov., in *Bergey's Manual of Determinative Bacteriology*, 8th. ed, R.E. Buchanan and N.E. Gibbons, Eds. Williams & Wilkins, Baltimore. 269–273.

Grant, W.D. and H. Larsen. 1989. Extremely halophilic archaeobacteria, order *Halobacteriales* ord. nov, in *Bergey's Manual of Systematic Bacteriology*, J.T. Staley, M.P. Bryant, N. Pfennig, and J.G. Holt, Eds. Vol. 3. Williams & Wilkins, Baltimore. 2216–2233.

Javor, B. 1989. *Hypersaline Environments. Microbiology and Biogeochemistry.* Springer-Verlag, Berlin.

Lanyi, J. K. and G. Váró. 1995. The photocycles of bacteriorhodopsin. *Israel J. Chem.* 35:365–385.

Madigan, M.T. 1997. Extremophiles. *Sci. Am.*, April 1997:66–71.

Niemi, T., J. Gat, and Z. Ben-Avraham, Eds. 1997. *The Dead Sea.* Oxford University Press, Oxford.

Norton, C.F. 1992. Rediscovering the ecology of halobacteria. *ASM News* 58:363–367.

Oesterhelt, D. 1995. Structure and function of halorhodopsin. *Israel J. Chem.* 35:475–494.

Oesterhelt, D., C. Bräuchle, and A. Hampp. 1991. Bacteriorhodopsin – a biological model for information processing. *Quart. Rev. Biophys.* 24:425–478.

Ollivier, B., P. Caumette, J.L. Garcia, and R.A. Mah. 1994. Anaerobic bacteria from hypersaline environments. *Microbiol. Rev.* 58:27–38.

Oremland, R.S. and G.M. King. 1989. Methanogenesis in hypersaline environments, in *Microbial Mats. Physiological Ecology of Benthic Microbial Communities*, Y. Cohen and E. Rosenberg, Eds. American Society for Microbiology, Washington, DC. 180–190.

Oren, A. 1988. The microbial ecology of the Dead Sea, in *Advances in Microbial Ecology*, Vol. 10, K.C. Marshall, Ed. Plenum, New York. 193–229.

Oren, A. 1994. The ecology of the extremely halophilic archaea. *FEMS Microbiol. Rev.* 13:415–440.

Oren, A. 1998. Salts and brines, in *Ecology of Cyanobacteria: Their Diversity in Time and Space,* B.A. Whitton and M. Potts, Eds. Kluwer Academic, Dordrecht, in press.

Rainey, F.A., T.N. Zhilina, E.S. Boulygina, E. Stackebrandt, T.P. Tourova, and G.A. Zavarzin. 1995. The taxonomic status of the fermentative halophilic anaerobic bacteria: description of *Haloanaerobiales* ord. nov., *Halobacteroidaceae* fam. nov., *Orenia* gen. nov., and further taxonomic rearrangements at the genus and species level. *Anaerobe* 1:185–199.

Rodriguez-Valera, F., Ed. 1988. *Halophilic Bacteria.* Vols. I and II. CRC, Boca Raton.

Rodriguez-Valera, F., Ed. 1991. *General and Applied Aspects of Halophilic Microorganisms.* Plenum, New York.

Spudich, J.L., D.N. Zacks, and R.A. Bogomolni. 1995. Microbial sensory rhodopsins: photochemistry and function. *Israel J. Chem.* 35:495–513.

Tindall, B.J. 1993. The family *Halobacteriaceae,* in *The Prokaryotes. A Handbook of Bacteria: Ecophysiology, Isolation, Identification, Applications,* A. Balows, H.G. Trüper, M. Dworkin, W. Harder and K.-H. Schleifer, Eds. Springer-Verlag, New York. 768–808.

Ventosa, A. and Oren, A. 1996. *Halobacterium salinarum* nom. corrig., a name to replace *Halobacterium salinarium* (Elazari-Volcani) and to include *Halobacterium halobium* and *Halobacterium cutirubrum, Int. J. Syst. Bacteriol.* 46:361.

Ventosa, A., J.J. Nieto, and A. Oren. 1998. Moderately halophilic eubacteria. *Microbiol. Mol. Biol. Rev.,* in press.

Vreeland, R.H. and L.I. Hochstein, Eds. 1993. *The Biology of Halophilic Bacteria.* CRC, Boca Raton.

Vreeland, R.H., C.D. Litchfield, E.L. Martin, and E. Elliot. 1980. *Halomonas elongata,* a new genus and species of extremely salt-tolerant bacteria, *Int. J. Syst. Bacteriol.* 30:485–495.

Woese, C.R. and G.E. Fox. 1977. Phylogenetic structure of the prokaryotic domain: the primary kingdoms. *Proc. Natl. Acad. Sci. USA* 74:5088–5090.

Woese, C.R., O. Kandler, and M.L. Wheelis. 1990. Towards a natural system of organisms: proposal for the domains Archaea, Bacteria, and Eucarya. *Proc. Natl. Acad. Sci. USA* 74:5088–5090.

Section I

Diversity and Ecology

2 Diversity of Members of the Family *Halobacteriaceae*

Masahiro Kamekura

CONTENTS

2.1 Introduction 13
2.2 Recently Described Genera 14
    2.2.1 *Halorubrum* 14
    2.2.2 *Natrialba* 14
    2.2.3 *Halobaculum* 15
2.3 The Alkaliphilic Members 15
    2.3.1 Sequences of 16s rRNA Encoding Genes 16
    2.3.2 Sequences of Spacer Regions 16
    2.3.3 Proposal of *Natronomonas* and Two Transfers 16
2.4 The Confusion Concerning Different Strains of *Halorubrum trapanicum/Halobacterium trapanicum* 21
2.5 *Haloarcula mukohataei*, a New Species of the Genus *Haloarcula* 22
2.6 Are Halophilic Archaea Everywhere? 23
    2.6.1 Halophilic Archaeal Clones from Hypersaline Ponds 23
    2.6.2 Halophilic Archaeal Clones from Forest Soils? 23
Acknowledgments 24
References 24

2.1 INTRODUCTION

The extremely halophilic, aerobic Archaea are classified within the family *Halobacteriaceae*, which currently contains nine validly described genera (as of June 1997): *Halobacterium, Halococcus, Haloarcula, Haloferax, Halorubrum, Halobaculum, Natrialba*, and the two alkaliphilic genera *Natronobacterium* and *Natronococcus* (Grant and Larsen, 1989; Kamekura and Dyall-Smith, 1995; McGenity and Grant, 1995; Oren et al., 1995). These divisions were based largely on chemotaxonomic criteria, particularly the composition of membrane polar lipids. The presence of phosphatidylglycerosulfate (PGS) and different types of glycolipids were used to assign new isolates of halobacteria to existing genera. Until the proposal of the genus

Halorubrum in 1995, there was a tacit consensus that species of extreme halophiles of the same genus should contain the same glycolipids. The two alkaliphilic genera were no exceptions, as members of those genera were found to be devoid of any readily detectable amount of glycolipid.

2.2 RECENTLY DESCRIBED GENERA

2.2.1 *HALORUBRUM*

In 1995, two articles were published concerning reclassification of the genus *Halobacterium*. McGenity and Grant (1995) proposed to transfer four species, *Halobacterium saccharovorum*, *Halobacterium sodomense*, *Halobacterium lacusprofundi*, and *Halobacterium trapanicum* NRC 34021, to a novel genus *Halorubrum*. The authors argued that these species possessed three different kinds of glycolipids: S-DGD-1 *(Halorubrum saccharovorum)*, S-DGD-3 *(Halorubrum sodomense* and *Halorubrum lacusprofundi)*, and S-DGD-5 *(Halorubrum trapanicum* NRC 34021) (see Kates [1993] for the structures of these glycolipids). Tindall (1990) had, however, shown that the sulfated diglycosyl diether lipid from *Halobacterium lacusprofundi* co-migrated with the glycolipids from *Halobacterium saccharovorum* and *Halobacterium sodomense* in two-dimensional thin layer chromatography (TLC). The proposal of the new generic name was accepted and the genus *Halorubrum* was validly published in the IJSB validation list no. 56.

In a paper published at almost the same time, Kamekura and Dyall-Smith (1995) proposed to transfer *Halobacterium saccharovorum*, *Halobacterium sodomense*, *Halobacterium lacusprofundi*, and *Halobacterium distributum* to a novel genus *Halorubrobacterium*, and proposed a novel species *Halorubrobacterium coriense*. They unambiguously showed by one-dimensional TLC that all these species contained S-DGD-3. In this paper, the confusion regarding the identities of different strains of *Halobacterium trapanicum* was briefly discussed (see Section 4), but the authors refrained from proposing a new genus containing *Halobacterium trapanicum*. The proposed new genus name *Halorubrobacterium* is a later synonym of *Halorubrum* in the IJSB validation list no. 57 (see also Oren and Ventosa, 1996).

The genus *Halorubrum* has thus been the first exception to the tacit consensus in the taxonomy of *Halobacteriaceae*. Though we are certain that glycolipids of *Halorubrum saccharovorum* and *Halorubrum sodomense* have the same mobilities on TLC, as was demonstrated by Tindall (1990), Grant's group argues they have slightly different mobilities (W.D. Grant, personal communication). These problems should be solved in the near future.

2.2.2 *NATRIALBA*

The strains 172P1 and B1T, isolated in Japan and Taiwan, respectively, have long been shown to contain a novel glycolipid, and its structure was finally determined as S_2-DGD-1 (2,3-di-*O*-phytanyl-or phytanyl-sesterterpenyl-1-*O*-[2,6-(HSO_3)-2-α-mannopyranosyl-(1→2)-α-glucopyranosyl]-*sn*-glycerol (Kates, 1993). Moreover, unlike other members of *Halobacteriaceae*, neither is red-pigmented; their content of bacterioruberins was less than 0.1% of that of *Halobacterium salinarum* NRC 34001 as estimated by absorption at 500 nm of the neutral lipid fractions. Cultivation under illumination does not induce pigment formation in these strains, whereas the intensity of pigmentation of strain GSL11 and *Halobacterium trapanicum* JCM 9743 was enhanced upon illumination during cultivation on agar slopes.

Strains 172P1 and B1T form a distinct group, as strongly supported by bootstrap analysis. The two strains possess C_{20},C_{25} core diether lipids as shown by chemical analyses, and also by the double spots of PG and PGP-Me (phosphatidyglycerophosphate methyl ester) on TLC. Thus the two strains 172P1 and B1T deserve a novel generic status based on possession of the novel glycolipid S_2-DGD-1 and on the sequences of 16S rRNA genes. Kamekura and Dyall-Smith (1995) proposed the generic name *Natrialba*. DNA–DNA hybridization experiments showed that they have more than 80% binding, indicating that they constitute a single species. The species was named *Natrialba asiatica*, referring to the geographical region from which the strains were isolated.

Recently, a variety of extreme halophiles were isolated from hypersaline ponds in northern China. Two of these strains, designated OF8 and QX1, have been suggested to belong to the genus *Natrialba*, as judged from the glycolipid compositions and 16S rRNA gene sequences (Kamekura et al., unpublished data).

2.2.3 *HALOBACULUM*

A mass bloom of halophilic Archaea developed in the Dead Sea in the summer of 1992. In the past, enrichment cultures in which Dead Sea water or sediment was used as inoculum have yielded three novel halophiles, *Haloferax volcanii, Haloarcula marismortui*, and *Halorubrum sodomense*. S-DGD-1 was found as the sole glycolipid in the polar lipids extracted from the biomass in the Dead Sea in 1992, suggesting that the dominant organisms were relatives of the genus *Haloferax*. A pleomorphic rod-shaped bacterium isolated from the biomass was further analyzed extensively by Oren's group. On the basis of 16S rRNA gene sequence data, they concluded that the isolate was sufficiently different from the previously described members of the *Halobacteriaceae* to warrant the creation of a new species, and they named the isolate *Halobaculum gomorrense* gen. nov., sp. nov. (Oren et al., 1995). This is the third group of extreme halophiles that possess S-DGD-1 as the membrane polar lipid (*Haloferax* spp., *Halococcus* spp., and *Halobaculum* sp.).

2.3 THE ALKALIPHILIC MEMBERS

The alkaliphilic members of the *Halobacteriaceae* form a distinct physiological group, as they require high NaCl concentrations, high pH, and low Mg^{2+} concentrations for growth. Microscopically, the initial isolates consisted of rods and cocci, and were accordingly separated into two genera, *Natronobacterium* and *Natronococcus* (Tindall et al., 1984). *Natronobacterium* currently contains four recognized species, *Natronobacterium gregoryi, Natronobacterium magadii, Natronobacterium pharaonis*, and *Natronobacterium vacuolatum*, with *Natronobacterium gregoryi* as the type species. Glycolipid analysis has had little impact in the classification of the natronobacteria, as these lack major amounts of glycolipids in their membranes. On the other hand, DNA-DNA hybridization experiments of the four species have indicated that they share very little sequence (31-38% binding only) (Mwatha and Grant, 1993).

2.3.1 Sequences of 16S rRNA Encoding Genes

The 16S rRNA genes from four natronobacteria (*Natronobacterium gregoryi*, *Natronobacterium pharaonis*, *Natronobacterium vacuolatum*, and strain SSL1) were sequenced (Kamekura et al., 1997) and compared with the previously determined sequences of *Natronobacterium magadii*, two species of the genus *Natronococcus*, and other closely related species. Surprisingly, the similarity values of the four recognized members of the genus *Natronobacterium* were below 92% (the highest similarity was observed between *Natronobacterium gregoryi* and *Natronobacterium magadii*, 91.9%). Devereux et al. (1990) and Fry et al. (1991) have proposed that a 16S rRNA similarity value of less than 93%-95% indicates that two organisms may belong to different genera. This implies that the four species of the genus *Natronobacterium* may be sufficiently different to warrant classification in different genera.

The sequences were aligned with those previously reported and used to construct phylogenetic trees, using *Methanospirillum hungatei* as the outgroup. Positions with any deletions or of uncertain alignment were removed and the remaining 1371 positions were used to construct the tree. A variety of algorithms were utilized (maximum likelihood, maximum parsimony, distance matrix), which gave very similar topologies. A representative example is shown in Figure 2.1, derived using maximum likelihood.

As anticipated from the similarity values, the four recognized *Natronobacterium* species and strain SSL1 do not form a monophyletic group. Using the aligned sequences, a number of signature sequences (Winker and Woese, 1991) were revealed that were specific for each genus. These bases are listed in Table 2.1. The sequence numbers refer to the refined alignment (1 to 1483) reported previously (Kamekura and Seno, 1993).

2.3.2 Sequences of Spacer Regions

Spacer regions between the 16S rRNA and 23S rRNA genes from the three *Natronobacterium* species and the five *Halorubrum* species were amplified by PCR. The sense and antisense primers were designed based on the sequence from nt 2332 to 2356, and from nt 2959 to 2981 of the *Halobacterium salinarum* rRNA operon. Both primers are situated in highly conserved regions. PCR products, which were expected to overlap by 51 bp with the 3'-termini of the 16S rRNA genes and by 71 bp with the 5'-termini of the 23S rRNA genes, were cloned and sequenced.

2.3.3 Proposal of *Natronomonas* and Two Transfers

We have proposed the following changes in the taxonomy of the genus *Natronobacterium* as described below (Kamekura et al., 1997):

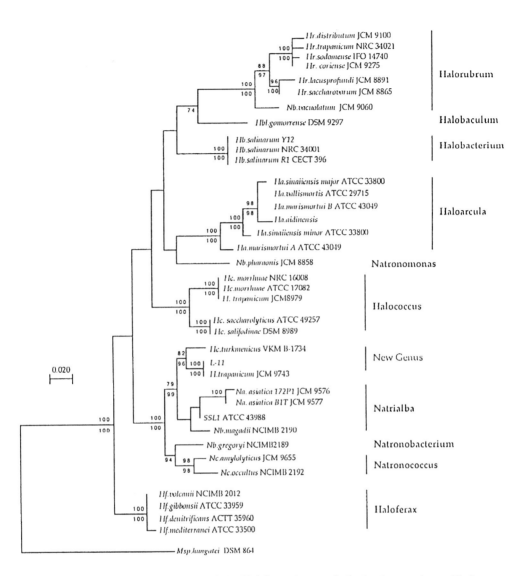

FIGURE 2.1 Phylogenetic tree of the *Halobacteriaceae*, derived using maximum likelihood. Branches with poorly supported lengths have been collapsed. Bootstrap values (100 replicates) shown at the nodes were obtained by parsimony (above) and distance matrix (below) methods and added to the maximum likelihood tree. Only bootstrap values above 70% are shown. The scale bar represents 0.02 expected changes per site.

1. *Natronobacterium gregoryi*, which clustered tightly with the two species of the genus *Natronococcus*, is the type species of the genus *Natronobacterium*.
2. *Natronobacterium pharaonis* was only distantly related to the other taxa. Similarities were less than 89.9%. We have proposed to transfer *Natronobacterium pharaonis* to *Natronomonas* gen. nov.

TABLE 2.1
Summary of Information on Species of 10 Genera of Halophilic Archaea

Genus, Species	Characteristic Glycolipids and Phospholipids	Accession numbers of sequences of:		
		16S rRNA Gene	Spacer Region	23S rRNA Gene
Halobacterium				
salinarum (halobium) R1	S-TGD, S-TeGD	M38280 & M11583	X03407	X03407
sp. strain Y12	S-TGD, S-TeGD	D14127		

Signature bases: 77C, 106G, 128T, 213C, 222T, 407A, 1383T[a]

Haloarcula				
vallismortis ATCC 29715	TGD-2	U17593		
vallismortis IFO14741		D50851		
marismortui ATCC 43049[b]	TGD-2	(A) X61688 (B) X61689	X13738	
hispanica ATCC 33960	TGD-2	U68541		
argentinensis JCM 9737	TGD-2	D50849		
"*sinaiiensis*" ATCC 33800	TGD-2	(major) D14129 (minor) D14130		
"*aidinensis*"	`TGD-2	AB000563		

Signature bases: 235T, 255A, 665A, 670G, 675C, 883C, 937T, 968C, 972T, 1173A, 1200T, 1202T, 1243- (-means a gap), 1250A, 1458A

japonica JCM 7785	TGD-2	D28872		
mukohataei JCM 9738	(S-DGD-1?, DGD-1?)	D50850		
Haloferax				
volcanii NCMB 2012	S-DGD-1	K00421	M19341	RDP[c]
mediterranei ATCC 33500	S-DGD-1	D11107		
gibbonsii ATCC 33959	S-DGD-1	D13378		
denitrificans ATCC 35960	S-DGD-1	D14128		

Signature bases: 1109A, 1218T, 1220T, 1227A

Halococcus				
morrhuae ATCC 17082	S-DGD-1	X00662	X72588	X72588
morrhuae NRCC 16008	S-DGD-1	D11106		
saccharolyticus ATCC 49257	S-DGD-1	AB004876		
salifodinae DSM 8989	S-DGD-1	AB004877		

Signature bases: 25T, 46C, 90C, 200C, 211C, 215A, 495A, 499A, 1246C, 1271T, 1280A

TABLE 2.1 (CONTINUED)
Summary of Information on Species of 10 Genera of Halophilic Archaea

Genus, Species	Characteristic Glycolipids and Phospholipids	Accession numbers of sequences of:		
		16S rRNA Gene	Spacer Region	23S rRNA Gene
Halobaculum				
gomorrense DSM 9297	S-DGD-1	L37444		

Signature bases: 106T, 213A, 246T, 521T, 603A, 690T, 702C, 706A, 777C, 943T

Natronobacterium				
gregoryi NCIMB 2189	PL1, PL3	D87970	AB003410	
Natronococcus				
occultus NCMB 2192	PL1, Pl2		Z28378	
amylolyticus JCM 9655	PL1	D43628		

Signature bases (together with the genus *Natronobacterium*): 1312C, 1313G, 1324C, 1325G

Natrialba				
asiatica JCM 9576	S2-DGD	D14123		
asiatica JCM 9577	S2-DGD	D14124		

Signature bases: 385G, 502G, 1108A

magadii NCMB 2190	PL3	X72495	X72495	X72495
sp. SSL1 ATCC 43988	DGD-4	D88256		
Halorubrum				
saccharovorum JCM 8865	S-DGD-3	U17364	AB003405	
		X82167		
NCIMB 2081	S-DGD-1			
trapanicum NRC 34021	S-DGD-5	X82168		
lacusprofundi JCM 8891	S-DGD-3	U17365	AB003406	
ACAM 34	S-DGD-3	S-DGD-3	X82170	
sodomense IFO 14740	S-DGD-3	D13379	AB003404	
ATCC 33755	S-DGD-3	X82169		
coriense JCM 9275	S-DGD-3	L00922	AB003408	
distributum JCM 9100	S-DGD-3	D63572	AB003407	
vacuolatum JCM 9060	PL3, PL4	D87972	AB003411	

Signature bases: 114T, <u>179A</u>, 189C, 206A, 611G, 683C, 783C, 794G, 795G, 818C, 1070C, 1087G, 1101G, 1105G, <u>1115C</u>, 1121G, 1127C, 1208A, <u>1274C</u>

Natronomonas pharaonis	JCM 8858	PL1	D87971	AB003409

Signature bases: 55G, 64-, 65-, 71-, 72-, 73-, 85T, 109G, 381A, 477C

H. trapanicum JCM 8979 (coccoid)	S-DGD-1	D63786		

TABLE 2.1 (CONTINUED)
Summary of Information on Species of 10 Genera of Halophilic Archaea

Genus, Species	Characteristic Glycolipids and Phospholipids	Accession numbers of sequences of:		
		16S rRNA Gene	Spacer Region	23S rRNA Gene
H. trapanicum ATCC 43102 (coccoid)	S-DGD-1			
H. trapanicum JCM 9743 (rod)	unidentified	D14125		
Strain GSL-11	unidentified	D14126		
Halococcus turkmenicus VKM B-1734	S2-DGD	AB004978		

Note: Bases that are confined to the six species, except *H. vacuolatum*, are underlined.

[a] Nucleotide numbers of signature bases refer to the refined alignment (1 to 1483) of all sequences cited in the table.

[b] There is a doubt about the authenticity of this strain (B.J. Tindall, personal communication).

[c] RDP: the sequence is available from the Ribosomal Database Project.

3. *Natronobacterium vacuolatum* showed a 16S rRNA sequence similarity of 95.6% to *Halorubrum saccharovorum*, the type species of the genus *Halorubrum*, indicating that they may belong to the same genus. Sequences of the six species of *Halorubrum* possess 19 genus-specific signature bases. The *Natronobacterium vacuolatum* sequence shares 16 of these 19 bases. Sequences of spacer regions from the five recognized species of *Halorubrum* were easily aligned with *Natronobacterium vacuolatum*, and the alignment showed long stretches of conserved sequence. If sequences from members of the other genera were introduced, the alignment was severely disrupted. Altogether, these data seem to support the view that *Natronobacterium vacuolatum* is related to members of the genus *Halorubrum*. We have proposed to transfer *Natronobacterium vacuolatum* to the genus *Halorubrum*, and the strain is to be called *Halorubrum vacuolatum* (comb. nov.).

   Whereas most of the species of the genus *Halorubrum* are characterized by the presence of glycolipid S-DGD-3 (Kamekura and Dyall-Smith, 1995), no glycolipids have been detected in *Natronobacterium vacuolatum* (Mwatha and Grant, 1993), but an exception has been reported, namely S-DGD-5 for *Halorubrum trapanicum* NRC 34021 (McGenity and Grant, 1995). We cannot exclude the possibility that a trace amount of glycolipid is present in *Halorubrum vacuolatum*.

4. *Natronobacterium magadii* showed a high sequence similarity of its 16S rRNA gene to that of strain SSL1 (95.9%), and to those of *Natrialba asiatica* 172P1T (type strain) and *Natrialba asiatica* B1T (93.3% and

93.7%, respectively). These data support the view that *Natronobacterium magadii* and the strain SSL1 are members of the genus *Natrialba*, and we have proposed to transfer *Natronobacterium magadii* to the genus *Natrialba* as *Natrialba magadii* and to call the strain SSL1, *Natrialba* sp. strain SSL1.

The lipid analyses again do not support the close 16S rRNA sequence similarities. No glycolipid has been detected in *Natrialba magadii*, but strain SSL1 has a minor amount of glycolipid DGD-4 (glucose 1→6 glucose-glycerol diether) (Kates, 1993; Upasani et al., 1994). *Natrialba asiatica* has S_2-DGD (mannose-2,6-disulfate →glucose-glycerol diether) as its glycolipid.

2.4 THE CONFUSION CONCERNING DIFFERENT STRAINS OF *HALORUBRUM TRAPANICUM*/*HALOBACTERIUM TRAPANICUM*

The history of *Halobacterium trapanicum* has been reviewed extensively by Tindall (1992). The 7th edition of *Bergey's Manual of Determinative Bacteriology* described detailed characteristics (Elazari-Volcani, 1957), and the 8th edition designated the type strain as [EV6.32.1 (Delft); NRC 34021]. As pointed out by Tindall (1992) and by Kamekura and Dyall-Smith (1995), however, there have been serious discrepancies between different laboratories when reporting the morphology and glycolipids of the type strains. According to the 7th edition of *Bergey's Manual*, cells are nonmotile rods, and colonies are light orange to very faintly colored. The 8th edition, and Vol. 1 and 3 of *Bergey's Manual of Systematic Bacteriology* also quoted similar descriptions. However, Ross and Grant (1985) reported that cells of NRC 34021 were motile pleomorphic rods. Later, Tindall (1992) stated that both he and R. Latta (of the NRC) had noticed that strains NRC 34021, NCIMB 767 and NRC 2856 all had coccoid morphology, thus casting doubt on the authenticity of the type strains.

On the other hand, Grant and a group of Italian chemists jointly reported that the structures of glycolipids from *Halobacterium sodomense* (Trincone et al., 1990) and *Halobacterium trapanicum* (Trincone et al., 1993) were S-DGD-3 and S-DGD-5, respectively (see Kates [1993] for detailed structures). At about the same time, Tindall (1992) examined the lipids of three strains (NRC 34021, NCIMB 767 and NRC 2856). All three were coccoid and had polar lipid compositions similar to members of the genus *Halococcus* that contains S-DGD-1.

At present, there are three strains designated *Halobacterium trapanicum*: JCM 8979 (derived from NCIMB 767 in 1994), JCM 9743 (derived from NCMB 767 in 1982 and kept by M. Kamekura since 1984, then deposited with JCM in 1995), and ATCC 43102 (obtained in 1995). Recently, the 16S rRNA gene sequence of *Halobacterium trapanicum* JCM 8979 was determined and found to be almost indistinguishable from those of two *Halococcus morrhuae* strains (98.6%-98.9% sequence similarity). Microscopic examination of strain JCM 8979 clearly showed these cells to be coccoid, often forming tight clusters. Colonies were bright red, not light orange as originally described. Cells of ATCC 43102 also showed coccoid morphology. On the other hand, the cells of strain JCM 9743 were long rods, and colonies were light orange in color. The 16S rRNA gene of *Halobacterium trapanicum* JCM 9743 has been sequenced, and found to be closely related to the strain GSL11 (Kamekura and Seno, 1993).

Lipids of JCM 8979 could not be extracted with a mixture of chloroform and methanol without prior sonication or grinding with quartz sand, as is the case with halococci, while JCM 9743 lipids were easily extractable. Thin-layer chromatography of the extracted polar lipids clearly showed that the glycolipid of JCM 8979 was S-DGD-1, the same as that of *Halococcus morrhuae*, while JCM 9743 gave a characteristic glycolipid pattern, as reported previously (Kamekura and Dyall-Smith, 1995).

Recently, it was found that the two strains of *Halorubrum trapanicum* NRC 34021 and NRC 2856 at NRC, Ottawa, had died during storage (R. Latta, personal communication in 1995). The only culture of the type strain (NRC 34021) possibly survives in the laboratory of W.D. Grant. It is recommended that *Halorubrum* (*Halobacterium*) *trapanicum* be considered a species *incertae sedis* until the type strain becomes available once more and/or a comparative study of the existing strains is completed (Oren and Ventosa, 1997).

2.5 *HALOARCULA MUKOHATAEI*, A NEW SPECIES OF THE GENUS *HALOARCULA*

Lately, two new species of the genus *Haloarcula* were validly published, *Haloarcula argentinensis* and *Haloarcula mukohataei* (Ihara et al., 1997). The latter (type strain arg-2) is the first exceptional species in this genus that has membrane glycolipids (similar to S-DGD-1 and DGD-1) different from TGD-2 common to all other members of *Haloarcula*. The authors argued that the 16S rRNA gene of *Haloarcula mukohataei* was most closely related to those of the members of the genus *Haloarcula* (e.g. 90.6% similarity with the *Haloarcula marismortui* A gene) than to any other genera. We now do not place as much emphasis on the glycolipid pattern as before in the taxonomy of *Halobacteriaceae*, as discussed above. However, a careful examination of the 16S rRNA gene sequences shows that nine out of the 15 signature bases of the genus *Haloarcula* are replaced by other bases in *Haloarcula mukohataei*. It may therefore be justified to propose another novel genus to accommodate the type strain arg-2.

In this connection, remember that all members of the genus *Haloarcula* possess considerably heterogeneous 16S rRNA genes: *Haloarcula marismortui* A and B (94.4% similarity), "*Haloarcula sinaiiensis*" major and minor (97.3 %), two genes for *Haloarcula vallismortis* (96.2%), etc. We also have detected at least two heterogeneous genes in *Haloarcula hispanica* and in strain XA3-1, an isolate from northern China (Kamekura, unpublished data). Since at least three 16S rRNA genes were detected in the genome of *Haloarcula mukohataei*, sequencing of the other genes would give additional information relating to the affiliation of this species.

2.6 ARE HALOPHILIC ARCHAEA EVERYWHERE?

In recent years, microbial diversity in nature has been studied by analysis of 16S ribosomal RNA genes amplified by PCR from DNA extracted from environmental samples. Molecular phylogenetic studies indicate a much greater phylogenetic and probably physiological diversity of Archaea than previously assumed. In particular, many striking discoveries have been reported for the kingdom Crenarchaeota. Many novel phylotypes of phenotypically yet-to-be-described Crenarchaeota have been isolated from a hot spring in Yellowstone National Park, USA. On the other hand, Crenarchaeotal phylotypes, originally thought to consist solely of hyperthermophilic organisms, have been detected in non-thermophilic environments such as subsurface ocean waters, polar seas, soybean fields, and freshwater lake sediments, etc. (Schleper et al., 1997).

2.6.1 HALOPHILIC ARCHAEAL CLONES FROM HYPERSALINE PONDS

Archaeal aerobic halophilic bacteria are poor in diversity in comparison with the other branches of the Archaea (methanogens, sulfur-dependent thermophiles). Some efforts have been made to find novel phylotypes of halobacteria in natural environments (Benlloch et al., 1995). Two clones, HAC1 and HAC4 (1437 bp and 1435 bp, respectively; accession numbers X84084, X94331), obtained from a sample from a saltern crystallization pond in Alicante, Spain, were found to be significantly different from all other 16S rRNA genes reported so far. Twenty two bases that break the consensus among 41 halophiles sequences compiled so far are scattered throughout the whole sequences. Comparison with the consensus sequences of methanogens, including those of the extremely halophilic methanogens *Methanohalophilus* and *Methanohalobium* spp., suggested that HAC1 and HAC4 were not members of methanogens. They might represent members of a novel family within the order *Halobacteriales*. Possibly Rodriguez-Valera's group will be successful in the isolation of those microorganisms harboring the HAC1 or HAC4 sequences. Such organisms might shed light on the evolutionary relationship between extreme halophiles and methanogens within the kingdom Euryarchaeota.

2.6.2 HALOBACTERIA IN FOREST SOILS?

Jurgens et al. (1997) sequenced nine 16S rRNA genes amplified by PCR with Archaea-specific primers from mixed-population DNA extracted from forest soil in northern Finland. They reported that those sequences may belong to a previously undescribed terrestrial group within the kingdom Crenarchaeota. The 869 bp sequences of other clones, FFSB12 (X96687, Schleper et al., 1997) and FFSB 9 (X96686, G. Jurgens, personal communication) are quite extraordinary. They show only 6 bp differences with the *Halobacterium salinarum* sequence. Jurgens' group has obtained more of this type sequences (personal communication). Do halobacteria indeed survive in non-salty environments?

ACKNOWLEDGMENTS

The author wishes to express his sincere thanks to collaborators M. Kates (Ottawa, Canada), M.L. Dyall-Smith (Melbourne, Australia), A. Ventosa (Sevilla, Spain), T. Ito (Riken Institute, Japan), and V. Upasani (M.G. Science Institute, India).

REFERENCES

Benlloch, S., A.J. Martinez-Murcia, and F. Rodriguez-Valera. 1995. Sequencing of bacterial and archaeal 16S rRNA genes directly amplified from a hypersaline environment. *Syst. Appl. Microbiol.* 18:574–581.

Devereux, R., S.-H. He, C.L. Doyle, S. Orkland, D.A. Stahl, J. LeGall, and W.B. Whitman. 1990. Diversity and origin of *Desulfovibrio* species: phylogenetic definition of a family. *J. Bacteriol.* 172:3609–3619.

Elazari-Volcani, B. 1957. Genus XII. *Halobacterium*, in *Bergey's Manual of Determinative Bacteriology*, 7th ed, R.S. Breed, E.G.D. Murray, and N.R. Smith, Eds. Williams & Wilkins, Baltimore. 207–212.

Fry, N.K., S. Warwick, N.A. Saunders, and T.M. Embley. 1991. The use of 16S ribosomal RNA analyses to investigate the phylogeny of the family *Legionellaceae*. *J. Gen. Microbiol.* 137:1215–1222.

Grant, W.D. and H. Larsen. 1989. Group III. Extremely halophilic archaeobacteria, in *Bergey's Manual of Systematic Bacteriology*, Vol. 3, J.T. Staley, M.P. Bryant, N. Pfennig, and J.G. Holt, Eds. Williams & Wilkins, Baltimore. 2216–2233.

Ihara, K., S. Watanabe, and T. Tamura. 1997. *Haloarcula argentinensis* sp. nov. and *Haloarcula mukohataei* sp. nov., two new extremely halophilic archaea collected in Argentina. *Int. J. Syst. Bacteriol.* 47:73–77.

Jurgens, G., K. Lindstrom, and A. Saano. 1997. Novel group within the kingdom Crenarchaeota from boreal forest soil. *Appl. Environ. Microbiol.* 63:803–805.

Kamekura, M. and M.L. Dyall-Smith. 1995. Taxonomy of the family *Halobacteriaceae* and the description of two new genera *Halorubrobacterium* and *Natrialba*. *J. Gen. Appl. Microbiol.* 41:333–350.

Kamekura, M. and Y. Seno. 1993. Partial sequence of the gene for a serine protease from a halophilic archaeum *Haloferax mediterranei* R4, and nucleotide sequences of 16S rRNA encoding genes from several halophilic archaea. *Experientia* 49:503–513.

Kamekura, M., M.L. Dyall-Smith, V. Upasani, A. Ventosa, and M. Kates. 1997. Diversity of alkaliphilic halobacteria: proposals for the transfer of *Natronobacterium vacuolatum*, *Natronobacterium magadii*, and *Natronobacterium pharaonis* to the genus *Halorubrum*, *Natrialba*, and *Natronomonas* gen. nov., respectively, as *Halorubrum vacuolatum* comb. nov., *Natrialba magadii* comb. nov., and *Natronomonas pharaonis* comb. nov., respectively. *Int. J. Syst. Bacteriol.* 47:853–857.

Kates, M. 1993. Membrane lipids of extreme halophiles: biosynthesis, function and evolutionary significance. *Experientia* 49:1027–1036.

McGenity, T.J. and W.D. Grant. 1995. Transfer of *Halobacterium saccharovorum*, *Halobacterium sodomense*, *Halobacterium trapanicum* NRC 34021 and *Halobacterium lacusprofundi* to the genus *Halorubrum* gen. nov., as *Halorubrum saccharovorum* comb. nov., *Halorubrum sodomense* comb. nov., *Halorubrum trapanicum* comb. nov., and *Halorubrum lacusprofundi* comb. nov. *Syst. Appl. Microbiol.* 18:237–243.

Mwatha, W.E. and W.D. Grant. 1993. *Natronobacterium vacuolatum*, a haloalkaliphilic archaeon isolated from Lake Magadii, Kenya. *Int. J. Syst. Bacteriol.* 43:401–404.

Oren, A. and A. Ventosa. 1996. A proposal for the transfer of *Halorubrobacterium distributum* and *Halorubrobacterium coriense* to the genus *Halorubrum* as *Halorubrum distributum* comb. nov. and *Halorubrum coriense* comb. nov., respectively. *Int. J. Syst. Bacteriol.* 46:1180.

Oren, A. and A. Ventosa. 1997. International committee on systematic bacteriology. Subcommittee on the taxonomy of *Halobacteriaceae*. Minutes of the meetings, 19 and 20 August 1996, Jerusalem, Israel. *Int. J. Syst. Bacteriol.* 47:595–596.

Oren, A., P. Gurevich, R.T. Gemmell, and A. Teske. 1995. *Halobaculum gomorrense* gen. nov., sp. nov., a novel extremely halophilic archaeon from the Dead Sea. *Int. J. Syst. Bacteriol.* 45:747–754.

Ross, H.N.M. and W.D. Grant. 1985. Nucleic acid studies on halophilic archaebacteria. *J. Gen. Microbiol.* 131:165–173.

Schleper, C., W. Holben, and H.-P. Klenk. 1997. Recovery of crenarchaeotal ribosomal DNA sequences from fresh water-lake sediments. *Appl. Environ. Microbiol.* 63:321–323.

Tindall, B.J. 1990. Lipid composition of *Halobacterium lacusprofundi*. *FEMS Microbiol. Lett.* 66:199–202.

Tindall, B.J. 1992. The family *Halobacteriaceae*, in *The Prokaryotes. A Handbook on the Biology of Bacteria: Ecophysiology, Isolation, Identification, Applications*. 2nd ed., Vol. I, A. Balows, H.G. Trüper, M. Dworkin, W. Harder, and K.-H. Schleifer, Eds. Springer-Verlag, New York. 754–808.

Tindall, B.J., H.N.M. Ross, and W.D. Grant. 1984. *Natronobacterium* gen. nov. and *Natronococcus* gen. nov., two new genera of haloalkaliphilic archaebacteria. *Syst. Appl. Microbiol.* 5:41–57.

Trincone, A., B. Nicolaus, L. Lama, M. De Rosa, A. Gambacorta, and W.D. Grant. 1990. The glycolipid of *Halobacterium sodomense*. *J. Gen. Microbiol.* 136:2327–2331.

Trincone, A., E. Trivellone, B. Nicolaus, L. Lama, E. Pagnotta, W.D. Grant, and A. Gambacorta. 1993. The glycolipid of *Halobacterium trapanicum*. *Biochim. Biophys. Acta* 1210:35–40.

Upasani, V.N., S.G. Desai, N. Moldoveanu, and M. Kates. 1994. Lipids of extremely halophilic archaeobacteria from saline environments in India: a novel glycolipid in *Natronobacterium* strains. *Microbiology* 140:1959–1966.

Winker, S. and C.R. Woese. 1991. A definition of the domains Archaea, Bacteria, and Eucarya in terms of small subunit ribosomal RNA characteristics. *Syst. Appl. Microbiol.* 14:305–310.

3 Contribution of Molecular Techniques to the Study of Microbial Diversity in Hypersaline Environments

Francisco Rodríguez-Valera, Silvia G. Acinas, and Josefa Antón

CONTENTS

3.1 Introduction 27
3.2 The Multi-Pond Saltern 28
3.3 Fingerprinting of the Bacterial and Archaeal Communities
  Inhabiting the Different Salinity Domains 29
3.4 Phylotypes in the Crystallizer Ponds 31
3.5 What is Retrieved by Cultivation? 33
3.6 Conclusion 36
Acknowledgments 36
References 36

3.1 INTRODUCTION

The recovery by PCR of phylogenetically informative molecules (typically 16S rDNA) from natural samples has allowed new insights into prokaryotic biodiversity (for a recent review see Pace, 1997). One general conclusion is that traditional pure culture techniques retrieve only a small fraction of the actual diversity present in natural environments, including water and soil. Although the consequences in terms of opening new perspectives for future work in microbiology are extremely important, relatively little ecologically relevant information (or for that matter of any other type) has been obtained. To illuminate the "black box" that microorganisms represent in natural environments would be one of the three wishes that any microbial ecologist would demand of the genie in the lamp. Furthermore, the potential harvest of this hidden biodiversity, in terms of biotechnologically useful products and processes, could be enormous, to the point of dwarfing that of the tropical rain forest.

Most of the problem resides in the fact that the sequence of 16S rRNA gives very little information about the phenotype of Bacteria or Archaea unless a very high match with a cultivated well known organism is found. This is highly improbable, even for organisms retrieved by standard culture technique (Benlloch et al., 1996; Ward et al., 1990). Moreover, there is no rule to correlate different 16S rRNA sequences with whole genome (or phenotypic) distance, particularly when similarity values are higher than 95%. Therefore, it is essential to devise new methods to get environmentally (or otherwise) relevant microorganisms in pure culture, or find alternative ways to increase and exploit the knowledge of their existence. Working with extreme environments, which are characterized by low diversity, can provide useful models of how to approach the biodiversity problem in other "high diversity" environments. In fact some of the pioneering work of the so-called molecular microbial ecology has been carried out in extreme thermal environments (Reysenbach et al., 1994; Ward et al., 1990; Weller et al., 1991).

We would like to describe our efforts to apply molecular techniques to the study of hypersaline waters, particularly NaCl-saturated (crystallizer) ponds of a multi-pond solar saltern, but also the salinity gradient that develops in this kind of system. We have used two molecular tools, both based on PCR amplification of 16S rDNA from the biomass present in the water. In one, the product of the amplification is digested with one or two restriction enzymes (four cutters) and the fragments obtained are run in a sequencing polyacrylamide gel. This technique has been called ARDRA (Amplified Ribosomal DNA Restriction Analysis) and is useful in obtaining a fingerprint of the prokaryotic diversity present in one sample (Martínez-Murcia et al., 1995; Massol-Deya et al., 1995). Obviously, only the most abundant phylotypes will be detected by this technique, but it has the advantages of being fast and cheap, and can be used to compare up to 30 or 40 samples (environments) simultaneously.

The second methodology is the more classical approach and consists simply of cloning and sequencing individual 16S rDNA molecules amplified from the environment. It is much more refined and informative, but is obviously limited by the sequencing step. In a high-diversity environment hundreds, perhaps thousands, of clones have to be sequenced to get an exhaustive image of the diversity present. We have combined both tools to define biodiversity domains throughout the salinity gradient from seawater to NaCl saturation. We have checked the stability of their diversity through time (different seasons and years), and the predominant phylotypes present in NaCl-saturated waters have been established. Finally, the effect of some factors of growth medium and incubation for halophilic Archaea on phylotype recovery have been analyzed.

3.2 THE MULTI-POND SALTERN

The operation, characteristics, environmental factors and general biota present in a number of multi-pond salterns have been described (for reviews see Javor, 1989; Oren, 1993; Rodríguez-Valera, 1988, 1993). Specifically, the saltern we studied has been described previously and has been used as a subject of classical (culture isolation–phenotypic characterization) biodiversity studies since the middle 1970s (Rodríguez-Valera, 1986; Rodríguez-Valera et al., 1981, 1985).

The sequential changes in the physico-chemical conditions and phytoplankton groups along the salinity gradient have been repeatedly described, and we refer the readers to the abundant literature (Javor, 1983; Oren, 1990; Rodríguez-Valera et al., 1985).

Over the salinity spectrum, three or four characteristic and different biota were observed and have been described (Calderón-Paz, 1997). The 3.5%-10% salinity range was characterized by a strong influence of marine biota and gradual disappearance of large forms and/or marine organisms. These were replaced by more euryhaline types, although "bona fide" halophilic organisms are almost absent within this range, and microbial mats are often found.

The primary producers included cyanobacteria, green algae, and diatoms, and even marine phanerogames at the lower end of the range. The 10%-20% salinity range still has a relatively high microbial biodiversity. The primary producers that are often present in mats and are very diverse at the beginning of the range are gradually substituted by planktonic *Dunaliella* species. At the end of this range, gypsum (CaSO · 2H$_2$O) precipitation precludes the formation of complex mat communities. Halophilic organisms requiring high salt concentrations to grow optimally are found in high numbers throughout this range. The 20%-30% salinity range marks the shift toward highly specialized halophilic habitats and few non-halophilic or non-halotolerant organisms are found. Primary production is carried out mostly by planktonic *Dunaliella salina*, which can reach very high densities, but are often kept in check by populations of the brine shrimp *Artemia salina*. Halophilic Archaea, as well as halophilic Bacteria of different physiological groups, are often isolated throughout this range. Over 30% salinity is the range in which no primary productivity (at least based on chlorophyll *a* determinations) is detected. The dense (up to 10^8 cells ml^{-1}) halophilic archaeal populations found here with doubling times recently established to be somewhere between two and 50 days, probably live on the allochtonous organic matter carried over from lower salinities (Guixa-Boixareu et al., 1996).

3.3 FINGERPRINTING OF THE BACTERIAL AND ARCHAEAL COMMUNITIES INHABITING THE DIFFERENT SALINITY DOMAINS

Figure 3.1 shows the dendrograms comparing the ARDRA fingerprints of different saltern ponds at different sampling times for Bacteria and Archaea (data from Martínez-Murcia et al., 1995, and from Acinas et al., in preparation). These dendrograms represent the similarity level between the ARDRA patterns obtained from the studied systems (e.g., the different ponds in the multi-pond saltern). In the cases of both Bacteria and Archaea, the populations are grouped into salinity domains. For Bacteria (Figure 3.1A) there are two major clusters corresponding to very high salt (>30%) and medium to high salt (8%-23%) that are clearly different from each other. The medium- to high-salt cluster is itself formed by two groups of populations inhabiting lower- (preparation ponds, PR, <12%) and higher-salinity ponds (concentration ponds, marked below as CO, 12%-23%). It is also noteworthy that temporal variation is higher in low-salinity environments.

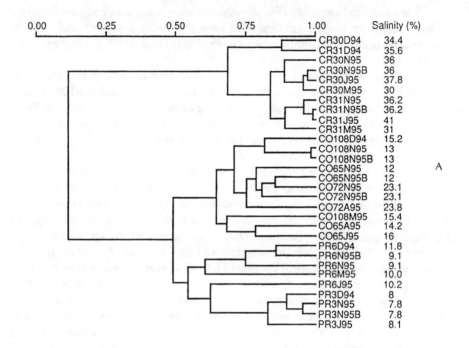

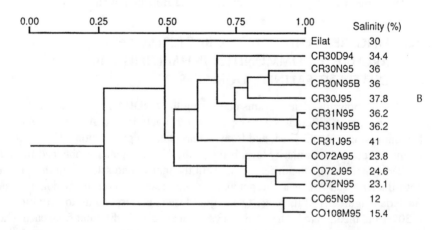

Archaeal communities (Figure 3.1B), as expected, are more diverse in very high-salt ponds. In fact, no amplification of archaeal 16S rDNA was obtained for the PR (8%-12% NaCl). For the CO (12%-15.4% NaCl) the number of bands in the ARDRA pattern was much lower than in the very high salinity (CR, >30% NaCl) ponds (data not shown). It seems reasonable to accept a correlation between the number of detected bands in a sample and its diversity (the more diverse a sample is, the more bands in the ARDRA pattern). Thus, our results show that the archaeal population is more diverse in very high-salinity ponds. Besides, as with Bacteria, the archaeal types that inhabit the salinity domains are markedly different. In fact, it is possible that the archaeal 16S rDNA amplified from ponds CO65 and CO108 (12% and 15.4% salinity, respectively) could belong to groups other than haloarchaea (e.g., methanogens). On the other hand, the population found in the crystallizers are relatively similar, with little temporal variation. The most significant variation was found for a sample of very high salinity and probably very high Mg^{2+} concentration (CR31J95), and for a single sample from a saltern in Eilat (Israel).

3.4 PHYLOTYPES IN THE CRYSTALLIZER PONDS

We have cloned and sequenced 16S rDNAs directly amplified from crystallizer water. In a first study (Benlloch et al., 1995) clones were retrieved by PCR from CR30. The sequence between positions 100–300 (*Halobacterium salinarum* numbering, a highly variable region in the cultivated species of haloarchaea) was determined. The first 12 sequenced clones were identical and different from any sequence of cultivated strains. Therefore, we stopped sequencing additional clones and concentrated on the analysis of this apparently predominant phylotype. The analysis of another variable region (positions 384–600) showed that six of the clones were identical and therefore can be confidently considered as representatives of the same species. Another two clones had nearly identical sequences but were slightly different (12 nucleotide substitutions) when compared with the predominant sequence; these could be representatives of a different, although closely related, species. Two clones, representatives of the two respective sequence classes, were fully sequenced, and showed 15 nucleotide differences, consistent with the two being closely related representatives of a new, yet

---

FIGURE 3.1 Similarity dendrograms of Bacteria (9A) and Archaea (9B) ARDRA. Bacterial and archaeal 16S rDNA fragments amplified from the salterns were digested with *Hinf*I and *Mbo*I. The products were separated by polyacrylamide gel electrophoresis. The patterns of each sample were compared, similarity coefficients were calculated, and a similarity dendrogram generated. Two preparation ponds (PR3 and PR6), three concentration ponds (CO65, CO72, and CO108) and two crystallizers (CR30 and CR31) were analyzed at different sampling times: December 94 (D94), May 95 (M95), June 95 (J95) and November 95 (N95). Thus CO72N95 refers to a sample from CO72 taken in November 95. "B" at the end indicates duplicate.

uncultivated, phylotype SPhT (designated "Susana's phylotype" after Susana Benlloch, probably with the taxonomic rank of genus, Figure 3.2). SPhT has a characteristic sequence for nucleotides 150–200, and signatures for every haloarchaeal genus can be found in this fragment. When, after two years, the experiment was repeated, six out of seven clones belonged to SPhT (Table 3.1). The seventh was very different and could correspond to yet another genus, but since the sequence has been retrieved only once, it could be an artifact.

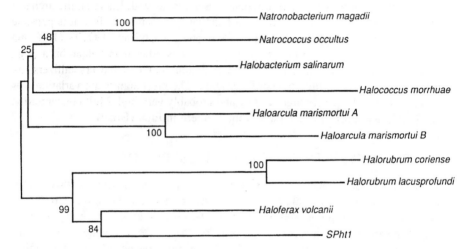

FIGURE 3.2 16S rRNA tree showing the phylogenetic relationships (neighbor-joining method) between SPhT and some cultured halobacteria.

The proposal of the three genera composing the non-alkaliphilic haloarchaea (*Halobacterium*, *Haloarcula* and *Haloferax*) was based mainly on data obtained from strains isolated from this saltern (*Haloferax mediterranei*, *Haloferax gibbonsii* and *Haloarcula hispanica*) and compared with culture collection strains. None of these genera are predominant here; if the PCR results represent the real diversity, they are rarities, and have been enriched by the culture media used. Furthermore, the sequencing of 15 clones from a sample of a saltern in Eilat gave the highest match (often high similarity values) to SPhT (Table 3.1). So the distribution of the organism represented by this sequence is remarkably global. Examples of other widely distributed phylotypes exist in the literature (Mullins et al., 1995).

The extreme saline environments that we have studied by the PCR technique show less 16S rDNA diversity than sea water or soil samples. However, as in high-biodiversity environments, high homologies to cultured organisms are very seldom found. The hypersaline waters of the crystallizer studied here show even less diversity by the direct rDNA amplification methodology than by culture isolation (Rodríguez-Valera et al., 1985). However, the organisms retrieved by one technique are markedly different from those obtained by the other, even when both techniques are applied to the same sample. This fits with the now widely accepted idea that culturing methods strongly bias the recovery of microorganisms that are often a minority in the environment.

TABLE 3.1
Clones and Isolates Recovered from Crystallizers

			% Similarity Best Match	Nucleotides
Solar Saltern (Alicante, 1996)				
Clone		HAC1 (SPhT)	*Haloferax volcani* (91%)	ALL
		HAC14 (SPhT)	*Haloferax volcani* (90.4%)	ALL
		CR30C1	*Halorubrum saccharovorum* (93%)	784
		CR30C2	SPhT (97%)	790
		CR30C3	SPhT (99.5%)	868
		CR30C4	SPhT (99%)	813
		CR30C5	SPhT (99%)	278
		CR31C1	SPhT (98%)	415
		CR31C2	SPhT (100%)	268
Solar Saltern (Eilat)				
Clone		C2	SPhT (93%)	218
		C3	SPhT (90%)	256
		C5	SPhT (96%)	312
		C10	SPhT (95%)	241
		C11	SPhT (99.5%)	177
		C15	SPhT (99%)	305
		C18	SPhT (96%)	300
		C19	SPhT (97%)	274
		C20	SPhT (95%)	267
		C26	SPhT (96%)	300
		C27	SPhT (96%)	234
		C28	SPhT (95%)	190
Isolates				
		B5[a]	*Halorubrum coriense* (96%)	264
		B19[b]	*Haloarcula marismortui* (99%)	324
		D3[b]	*Halorubrum trapanicum* (93%)	312
		D6[b]	*Halorubrum saccharovorum* (93%)	324
		E5[c]	*Halorubrum saccharovorum* (93%)	324
		E19[c]	*Haloarcula vallismortis* (94%)	139
		H4[d]	*Halorubrum coriense* (94%)	159
		H14[d]	*Halorubrum saccharovorum* (96%)	277

Culture conditions (see text): [a]yeast extract, high oxygen, [b]yeast extract, low oxygen, [c]fish meal, high oxygen, [d]fish meal, low oxygen.

3.5 WHAT IS RETRIEVED BY CULTIVATION?

What is wrong with the usual cultivation techniques that bias the retrieval of organisms to the ones that are not abundant in the environment? In a way, the answer is rather obvious. Only once in the history of microbiology, when pathogenic and saprophytic bacteria of humans were isolated, has a serious effort been made to

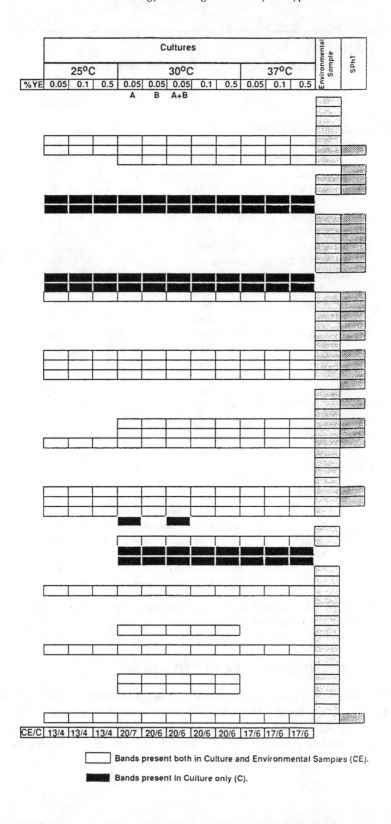

develop good culture media and cultivation techniques. In fact, the intestinal biota can be retrieved efficiently by cultivation as shown recently (Wilson and Blitchington, 1996). These organisms are generally copiotrophs and fast growers that compete poorly in natural (e.g. aquatic) habitats. Successful inhabitants of these will probably be specialized in growing slowly at the expense of low concentrations of dissolved organic matter or by direct hydrolysis of polymeric substances found in particulate matter. None would like the conditions found in the standard media used for organotrophs — e.g., peptones, yeast extract or chemically defined media. The high oxygen concentration on the surface of a Petri dish could also be inhibitory for many unknown microorganisms. On the other hand, liquid media favor fast growers and severely reduce the possibility of recovering organisms with low growth rates.

There are now a number of reports in which the organisms retrieved by culturing are compared with those obtained by direct PCR amplification of the 16S rDNA (Santegoeds et al., 1996; Torsvik et al., 1990; Wilson and Blitchington, 1996). DNA of four strains of haloarchaea isolated from the same sample on a standard haloarchaea synthetic medium was extracted and a part of the 16S rDNA sequenced. In our own work (Benlloch et al., 1995) all the strains isolated had the same sequence in the region 130–270 and showed the highest identity (97%) to *Haloarcula marismortui*, very probably belonging to a closely related species.

This experiment was repeated a couple of years later using different culture conditions: two complex carbon sources (yeast extract and fish meal) at a concentration of 1 g l^{-1} and two oxygen tensions obtained by inoculating plates with soft agar (high oxygen) or overlaying the soft agar with yet another layer of 3–4 mm of solid medium (low oxygen). Randomly picked colonies from the different plates were used for 16S rDNA sequencing. Most of the isolates (see Table 3.1; data from Antón et al., in preparation) bear high homology with cultured halobacteria, especially with the genus *Halorubrum*, as is shown by both the similarity values and the genus diagnostic sequence (Benlloch et al., 1995). Still, some of the isolates present very low homology with cultured microorganisms. In other words, what is retrieved by culture in terms of sequence is also diverse, but different from what is retrieved by PCR.

The question we addressed was what part of the population we are retrieving by cultivation. We first focused on plating techniques with fairly "standard" culture conditions (carbon source: yeast extract at 0.05, 0.1 and 0.5%, aerobic incubation at 25°C, 30°C and 37°C). To avoid bias caused by picking colonies that may look the same and yet be different, our approach was to collect all of the colonies, including microcolonies, that grew on a plate. 16S rDNA was amplified and subjected to ARDRA as explained above. In parallel, we analyzed the original sample (crystallizer water used as inoculum) that would represent the total population. Figure 3.3 shows the results of this study (data from Antón et al., in preparation). Bearing in mind the limitations of this technique, we can draw several conclusions from this figure. First, as was expected, there are bands in the cultures

---

FIGURE 3.3 Schematic representation (opposite) of an electrophoresis gel showing *Hinf*I/*Mbo*I double digestion or archaeal 16S rDNA amplified from colonies grown in different culture conditions. Each rectangle represents a band in the gel. A and B are different pools of colonies from the same plate. SPhT: digestion pattern of SPhT cloned 16S rDNA.

that are not present in the environmental sample. Again, this tells us that culturing over-represents minor components in the original population. Second, the carbon-source concentration does not seem to affect the nature of the cultured population, but temperature does. Interestingly, the temperature that gave the best yield (more bands in our case) is 30°C, not 37°C, which is the standard temperature for halobacterial cultivation. The third and last conclusion is that we are not able to culture SPhT using these techniques (none of the culture patterns included all the SPhT bands). This means that we did not miss it when picking colonies; we actually did not culture it.

3.6 CONCLUSION

In 1985, we proposed the classification of non-alkaliphilic halophilic Archaea in four genera: *Halobacterium, Haloarcula, Haloferax* and *Halococcus* (Rodríguez-Valera et al., 1986). Since then, at least two new genera have been added. But what if there are 100 genera of haloarchaea or, worse still, what if there is a gradient of diversity in which the sharp discontinuities that allow taxonomists to do their work cannot be found? And if this happens in saltern ponds, an extreme environment, imagine the situation in soil or sea water. That would explain why, even by standard culture techniques, isolates retrieved randomly in non-selective media rarely show enough similarity to described species (at least at the level of 16S rRNA sequence) to be classified into one of them. Generally speaking, it does not seem to be the case with pathogenic bacteria. But it is a possible scenario that could be described as the search for the "needle in the haystack." If this is the situation in the real world, many of our preconceptions about how to classify and study microorganisms will have to be reconsidered.

ACKNOWLEDGMENTS

While writing this review, work of the authors was supported by grants ENV4-CT96-0218 (European Commission) and PM95-0111-CO2-01 and UE96-0032 (Spanish Ministry of Education and Culture). Kathy Hernández and Stuart Ingham provided secretarial assistance and graphics. We are grateful to Miguel Cuervo Arango, owner of the saltern ponds, for his kind help.

REFERENCES

Benlloch S., A.J. Martínez-Murcia, and F. Rodríguez-Valera. 1995. Sequencing of bacterial and archaeal 16S rRNA genes directly amplified from a hypersaline environment. *Syst. Appl. Microbiol.* 18:574–581.

Benlloch, S., F. Rodríguez-Valera, S.G. Acinas, and A.J. Martínez-Murcia. 1996. Heterotrophic bacteria, activity and bacterial diversity in two coastal lagoons as detected by culture and 16S rRNA genes PCR amplification and partial sequencing. *Hydrobiologia* 329:3–17.

Calderón-Paz, J.I. 1997. *Ecology of Heterotrophic Bacteria in Planktonic Ecosystems.* Ph.D. thesis, Universitat de Barcelona.
Guixa-Boixareu, N., J.I. Calderón-Paz, M. Heldal, G. Bratbak, and C. Pedrós-Alió. 1996. Viral lysis and bacterivory as prokaryotic loss factors along a salinity gradient. *Aquat. Microb. Ecol.* 11:215–227.
Javor, B.J. 1983. Nutrients and ecology of the Western Salt and Exportadora de Sal saltern brines, in *Sixth International Symposium on Salt.* Vol. 1. The Salt Institute, Alexandria, VA. 195–205.
Javor, B. 1989. *Hypersaline Environments. Microbiology and Biogeochemistry.* Springer-Verlag, Berlin.
Martínez-Murcia, A.J., S.G. Acinas, and F. Rodríguez-Valera. 1995. Evaluation of prokaryotic diversity by restrictase digestion of 16S rDNA directly amplified from hypersaline environments. *FEMS Microbiol. Ecol.* 17:247–256.
Massol-Deya, A.A., D.A. Odelson, R.F. Hickey, and J.M. Tiedje. 1995. Bacterial community fingerprinting of amplified 16S and 16-23S ribosomal DNA gene sequences and restriction endonuclease analysis (ARDRA), in *Molecular Microbial Ecology Manual 3.3.2,* D.L. Akkermans, J.D. van Elsas, and F.J. de Bruijn, Eds. Kluwer Academic, Dordrecht. 1–8.
Mullins, T.D., T.B. Britschgi, R.B. Krest, and S.J. Giovannoni. 1995. Genetic comparisons reveal the same unknown bacterial lineages in Atlantic and Pacific bacterioplankton communities. *Limnol. Oceanogr.* 40:148–158.
Oren, A. 1990. Estimation of the contribution of halobacteria to the bacterial biomass and activity in solar salterns by the use of bile salts. *FEMS Microbiol. Ecol.* 73:41–48.
Oren, A. 1993. Ecology of extremely halophilic microorganisms, in *The Biology of Halophilic Bacteria,* R.H. Vreeland and L.I. Hochstein, Eds. CRC, Boca Raton. 25–53.
Pace, N.R. 1997. A molecular view of microbial diversity and the biosphere. *Science* 276:734–740.
Reysenbach, A.L., G.S. Wickham, and N.R. Pace. 1994. Phylogenetic analysis of the hyperthermophilic pink filament community in Octopus Spring, Yellowstone National Park. *Appl. Environ. Microbiol.* 60:2113–2119.
Rodríguez-Valera, F. 1986. The ecology and taxonomy of aerobic chemoorganotrophic halophilic eubacteria. *FEMS Microbiol. Rev.* 39:17–22.
Rodríguez-Valera, F., Ed.1988. Characteristics and microbial ecology of hypersaline environments, in *Halophilic Bacteria,* Vol. I. CRC, Boca Raton. 3–30.
Rodríguez-Valera, F. 1993. Introduction to saline environments, in *The Biology of Halophilic Bacteria,* R.H. Vreeland and L.I. Hochstein, Eds. CRC, Boca Raton. 1–23.
Rodríguez-Valera, F., A. Ventosa, G. Juez, and J.F. Imhoff. 1985. Variation of environmental features and microbial populations with salt concentrations in a multi-pond saltern. *Microb. Ecol.* 11:107–115.
Rodríguez-Valera, F., F. Ruiz-Berraquero, and A. Ramos-Cormenzana. 1981. Characteristics of the heterotrophic bacterial populations in hypersaline environments of different salt concentrations. *Microb. Ecol.* 7:235–243.
Rodríguez-Valera, F., M. Torreblanca, G. Juez, A. Ventosa, M. Kamekura, and M. Kates. 1986. A proposal for the classification of non-haloalkaliphilic halobacteria into three genera, in *Archaebacteria 85,* O. Kandler and W. Zillig, Eds. Gustav-Fisher-Verlag, Stuttgart. 411–412.
Santegoeds, C.M., S.C. Nold, and D.M. Ward. 1996. Denaturing gradient gel electrophoresis used to monitor the enrichment culture of aerobic chemoorganotrophic bacteria from a hot spring cyanobacterial mat. *Appl. Environ. Microbiol.* 62:3922–3928.

Torsvik, V., S. Käre, R. Sørheim, and J. Goksøyr. 1990. Comparison of phenotypic diversity and DNA heterogeneity in a population of soil bacteria. *Appl. Environ. Microbiol.* 56:776–781.

Ward, D.M., R. Weller, and M.M. Bateson. 1990. 16S rRNA sequences reveal numerous uncultured microorganisms in a natural community. *Nature* 345:63–65.

Weller, R., J.W. Weller, and D.M. Ward. 1991. 16S rRNA sequences of uncultivated hot spring cyanobacterial mat inhabitants retrieved as randomly primed cDNA. *Appl. Environ. Microbiol.* 57:1146–1151.

Wilson, K.H. and R.B. Blitchington. 1996. Human colonic biota studied by ribosomal DNA sequence analysis. *Appl. Environ. Microbiol.* 62: 2273–2278.

4 The Microbial Ecology of Solar Salt Plants

Carol D. Litchfield, Amy Irby, and Russell H. Vreeland

CONTENTS

4.1 Introduction: the Discovery of Microorganisms in Solar Salterns 39
4.2 Review of Published Microbial Ecological Studies of Solar Salt Plants 41
    4.2.1 Solar Salterns in Spain 41
    4.2.2 Western Hemisphere: Mexico and Southern California 43
4.3 Microbial Ecology of the Bonaire, Netherlands Antilles Saltern 44
4.4 Preliminary Studies on the Microbial Ecology of the
    Cargill Solar Salt Plant, Newark, CA 47
4.5 Conclusion 50
Acknowledgments 51
References 51

4.1 INTRODUCTION: THE DISCOVERY OF MICROORGANISMS IN SOLAR SALTERNS

Sodium chloride, a necessity for the chemical, food, and deicing industries, historically has been obtained by three methods, depending on the location of the salt producers: 1) the boiling of brine from springs at inland locations, 2) mining or quarrying of rock salt (as was common in India in pre-Roman times [Pliny, trans. 1963] and in Austria in prehistoric times [Gouletquer, 1974]), and is still practiced today; and 3) solar evaporation, especially along the sea coasts or at inland hypersaline lakes. Undoubtedly the first solar salt production and collection came from sea- or brine-spring sprays that dried on nearby rocks or twigs; this later gave way either to boiling the brine or using single pans for solar evaporation of the water. These operations probably provided sufficient salt for a single family or small tribe or clan.

    As commerce developed, it was recognized, possibly first by the Chinese (Forbes, 1968), that multiple pans containing increasing concentrations in the salt content resulted in purer salt. This salt was whiter and sweeter, and we now know that the improvement in taste resulted from the precipitation in the concentrators of the gypsum, iron, and other inorganic impurities prior to the precipitation of sodium

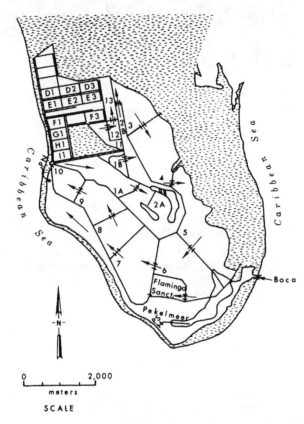

FIGURE 4.1 Plan of the Bonaire, Netherlands Antilles solar salt plant showing the directions for water flow. The concentrators have Arabic numbers and the crystallizers are designated by a letter and number. All samples were obtained at the outlet gates for the pans.

chloride. A typical solar salt plant (also called a saltern or saline), with its increasing salt concentrations, is shown in Figure 4.1. This type of system uses seawater and is called a thalassohaline system, while inland brine springs frequently have a different chemical composition from seawater and are athalassohaline waters.

Because microorganisms had not been discovered, none of the early salt-production methods reported their presence. The recognition of bacteria did not occur until 1683 with van Leeuwenhoek's description in his letter to the Royal Society (1684). By the latter part of the 19th century, however, microorganisms were generally recognized as playing important roles in both health and the environment. During the voyage of the *Beagle* in 1826, Darwin described the production of solar salt in an inland lake 23 km from El Carmen, Patagonia (Darwin, 1901). Although unfamiliar with bacteria, Darwin recognized that there was a "putrefying life-form" in the mud; that algae caused a green color on the lake; that flamingoes fed on the brine shrimp and worms; and that other "infusorial animalcula" were part of the salt-producing system.

However, his notations made little impact on the general impression that salterns were essentially sterile environments. Many visual observations of solar evaporation pans showed that fish could not live in them, and therefore it was assumed that nothing could survive in solar salterns. This view persisted despite mounting evidence that there were microorganisms in the solar salt pans. Payen in the mid-19th century had shown that *Artemia* did indeed survive, grow, and reproduce in brines (Payen, 1837). Even with the discovery of *Dunaliella salina* (Teodoresco, 1905) there was still considerable questioning about the existence of other microscopic life in salterns. This dispute was finally settled after Pierce (1914), Kellerman and Smith (1916), Brown (1922), and Clayton and Gibbs (1927) all successfully cultured bacteria from brines as well as the solar-produced salt. These investigators performed the first microbial ecology studies of solar salt works by simply showing that bacteria did indeed live in these "sterile" environments.

Since then, there has been a continuing interest in the halophilic bacteria, their metabolism, structure, taxonomy, and survival mechanisms. However, there have been comparatively few studies on the ecology of the microorganisms in solar salterns. This chapter will review some of these studies, present some current work, and describe several factors affecting the microbial population that seem to be common regardless of the geography of the solar salt plant.

4.2 REVIEW OF PUBLISHED MICROBIAL ECOLOGICAL STUDIES OF SOLAR SALT PLANTS

Although solar salterns have been in existence since prehistoric times, the microbial ecology of these operations has never been extensively studied, probably because it has been assumed by the general microbiological community that these hypersaline environments are highly selective and not much of microbiological interest is happening in them. The few investigators who have examined solar salterns would disagree. This section will discuss the studies conducted at salines in Spain, the Exportadora de Sal saline in Mexico, and the Western Salt Company in southern California.

4.2.1 SOLAR SALTERNS IN SPAIN

Solar salt is produced in Spain not only by the evaporation of seawater along the coast but also by the evaporation of athalassohaline waters from natural inland brine springs. These springs typically contain about 15–20% total salt solutions and normally have higher levels of magnesium, calcium, and potassium than are found in seawater (Ramos-Cormenzana, 1993). Bacterial enumerations from the saltern of La Malá near Granada, Spain were reported to be dependent on the salt content of the isolation medium (Del Moral et al., 1987). Six sites were sampled and the waters were plated on media with the same composition but at four different salt concentrations (3, 10, 18, and 25% w/v total salt). Because the source water was 18%, the salt concentrations in each pan were relatively constant. Thus, this was not a graduated solar salt plant as seen on the coast. However, even with this higher salt concentration, both eubacteria and halophilic archaea were found in all of the pans regardless of the salinity of the medium (Ramos-Cormenzana, 1993).

Meanwhile, a coastal saltern 22 km south of Alicante, Spain has been the subject of several studies by Rodriguez-Valera and coworkers. These investigators scored the colony growths according to the pigmentation of the cells and used this as an indicator of Bacteria vs. Archaea. They found nonpigmented organisms were more common in ponds containing up to 6–14% salt, while the red-pigmented colonies appeared at about 18% salt, peaked at about 27–30% NaCl and were barely present at 34.3% NaCl. In all cases, the total colony-forming units (TCFU) were substantially lower at the higher salt concentrations. The nonpigmented strains grew in 2–10% salt while the pigmented strains required 20% salt or higher at 38°C (Rodriguez-Valera et al., 1981). They examined 384 nonred-pigmented colonies and found they were generally halotolerant and the predominant genera were *Vibrio, Flavobacterium, Alcaligenes, Alteromonas*, and *Chromobacterium* (Ventosa et al., 1982).

In a later paper on the Alicante saltern, Rodriguez-Valera and co-workers reported the physical/chemical characteristics of various pans where microbiological samples had been taken. The data (Table 4.1) show higher concentrations of nitrogen and phosphate at the lower salt concentrations than had been reported by Javor (1983) (see below). However, the numbers of culturable bacteria do not appear to be much higher (Rodriguez-Valera et al., 1981).

TABLE 4.1
Inorganic Nutrient Concentrations at Selected Solar Salterns

	Percent NaCl in the Saltern Pans							
Location	5-6%	12-13%	15-17%	20-21%	22-25%	27-30%	32-34%	Source
Ammonia (mg l^{-1})								
ESSA	<0.2	<0.2	<0.2	<0.2	<0.2	0.2	1.0	Javor, 1995
WSC	0.2	0	0	0	0.4	<0.2	0.5	Javor, 1995
Alicante[a]	<1	3	3	6.5	—[b]	6	8	Rodriguez-Valera et al., 1985
Bonaire	0.45	0.3	0.2	0.45	0.26	—	0.67	This paper
Nitrate (mg l^{-1})								
ESSA	0	0	0	0	1.1	2.2	3.8	Javor, 1985
WSC	4.2	0	0	0	0	1.98	2.09	Javor, 1985
Alicante	—	—	—	—	—	—	—	
Bonaire	0.19	0.24	0.05	0.18	0.16	—	0.54	This paper
Phosphate (mg l^{-1})								
ESSA	0	0	0	0	0.16	0.4	1.59	Javor, 1985
WSC	40.71	0	0	0	0	0.32	0.32	Javor, 1985
Alicante	0.75	0.1	0.12	0.12	—	0.12	0.25	Rodriguez-Valera et al., 1985
Bonaire	0.02	0.04	0.04	0.06	0.04	—	0.05	This paper

[a] Kjeldahl nitrogen values were used for Alicante ammonia concentrations.
[b] Values not reported or not tested.

A frequent question concerns the source of the red Archaea in these solar salt plants. Are the red bacteria indigenous, induced or mutated halotolerant strains, or introduced from some unusual source? To answer this question, Rodriguez-Valera et al. (1979) isolated several strains of Archaea from seawater, but all belonged to the genus *Halococcus*. They did not find any pleomorphic, Gram-negative rods that they could identify as Archaea, though, so the question is still only partially answered.

Another saltern in Spain was investigated by Marquez et al. (1987). This saltern was on the Atlantic Ocean instead of the Mediterranean Sea, and the authors compared the types of isolates obtained from Huelva saltern with those from Alicante. Their isolates were grouped according to the salt requirements and included 154 out of the total 564 that could be classified as Gram-positive isolates. The Alicante studies had also shown a large proportion of nonhalophilic Gram-positive organisms (Rodriguez-Valera et al., 1981). Vibrios were again the predominant type in the Huelva saltern, but at about one-half the TCFU found in seawater, indicating some degree of selectivity in this saltern system. They also identified 145 halophilic Archaea with *Halobacterium salinarum* composing 59% of the archaeal isolates (Marquez et al., 1987). Unfortunately, since the authors did not report any of the physical/chemical characteristics of this saltern, the microbial numbers and the nutrient levels cannot be compared to see what effect these might have had on the bacterial distributions.

4.2.2 Western Hemisphere: Mexico and Southern California

A major salt-producing saline in Mexico, the Exportadora de Sal (ESSA), has been the subject of several microbial ecological studies by Javor. After measuring the major ions in the saline, Javor correlated the changing ionic compositions with the various types of microorganisms detected in the liquid phase at two different salines, ESSA and the Western Salt Company near San Diego, CA. She noted that the Western Salt saline had unidentified bacteria and various algae starting at 6–13 °Bé (4–14% NaCl), while the Exportadora de Sal waters were dominated by algae, cyanobacteria, brine shrimp, and insect larvae until about 20–30 °Bé (22–35% NaCl). In the latter case of the ESSA, however, the TCFU were much less dense. At the higher salt concentrations, the halophilic bacteria became the exclusive microorganisms in the ESSA saline, but Western Salt continued to have both the halophilic bacteria and *Dunaliella* in the high brine waters (Javor 1983, 1985).

More importantly, Javor also measured the inorganic nutrient concentrations of these waters and found that the Western Salt brines had higher levels of nitrate, reactive phosphate, and ammonium ions (Table 4.1), which she correlated with the concentration of biomass each saltern supported. In general there was little fluctuation in the concentration of the phosphate and nitrate, which were uniformly low, while the ammonia concentrations showed some variability; the concentrations of all three nutrients increased dramatically in the more saturated brines at ESSA (Javor, 1985).

4.3 MICROBIAL ECOLOGY OF THE BONAIRE, NETHERLANDS ANTILLES SALTERN

Over a four-year period, an extensive seasonal investigation was conducted of the factors affecting the distribution of microorganisms in a solar salt facility located in Bonaire, Netherlands Antilles (Figure 4.1) (Litchfield, 1977; Vreeland, 1976). Samples were collected from the inlet, through the various concentrators, and the final crystallizers. Besides plate counts of the microorganisms that would grow on different nutrient media prepared with different amounts of solar salt (3.5, 10, and 25%), nutrient chemistries were also determined: urea, ammonia, phosphate, nitrate/nitrite, and oxygen along with the temperature, pH, and density of the pans. During the latter two years of the study, the numbers of culturable photosynthetic organisms were also determined.

The results of these cultivations are shown in Figure 4.2, where the averages for the plate counts for the three major ecological zones have been plotted. Both total heterotrophic colony-forming units (TCFU) and the red-pigmented colonies (TPCFU) are shown. The ESWA/10 medium is the reduced nutrient medium of Litchfield et al. (1974), and HSC is the high-salt casein medium of Vreeland et al. (1980). All samples were surface-spread plated onto the media in quintuplicate and incubated for up to 21 days at 30°C. Dilutions were made in either artificial seawater, 10% solar salt, or 25% solar salt, as appropriate.

In general, the TCFU on 1/10 ESWA ranged from 10^5 to 10^7 per liter. There were no discernible seasonal fluctuations. The numbers of TPCFU followed the same pattern as the TCFU. On the ESWA medium containing 10% solar salt, there was inconsistent growth until concentrator 4, where the density ranged between 1.039-1.045 (5-7% salt). The maximum numbers to grow on this medium were found in concentrators 8 to 10 (13–18% salt) with decreases as the waters approached saturation with respect to NaCl. This resulted in a bell-shaped curve for the CFUs on this medium (data not shown). Growth on the HSC medium was always less, especially in the lower concentrators. However, red-pigmented organisms were consistently found only from concentrator 8 onward and were the dominant type of organism in the crystallizers. The high increase in concentrators 6, 7, 9, and 10 during the August sampling was likely due to nutrient enrichment that resulted from flooding the flamingo sanctuary located in the center of concentrator 6.

Because there was no apparent seasonal influence on the bacterial distribution, the nutrient data were examined to determine if there were any seasonal effects on their concentrations. The product moment correlation coefficients for the data pairs were calculated (Barr and Goodnight, 1972). The chemical and culture data were analyzed in three ways. First, all of the samples were analyzed in a single matrix to determine if one factor might be common to all of the samples; second, the entire facility was divided into five arbitrary zones based on density; and third, each individual concentrator was analyzed to see if different factors impacted each concentrator.

The only overall significant correlation within the entire saltern involved nitrogen and ammonia (0.992, $p = 0.0001$) indicating that ammonia is a major contributor to the nitrogen in the system. Statistical analyses of the combined microbiological and chemical data resulted in the recognition of three distinct ecological zones. The first

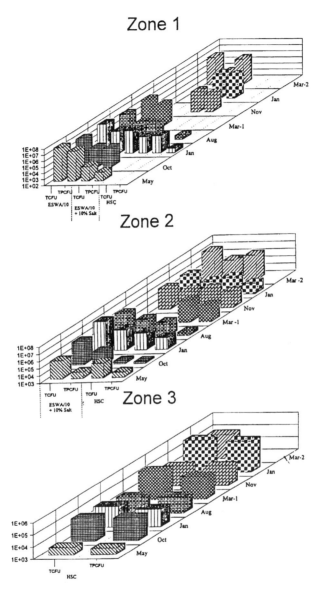

FIGURE 4.2 Averages for the plate counts for the three major ecological zones at the Bonaire solar salt plant. Each sample was plated in quintuplicate and the CFU for that sample averaged. All samples within the density range of the individual zones have been totaled and averaged for this figure.

is composed of pans containing salt densities from 1.026 (inlet) (3.5% salt) through 1.045 (7% or concentrator 9). This is the most diverse zone, not only in terms of bacteria, but also higher organisms. Bacterial fluxes here correlated with the amount of nitrogen (ammonia and urea) at $p = 0.0005$, and there was no correlation with phosphate, which was frequently not detectable in these waters.

The second zone consisted of density areas from 1.045 through 1.190 (7–25% salt) — all of the remaining concentrators until the crystallizers. The dominant organisms were those capable of growth on 10% solar salts media with bacteria able to grow on 25% salt media also routinely encountered, but they were not the dominant forms even in the higher salt concentrations. At the lower end of this zone, nitrite levels could be correlated (at $p = 0.001$) with the total colony-forming units, and as the salt concentration increased, this became the determining factor in the numbers of culturable bacteria.

The final ecological zone consisted of the crystallizers. There, red and green photosynthetic organisms were occasionally isolated, but the entire community was dominated by the red halophilic and nonpigmented halotolerant bacteria. Both oxygen and phosphate were the most limiting nutrients, being frequently nondetectable, while nitrogen levels remained fairly constant. Despite seasonal temperature and rainfall changes, neither of these was a factor in the microbial ecology, except for the one year with unusual rainfall that resulted in crystallizer densities falling below 1.2.

In all of these studies (Spain, Mexico, United States), nitrogen appears to be highly correlated with the bacterial populations throughout the saltern. Also, oxygen and phosphate are the limiting factors since these are frequently nondetectable.

A model food web for this facility has been developed and is shown in Figure 4.3. All components listed in the figure are active in the first ecological zone (inlet to 7% NaCl); organisms listed below line A are active in the second ecological zone (approximately 8–25% NaCl); while only those listed below line B are active in the third zone (crystallizers) (Vreeland, 1976).

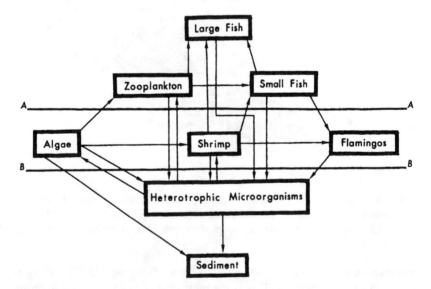

FIGURE 4.3 Theoretical food web for the three ecological zones at the Bonaire salt plant. The entire figure is active in Zone 1; those below line A are active in Zone 2; Zone 3 includes those organisms listed below line B.

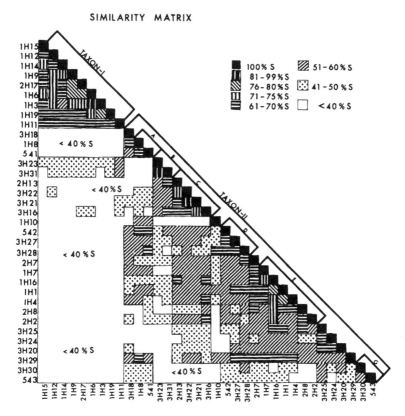

FIGURE 4.4 Similarity matrix for 35 halophilic bacteria. Taxon I contains the genus *Halomonas*, while the Archaea are grouped under Taxon II. Strains 541, 542, and 543 are *Halobacterium cutirubrum* (*salinarum*), *Halobacterium salinarum*, and *Halococcus morrhuae*, respectively.

Forty-seven bacterial strains were isolated, purified, and tested using standard microbiological procedures. The resulting data were analyzed using numerical taxonomy, with the results shown in Figure 4.4 (Colwell et al., 1979). By single-linkage Jaccard analysis, two major taxa were found. The first comprises the isolates later identified as the genus *Halomonas* (Vreeland et al., 1980). The second major taxon was actually composed of six subsets that included known isolates of the halophilic Archaea. From this study, the interrelationship of salt and temperature on the growth ranges of the genus *Halobacterium* was also noted (Colwell et al., 1979).

4.4 PRELIMINARY STUDIES ON THE MICROBIAL ECOLOGY OF THE CARGILL SOLAR SALT PLANT, NEWARK, CA

Current studies at the Cargill Solar Salt Plant in Newark, CA are designed to determine the similarity of the microbial flora of a more northerly saline to the

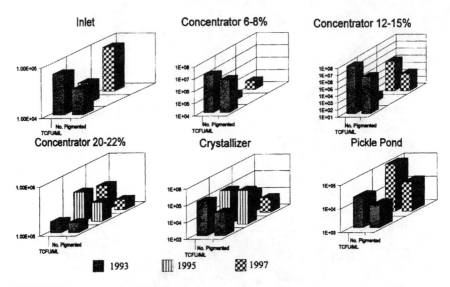

FIGURE 4.5 Averages of the plate counts on MCA medium taken over a three-year period at the Cargill Solar Salt Plant, Newark, CA. Each brine sample was plated in duplicate, and the plates incubated at room temperature, which approximates the *in situ* temperature.

microbial populations in salterns in southern California, the Caribbean, Mediterranean, and Israel. Three sets of samples have been obtained to date, and the resulting plate counts are shown in Figures 4.5 and 4.6. The Modified Casamino Acid (MCA) medium is similar to the HSC used in Bonaire, so the numbers of bacteria should be comparable. Indeed, except for the direct effects of rainfall on the Cargill TCFU (Figure 4.5), the microbial counts are similar to those from Bonaire (Figure 4.3).

A comparison of the results plotted in Figures 4.5 and 4.6 show that the culture medium makes a significant difference in the numbers of viable culturable bacteria recovered. The MR2A medium, which contains low levels of pyruvate, starch, glucose, and proteose peptone and was modified by supplemental magnesium sulfate, supported either about the same or an order of magnitude higher numbers than the low yeast extract MCA, even in the crystallizers. The numbers of pigmented colonies were also higher on the MR2A medium, indicating that the bacteria cultured on this medium have a broader range of metabolic capabilities than is normally considered.

Another factor to be taken into account in evaluating these data is that the northern California area had been under drought conditions for several years, and the drought broke following sample collection at the end of 1993. This resulted in a lowering of the total salt concentrations in the pans, which is reflected in the decreases in numbers during 1995 and February 1997. For example, viable bacterial numbers decreased from $7*10^7$ ml^{-1} in 1993 in the concentrator containing 12–15% salt to <100 ml^{-1} in the same concentrator in 1995. In general, the TCFU ml^{-1} are comparable to the standing crops of bacteria noted for the Spanish salines.

In an effort to gain a better understanding of the nutritional needs and metabolic diversity of the whole microbial community, samples were placed in BIOLOG™

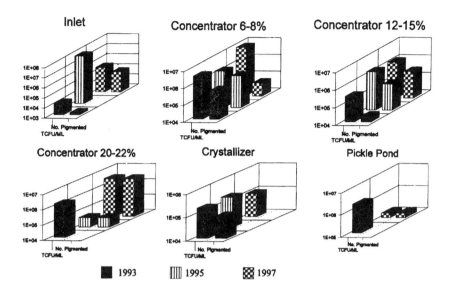

FIGURE 4.6 Averages of the plate counts on MR2A medium taken over a three-year period at the Cargill plant. Each brine sample was plated in duplicate, and the plates incubated at room temperature, which approximates the *in situ* temperature.

GN plates. These plates contain 95 separate substrates. Incubation of the water samples in the plates resulted in metabolic patterns that demonstrated that 69% of the samples could use those compounds listed as positive in Table 4.2. In 69–88% of the samples, the microbial community was unable to use those compounds listed as negative in Table 4.2. Interestingly, glycerol, found by Oren (1993) to be so important to the Dead Sea bacteria, is not generally used by the populations in this saltern; amino acids appear to be the most common class of organic nutrients generally consumed by the microbial community.

TABLE 4.2
Summary of the Carbon Compounds Most Used (>67%) and Those Generally Not Consumed (>67%) by the Whole Community in the Cargill Solar Salt Plant

Generally used carbon sources		Generally not used carbon sources	
α-cyclodextrin	D-glucuronic acid	β-methyl-D-glucose	succinamic acid
dextrin	propionic acid	D-psicose	L-leucine
glycogen	quinic acid	itaconic acid	D,L-carnitine
N-acetylglucosamine	L-alanine	α-ketovaleric acid	phenethylamine
D-fructose	L-asparagine	α-ketobutyric acid	2-aminoethanol
D-glucose	L-glutamic acid	glucuronamide	D,L-α-glycerolphosphate
sucrose	L-phenylalanine	alaninamide	p-hydroxyphenylacetic acid
citric acid	L-pyroglutamic acid		

4.5 CONCLUSION

As more data on the complexity of solar salterns are obtained, a few patterns begin to emerge. Although salt is made in salt lakes, there are distinct differences in the microbial ecology between solar salterns and hypersaline lake systems. Temperature and rainfall do not appear to be deciding factors in the distributions of microorganisms in salines, but are important in salt lakes such as the Dead Sea (Oren, 1988; Oren and Gurevich, 1993) and the Great Salt Lake (Post, 1977; Rushforth and Felix, 1982). Nutrient chemistries, especially the lack of phosphate, are more significant determinators of microbial populations for salines, where the gradual increase in salinity results in the precipitation of gypsum, and the bacteria and algae appear to compete for the scarce phosphate ions, especially in the higher-density concentrators and crystallizers.

Bacteria are found throughout the saltern system, while they seem to play a rather insignificant role in salt lakes. Conversely, halophilic Archaea do not appear to be limited to just 12–15% salt concentrations, but have been recovered from inlet and 4–6% salinity waters in Bonaire, at Cargill (data not presented), and in the Mediterranean Sea (Rodriguez-Valera et al., 1979).

As the seasonality in salterns is investigated, it becomes obvious that the salt concentrations are not constant in the different pans. There may be a 5–8% variation (Bonaire data, not shown, and Javor, 1983) so it should not be surprising that many of the bacteria demonstrate broad salinity tolerance. This adaptability may account for the widespread distributions of both halophilic Archaea and halotolerant Bacteria throughout the salt plant. This would explain the greater degree of microbial diversity in solar salterns than has been generally expected.

These studies have barely begun to clarify the complexities of solar salterns. Future studies will include the application of molecular techniques such as 16S rRNA and DNA fingerprinting to evaluate those microbes that have not been cultured. A few preliminary attempts have been initiated in this area (Martinez-Murcia et al., 1995), but these have not been long-term studies that could show fluxes in the microbial populations. Other approaches include the examination of the whole community lipid patterns and correlation of these data with predominant culture isolations (Oren and Gurevich, 1993). However, polar lipid patterns and pigments of bacteria are modified during cultivation, so it is difficult to make such direct comparisons. However, examination of whole community patterns do show changes with time, and this may reflect shifts in the types of dominant microbes (Litchfield et al., unpublished results).

In addition, examination of the importance of bacteriophages and halocins to control of the populations will also be important in understanding not only distributions but concentrations of bacteria in different-density ponds. Finally, further work on the nutrient chemistries and the identity of community properties, as well as the individual microbial isolates, will help us to understand the microbial population fluxes in these very dynamic systems.

ACKNOWLEDGMENTS

The authors thank the personnel at the Antilles Salt Company, Bonaire, Netherlands Antilles. L.A. Kiefer performed the phosphate analyses and assisted in the Bonaire sampling. Assistance with the Cargill cultures has been provided by F. Al-Hothali, E. Fulnecky, L. Kristianson, K. Fitzgerald, M. Bannasch, S. Vinayak, and V. Chandhoke. We also thank the United States–Israel Binational Science Foundation (BSF, Jerusalem) for supporting portions of this work under grant no. 95-00027, and the International Salt Company for supporting the Bonaire studies.

REFERENCES

Barr, A.J. and J.H. Goodnight. 1972. *Statistical Analyses System*. North Carolina State University, Raleigh.

Browne, W.W. 1922. Halophilic bacteria. *Proc. Exper. Biol. Med.* 19:321–322.

Clayton, W. and W.E. Gibbs. 1927. Examination for halophilic micro-organisms. *Analyst 52*: 395–397.

Colwell, R.R, C.D. Litchfield, R.H. Vreeland, L.A. Kiefer, and N.E. Gibbons. 1979. Taxonomic studies of red halophilic bacteria. *Int. J. Syst. Bacteriol.* 29:379–399.

Darwin, C. 1901. Journal of Researches, in *A Library of Universal Literature in Four Parts, Part One - Science*, P.F. Collier and Son, New York. 81–83.

Del Moral, A., E. Quesada, and A. Ramos-Cormenzana. 1987. Distribution and types of bacteria isolated from an inland saltern. *Ann. Inst. Pasteur/Microbiol.* 138:59–66.

Forbes, R.J. 1968. Zoutzieden door de tijden, in *Het Zout der Aarde*, R.J. Forbes, Ed. NV Koninklijke Nederlandsche Zoutindustrie, Hengelo. 191–220.

Gouletquer, P.L. 1974. The development of salt making in prehistoric Europe. *Essex J.* 8:2–14.

Javor, B.J. 1983. Planktonic standing crop and nutrients in a saltern ecosystem. *Limnol. Oceanogr.* 28:153–159.

Javor, B.J. 1985. Nutrients and ecology of the Western Salt and Exportadora de Sal Saltern brines, in *Sixth International Symposium on Salt*, Vol. 1, B.C. Schreiber and H.L. Harner, Eds. The Salt Institute, Alexandria, VA. 195–205.

Kellerman, K.F. and N.R. Smith. 1916. Halophilic and lime precipitating bacteria. *Centralbl. Bakteriol.*, II Abt. 45:371–372.

Litchfield, C.D. 1977. *Final Report on a Study of the Factors Controlling the Distribution of Halophilic Bacteria in The Antilles International Salt Co, N. N. Facility on Bonaire, Netherlands Antilles*. 12 pp.

Litchfield, C.D., J.B. Rake, J. Zindulis, R.T. Watanabe, and D.J. Stein. 1974. Optimization of procedures for the recovery of heterotrophic bacteria from marine sediments. *Microb. Ecol.* 1:219–233.

Marquez, M.C., A. Ventosa, and F. Ruiz-Berraquero. 1987. A taxonomic study of heterotrophic halophilic and nonhalophilic bacteria from a solar saltern. *J. Gen. Microbiol.* 133:45–56.

Martínez-Murcia, A.J., S.G. Acinas, and F. Rodriguez-Valera. 1995. Evaluation of prokaryotic diversity by restrictase digestion of 16S rDNA directly amplified from hypersaline environments. *FEMS Microbiol. Ecol.* 17:247–256.

Oren, A. 1988. The microbial ecology of the Dead Sea, in *Advances in Microbial Ecology,* Vol. 10, K.C. Marshall, Ed. Plenum, New York. 193–229.

Oren, A. 1993. Availability, uptake, and turnover of glycerol in hypersaline environments. *FEMS Microbiol. Ecol.* 12:15–23.

Oren, A. and P. Gurevich. 1993. Characterization of the dominant halophilic archaea in a bacterial bloom in the Dead Sea. *FEMS Microbiol. Ecol.* 12:249–256.

Payen, A. 1837. Note sur les causes de la coloration en rouge des marais salans. *Annal. Chim. Phys. Ser.* 2 65:156–169.

Pierce, G.J. 1914. The behavior of certain micro-organisms in brine. *Carnegie Inst. Washington. Publ.* 193:49–69.

Pliny the Elder. 1963. *Natural History,* translated by W.H.S. Jones, Vol. VIII, Book 31, 431. Harvard University Press, Cambridge, MA.

Post, F.J. 1977. The microbial ecology of the Great Salt Lake. *Microb. Ecol.* 3:143–165.

Ramos-Cormenzana, A. 1993. Ecology of moderately halophilic bacteria, in *The Biology of Halophilic Bacteria,* R.H. Vreeland and L.I. Hochstein, Eds. CRC, Boca Raton. 55–86.

Rodriguez-Valera, F., F. Ruiz-Berraquero, and A. Ramos-Cormenzana. 1979. Isolation of extreme halophiles from seawater. *Appl. Environ. Microbiol.* 38:164–165.

Rodriguez-Valera, F., F. Ruiz-Berraquero, and A. Ramos-Cormenzana. 1981. Characteristics of the heterotrophic bacterial populations in hypersaline environments of different salt concentrations. *Microb. Ecol.* 7:235–243.

Rodriguez-Valera, F., A. Ventosa, G. Juez, and J.F. Imhoff. 1985. Variation of environmental features and microbial populations with salt concentrations in a multi-pond saltern. *Microb. Ecol.* 11:107–115.

Rushforth, S.R. and E.A. Felix. 1982. Biotic adjustments to changing salinities in the Great Salt Lake, Utah, USA. *Microb. Ecol.* 8:157–161.

Teodoresco, E.C. 1905. Organisation et développement du *Dunaliella,* nouveau genre de Volvocacée-Polybléphariée. Beiheft 7. *Botan. Centralbl.* 18 (Abt. 1):215–232.

van Leeuwenhoek, A. 1684. Microscopical observation about animals in the surface of the teeth. *Phil. Trans. R. Soc. London* 14:568–574.

Ventosa, A., E. Quesada, F. Rodriguez-Valera, F. Ruiz-Berraquero, and A. Ramos-Cormenzana. 1982. Numerical taxonomy of moderately halophilic gram-negative rods. *J. Gen. Microbiol.* 128: 1959–1968.

Vreeland, R.H. 1976. *Microbial Ecology of a Solar Salt Facility.* M.Sc. thesis, Rutgers–The State University, New Brunswick, NJ. 104 pp.

Vreeland, R.H., C.D. Litchfield, E.L. Martin, and E. Elliot. 1980. *Halomonas elongata,* a new genus and species of extremely salt tolerant bacteria. *Int. J. Syst. Bacteriol.* 30:485–495.

5 Considerations for Microbiological Sampling of Crystals from Ancient Salt Formations

Russell H. Vreeland and Dennis W. Powers

CONTENTS

5.1 Introduction 53
5.2 Isolation Over Geological Time 56
    5.2.1 Age of Salado Formation 56
    5.2.2 Permeability of Evaporites 57
    5.2.3 Origin of Salado Rocks 57
5.3 Potential Trapping Mechanisms in Evaporites 58
5.4 Inclusion Selection Criteria 61
5.5 Changes in Rock and Inclusion Conditions Over Geological Time 62
5.6 Summary of Geological Isolation Factors 63
5.7 Microbiological Considerations 63
    5.7.1 Isolation/Sterility Following Field Sampling 64
    5.7.2 Isolation During Culturing 65
    5.7.3 Isolation Media and Growth Conditions 68
5.8 Conclusion 70
Acknowledgments 71
References 71

5.1 INTRODUCTION

Debates about the significance of the presence of bacteria in ancient geological salt formations have been ongoing for several years. Some scientists believe these bacteria represent actual relics of bygone eras (Cano and Borucki, 1995; Dombrowski, 1963; Norton et al., 1993; Reiser and Tasch, 1960; Tasch, 1963a, b). Others believe such occurrences are nothing more than instances of recent contamination (Nehrkorn, 1966; Nehrkorn and Ritzerfeld, 1966). Scientists who support the antiquity argument point to the use of accepted aseptic sampling techniques and the use of crystals, freshly brought from underground, as support for their position (Dombrowski, 1963;

Tasch, 1963a, b). However, other authors note that the bacteria that have been isolated from these samples grow well in media without any added NaCl and appear to be similar to recent surface isolates at molecular levels (Nehrkorn and Ritzerfeld, 1966).

Recently, other investigative groups have taken a different, more fruitful, approach to this microbiological problem in that they have concentrated their efforts on isolating extremely halophilic bacteria from the domain Archaea. The first recent work in this area was reported by Huval and Vreeland (1991), who described several halophiles that had been isolated directly from subterranean brines. In one important instance, the concentrated brines sampled were formed from salt that had dissolved in fresh water (< 1% NaCl) that had been accidentally introduced and trapped in the underground formation (Gold, 1981). The bacterium isolated from this brine initially required at least 20% NaCl for survival and grew optimally in 30% NaCl. It could not have survived in the fresh water that initially entered the cavern (Huval and Vreeland, 1991). In addition, Norton et al. (1993) and Denner et al. (1994) have reported isolating extreme halophiles from three European salt mines. However, the crystals used by these scientists were described as either "recrystallized salt" or simply "rock salt" (without any qualifiers) from a formation that was created 195–270 million years before the present era.

Unfortunately, these studies suffered from basic problems. Claims that an ancient bacterium has been isolated cause both excitement and increased scrutiny. This research involves complex geological systems as well as complicated laboratory procedures because of the issues of contamination. Much of the work to date has not been completely systematic in attacking the whole problem, from rock to laboratory. Instead, all of us looked at underground salt formations as if they were nothing more than a hypersaline lake without the overlying water.

Microbiological approaches to sampling subsurface environments have generally been based on either drill cores, or samples selected at random from mines. Throughout most of the previous studies, little consideration has been given to the nature of the samples and even to the types of microorganisms that may be present in the samples.

To put the reader into perspective, a microorganism isolated from a Permian-age salt formation would have been encased in that salt for 100 million years by the time the oldest of the dinosaurs set foot on land. Further, many researchers have assumed (again incorrectly) that standard microbiological techniques that work well above ground, on smooth hard surfaces, are sufficient for working with salt crystals. The purpose of this chapter is to basically take a step backward, to examine the previous research from the advantage of 20/20 hindsight and to discuss the considerations that should now be applied to these studies.

As a part of testing for viable organisms in rocks, it is critical to establish that the organisms have remained isolated for a long period rather than being introduced by recent contamination. Rock salt is one of the rock types considered a possible preserver of bacteria in isolation for geological periods of time. There are specific features of rocks that help to assure that any organisms found in the rock did not become trapped there as a consequence of more-recent geological events unrelated to the original deposition.

In this chapter, we focus on both the geological and microbiological sciences that must be carefully applied when conducting underground sampling. The primary examples used throughout the discussion have been taken from our studies conducted primarily in the Salado Formation of southeastern New Mexico and west Texas, a salt deposit that is an excellent candidate because of access through mining and drilling operations. It is also one of the primary sampling targets for research by R.H. Vreeland. These beds reveal a panoply of features that indicate isolation from the biosphere for hundreds of millions of years. While some of the features are specific to evaporite beds, they also illustrate some important criteria for determining isolation.

During the previous microbiological studies, none of the researchers were concerned with determining the geological history of the formations that yielded their samples (Tseng et al., 1996). First, the use of simple descriptions such as "rock salt" is inadequate since it provides no information about the quality of the crystal used for the sample. Second, "recrystallized salt" can mean several different things, all of which signify that the salt crystallized in a new form at some time since the original deposition of the formation. Salt can recrystallize under heavy loading by burial. Salt can recrystallize by interaction with pore fluids. Some salt characterized as "recrystallized" precipitated from ground water that entered mines, dissolved some salt, and then evaporated upon reaching mine surfaces. Such salt crystals may actually be only a few days old. Several of the studies have also reported sampling brines that have been flowing within the formation (Huval and Vreeland, 1991; Norton et al., 1993). Brine samples, while being more likely to provide living material, are not suitable for determining whether cells were preserved for geological periods of time, since the brine origin and history may not be sufficiently constrained. If the brine history can be demonstrated, however, some brine sources may be appropriate "collection" points.

Crystals that have been buried for long periods can be subjected to a wide variety of forces and stresses that can affect their integrity and value as a microbiological sample. Many of these forces, and their effect on the morphology of the crystal, have been documented and photographed by Roedder (1984). Primary crystals, those that formed at the time the sediment or rock was being deposited, often have diagnostic features that indicate they formed at the time of deposition. As an example, salt crystals that grow on the sediment surface under brine commonly show features that are diagnostic of growth in that environment. Crystals showing the preservation of the unaltered features and textures are among the most secure evidence of isolation over geological time. The material contained within these unaltered inclusions has been considered material undisturbed since the original depositional events occurred (Das et al.,1990; Horita, 1990; Horita et al., 1991, 1996; Land et al., 1995; Lazar and Holland, 1988; Roedder, 1984, 1990; Roedder et al., 1987). Many rocks are altered by a variety of processes after deposition and new crystals form as a consequence of changing pressure, temperature and fluid chemistry. Some of these rocks are very likely to be excellent sources of samples from different periods in the history of the rock, if it can be demonstrated that viable organisms are indeed preserved over geological time. Another area in which previous research could be considered lacking is that the work was done without regard to the stratigraphy of the sample and the information the sedimentologic studies could provide about depositional environments and isolation from the biosphere.

We believe that the true scientific analysis of the possibility that halophilic bacteria have survived, trapped in salt for hundreds of millions of years, has just begun. The considerations and discussions in this chapter are not meant to invalidate the previous work. Rather, the basic rules laid out here should, when followed, provide sufficient supporting evidence to justify a claim that a particular culture was entrapped in a rock for millions of years.

5.2 ISOLATION OVER GEOLOGICAL TIME

In this section, several types of geological information that should be involved in selecting a study site are considered. Many of the parameters would be available for well-characterized geological formations or can be readily bounded by reconstruction of the geological history. While these types of studies would not normally be included in a direct microbiological study, they should be available in supporting documentation and their import should not be overlooked.

The salt beds of the Salado Formation of southeastern New Mexico and west Texas are used as an example. The comments have not been generalized for all rocks but the parameters discussed represent approaches that should be taken.

5.2.1 AGE OF SALADO FORMATION

Geologists spend much time and energy determining sequences of geological events and directly dating or bounding the ages of rocks and the events. The tools available include determining ages through radiometric techniques and fossil correlations, bounding ages where rocks above or below can be directly dated, and using a variety of other methods (such as stable isotope values or magnetic polarity of the rock) to refine age estimates. These techniques cover the range of geological history for the earth with varying precision and applicability, depending on the rock composition and history. Here, the main concern is to assure a period of isolation from contamination by recent geological events. This need will obviously change with a successful demonstration that viable organisms survive within rocks over geological time. The Salado is considered to be latest Permian in age (about 250 million years old) principally because beds below and above the evaporite sequence bear fossils that are believed to bracket formation of the Salado to this time period. The Salado itself has not yielded any fossils.

Within the Salado, there are minerals with sufficient potassium and rubidium to attempt direct radiometric dating of the mineral formation. Brookins and his colleagues (Brookins, 1980; Brookins et al., 1980) applied K-Ar and Rb-Sr techniques to different minerals and achieved results consistently in the range of 200–225 Ma (million years), with most ages between about 210 and 215 Ma. A more recent study using Ar–Ar techniques (Onstott et al., 1995) yielded radiometric ages of about 210 Ma.

There are several possible explanations for the difference between geological and radiometric assessments of the age of the Salado (or any formation with such differences). The geological age based on bounding rocks may be too old and based on too limited a data set. The minerals may have formed well after deposition, or the minerals may not have sealed to loss of daughter products until after deeper burial. In cases where the radiometric ages are older than the geological age calculated from other information, the rock may have inherited older minerals that were included in the analyses. While some research is in progress to further test the radiometric ages of the Salado minerals, such differences between methods should not be critical for now, where, by all reckoning, minerals have been isolated for geological periods of time.

The radiometric ages are, however, strong evidence that a system has been sealed to external sources of water or brine and loss of daughter products for a period in excess of 200 million years. Many other rocks can be tested in similar ways.

5.2.2 Permeability of Evaporites

Casas and Lowenstein (1989) showed that the porosity of halite beds in modern salt pans may be very high near the surface but declines rapidly with depth. At 10 m, porosity is less than 10%. At 45 m depth, porosity is no longer visible. Except for inclusions, the same is true for samples of the "undisturbed" Salado. Sophisticated *in situ* tests of Salado halite beds at a depth of about 650 m yield permeabilities ranging from about 5.1^*10^{-20} to 2.9^*10^{-19} m^2 (Beauheim et al., 1993). These are exceedingly low values, five to 10 orders of magnitude below that of rocks commonly measured. While such tests indicate the theoretical possibility of very slight fluid movement within these rocks, there is no evidence consistent with fluids passing through the formation over geological time.

Microscopic and physical testing of bedded evaporites, including the Salado, consistently demonstrate the difficulty of moving fluids through the formation after burial to shallow depths. This is strong supporting evidence that these rocks have been isolated from the biosphere since approximately the time of deposition.

Some other sedimentary rocks have low permeabilities, but evaporites are at the low end of the spectrum. The oil and gas businesses are based on sedimentary rocks with permeabilities that allow fluids to move readily from the rock to the well. While these other rocks were deposited in environments with a great deal of biological activity, and might be expected to preserve organisms, it will take greater effort to convince ourselves and reviewers of isolation from the biosphere for geological periods of time if these rocks are to be tested for viable bacteria.

5.2.3 Origin of Salado Rocks

Many of the bedded evaporites around the world were deposited from relatively shallow waters. Others might have been deposited in deeper water. The depth of water during formation of the well known Castile Formation, which underlies the Salado, is still being determined. Some evaporites (e.g. Louann salt of the Gulf Coast) have been deeply buried and are now rising through the overlying sediments as intrusions (diapirs) because of the buoyancy of the salt relative to other rocks and pressures at depths. Such evaporites retain little or no evidence of primary characteristics.

The Salado was deposited in a very large, shallow basin from a series of cycles of flooding and desiccation (Gard, 1968; Jones, 1972; Lowenstein, 1988; Holt and Powers, 1990a, b). As the initial flooding of the basin occurred for the cycle, sulfate minerals (generally gypsum, $CaSO_4 \cdot 2H_2O$) were deposited; major cycles formed thicker sulfate beds that are continuous over large areas, are distinctive, and have been designated marker beds. Most are less than 1 m thick, though a few are considerably thicker. As sodium and chloride reached saturation in the brine, thinner beds (commonly 10–20 cm thick) of halite (NaCl) formed, mainly from crystals growing on the sediment surface. These halite beds include few impurities (< 1%), most of which are sulfate minerals. Clay minerals normally increase in proportion upward in the cycle, as the basin dries up. The clay is believed to be blown in as dust, as well as distributed by occasional flooding or wash from precipitation.

As evaporation continued, the water table dropped below the surface of the sediment. The salt and clay cracked, and rainwater dissolved pits and pipes in the argillaceous salt down to the level of the water table (Holt and Powers, 1990a, b; Powers and Hassinger, 1985). Renewed flooding raised the water table. Brines in the sediment precipitated coarse, clear halite in the pits and pipes (Figures 5.1, 5.2). As flooding inundates the basin with fresher sulfate-rich water, a new cycle of deposition begins.

5.3 POTENTIAL TRAPPING MECHANISMS IN EVAPORITES

Salt crystals can grow very rapidly in warm brines. One of the consequences of this rapid growth is that the crystals trap brine as fluid inclusions along the crystal growth surface (see, e.g., Roedder, 1984, for an extensive review of fluid inclusions). Small pores can be trapped where crystals grow together. Most likely these pores have been isolated from the biosphere as effectively as the inclusions within the crystals. Nevertheless, fluid inclusions trapped within salt crystals are more reasonably demonstrated to have been isolated and should be the main target for now. Because evaporites are such "tight" rocks, targets such as intercrystalline brine-filled pores are legitimate, but focusing on the intracrystalline fluids supports a more conservative early strategy.

In many primary halite crystals, fluid inclusions are less than 1 μm across, and are so numerous that the crystals are cloudy rather than clear. Still others have well-defined bands parallel to the crystal growth surface. The fabrics created are called chevron or cornet, depending on the orientation of the original crystal on the sediment surface (Figure 5.3). Fluid-inclusion bands along the crystal surfaces of salt are taken as evidence that the crystal has not been dissolved or recrystallized since it was deposited. Fluid-inclusion bands along the halite growth surfaces are taken as evidence that the crystal is primary — it has not been dissolved or recrystallized since it was deposited.

Considerations for Microbiological Sampling of Crystals 59

FIGURE 5.1 Photomosaic of a complete Salado depositional cycle showing halite grown from standing brine (stratified mud-poor halite, SMPH) as well as pits and pipes (outlined in photo) developed during Permian time when the surface was exposed after the standing brine evaporated and the water table dropped below the surface. See Figure 5.2, lithofacies corresponding to the photo. HM: halitic mudstone; DMRH: "dilated" mud-rich halite; PMH: "podular" muddy halite (from Holt and Powers 1990a, b. With permission).

Still other crystals of halite in the Salado and similar salt deposits are large, clear, and contain only a few large fluid inclusions. Some may reach several cm across, though these are rare. Inclusions of 1 to 5 mm across are more common.

The common excuse for rejecting these crystals is that large, clear halite crystals have been recrystallized. The time of such recrystallization has generally not been specified, but, if the crystal formed by recrystallization in the presence of a fluid, that might be considered as reasonable doubt of isolation and cause for rejection. For much of the Salado, however, studies show that such large crystals are primary sediments formed during the Permian, and are not recrystallized halite (Holt and Powers, 1990a, b). Having these types of data on hand would reclaim the usefulness of crystals with large brine-filled inclusions, since these rocks could now be recognized as unaltered since deposition of the formation.

Holt and Powers (1990a, b) found that coarse, clear halite crystals are the main component in the pits and pipes that formed during water table fluctuations as the basin dried up during deposition. The pits and pipes are clearly overlain by undisturbed bedded units, which is proof that the feature formed before the next layer

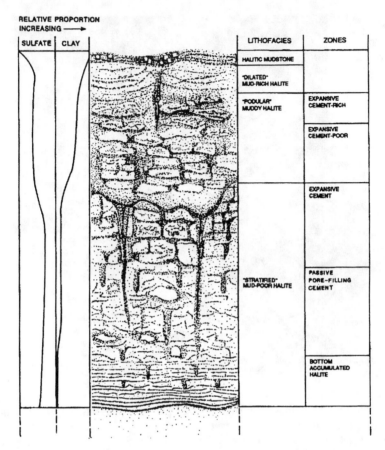

FIGURE 5.2 Idealized sequence of halitic rock types in a depositional cycle of the Salado Formation. Lower part is halite that grew in standing brine and commonly shows crystal features unaltered since deposition (see Figure 5.3). The upper part shows the effects of the depositional basin drying up frequently, forming pits and pipes that are filled in many cases with large clear crystals that are the same age as the formation (from Holt and Powers 1990a, b. With permission).

was deposited. The halite shows crystal boundaries characteristic of growth in an open space in brine, until the space was filled with crystals. One characteristic of rocks recrystallized under pressure is that crystal faces tend to intersect along boundaries with triple junctions (at about 120 degrees). Since these halite crystals don't display such boundaries, they have not been recrystallized since deposition.

At least two different populations of fluid inclusions in the Salado can be demonstrated to have survived since deposition without recrystallization. They offer different sizes to work with, and they also represent two somewhat different environments.

Crystals growing on the sediment trapped fluid from a standing body of brine saturated with respect to halite. Fluid inclusion bands in these crystals are generally small, and the chemistry would be similar to the original brine body. In contrast, the coarse halite from pits and pipes represents a later stage in the depositional cycle.

FIGURE 5.3 Halite crystals growing vertically on sediment surface in brine compete for space. Chevron crystals, growing from a seed crystal along the diagonal, will crowd out cornets, growing vertically perpendicular to the cubic crystal side. Shearman (1970) illustrated similar crystal growth relationships. The parallel lines in the diagram represent relative crystal growth. Halite crystals often have bands of fluid inclusions that show this same pattern. These crystal shapes, growth relationships, and fluid-inclusion bands demonstrate that the crystals have not changed since the rock formed.

Holt and Powers (1990a, b) associated this desiccation stage with some potash minerals, and the concentrations of some solutes (especially K and Mg) could have been much higher as these crystals were forming. At this stage in the cycle, the chemical environment for halophiles likely differed from that early in the depositional cycle.

Holt and Powers (1990a, b) also interpreted other halite forms that likely have remained intact since deposition, but the relationships are not as secure as the two forms already described.

Many crystals (besides halite) are precipitated from a fluid medium and trap fluid inclusions. An extensive geological literature exists (e.g. Roedder, 1984) in which either the properties of the fluid are carefully measured to infer the chemistry and temperature of the fluid medium from which the crystal formed, or microtechniques are used for direct analysis of fluid composition. Salt is the simplest evaporite mineral to deal with, but other minerals such as gypsum might be considered in the future.

5.4 INCLUSION-SELECTION CRITERIA

While the geological context is important in deciding that a rock has remained isolated for long periods of time, fluid-inclusion features add confidence in selecting particular specimens for analysis. Roedder (1984) has reviewed criteria by which fluid inclusions have been judged to be primary, and some of the criteria are less than satisfactory.

Though most inclusions in Salado salt are filled with only a liquid phase, some inclusions have a gaseous or mineral phase. (Inclusions from some environments may even contain immiscible liquids such as brine and hydrocarbons.) The gaseous or mineral phases generally separate when temperature or pressure changes occurring after mineral formation affect solubilities.

The gaseous phase may also form from leakage if the crystal cracks as stress is relieved. Weaker minerals such as halite are more subject to this problem.

A conservative strategy that could be used during exploratory phases of research to detect viable organisms in ancient salts is to reject samples in which gas bubbles are abundant. This strategy will, however, eliminate some samples that have been isolated since deposition. Samples that have cracks through sampled inclusion zones should also be eliminated, even if there are no gas bubbles evident in the fluid inclusions.

5.5 CHANGES IN ROCK AND INCLUSION CONDITIONS OVER GEOLOGICAL TIME

Some evaporites have undergone severe physical changes in pressure and temperature since deposition. Diapiric or dome salt, which occurs in many parts of the world, pierces overlaying rock layers as it rises from deep burial. Some chemical changes can also occur over geological time. The Salado and other bedded evaporites have been subject to lesser changes, but they should be evaluated beforehand as part of the strategy of selecting samples for exploratory biological testing. Rocks that have remained closer to the original conditions of formation would seem more likely to have preserved organisms than rocks that have been highly deformed, heated, or otherwise subjected to extreme conditions. It is usually possible to reconstruct the geological history of most formations sufficiently to estimate the range of physical conditions that have affected the rock.

Powers and Holt (1995) estimated the maximum possible burial depth of the Salado over geological time since the formation was deposited. For the lower Salado, which is directly accessible through mining now at a depth of 650 m, the maximum was slightly more than double the present depth and likely occurred during the late Cretaceous, about 80 Ma. The lithostatic load or pressure (assuming an average rock density of 2.5 g cm^{-3}) would be about $3.3*10^6$ kg m^{-2}. If a hydrostatic loading was assumed, the pressure (with a brine column with a density of 1.25 g cm^{-3}) would be half the lithostatic load, or about $1.65*10^6$ kg m^{-2}.

The thermal history of most rock units can be estimated as well. Natural shallow brines can exceed 70°C through solar heating. Roedder and Belkin (1977) reported that fluid inclusions from the Salado show that most fluids were trapped at temperatures between about 25 and 45°C. *In situ* temperature at a depth of about 650 m in the Salado is about 26–27°C. If these rocks were buried about twice as deep, the temperature would have increased by only some 5–6°C; the current thermal gradient in the Salado evaporites is about 0.8°C/100 m depth (Mansure and Reiter, 1977).

The physical environment and even chemistry of the rocks can change greatly over geological time, even in a closed system. These are factors that should be assessed as part of the design of experiments to search for viable organisms in rocks.

Inclusion-fluid chemistry can also change with time. As the halite crystal is buried, the temperature might decline sharply, from, for example, 70°C at the brine-sediment interface to around 20°C at shallow depths. Some minerals could precipitate at lower temperatures, drastically altering the fluid chemistry.

Considering the thermotolerance of many halophilic enzymes (Keradjopoulos and Holldorf, 1977) these conditions would certainly not preclude the survival of trapped bacteria even under the most extreme conditions experienced in a bedded formation such as the Salado. Many other formations have similar low-temperature histories.

5.6 SUMMARY OF GEOLOGICAL ISOLATION FACTORS

Three separate isolation factors need to be considered while examining rock samples for viable organisms. Did the rock remain isolated from the biosphere during a relevant (geological) period of time? Was the rock contaminated during sampling? Did laboratory techniques maintain sterile conditions during analysis? Geological evidence can be used to address the first factor and will contribute to the second.

The geological history of a rock body can be reconstructed to estimate whether physical and chemical conditions remained within the range that is reasonable for organisms. As a first approximation, formations that show evidence of having been exposed to extreme conditions should be avoided unless one is prepared to conduct more-extensive studies to document that the formation isolated organisms after the extreme conditions were modified. Tseng et al. (1996) provide an excellent example of detailed studies that show how a formation that had been heated to temperatures of 120–210°C subsequently trapped organisms at much lower temperatures. Details of the rock formation, and subsequent history, indicate the degree of isolation. Rock ages determined by radiometric techniques can be used to infer the length of time that a system has been closed. From a geological point of view it is preferable, but not necessary, that radiometric ages be the same as those determined by other geological techniques (e.g., fossils). Evaporites are useful for initial experiments for several reasons. The potential for fluid movement is very limited by the rock permeabilities. Rock features show relationships that can only have formed at the time of deposition, and crystals include fluids in patterns diagnostic of primary crystals (unmodified since deposition). In addition, these fluid inclusions can be examined for evidence of closure during the sampling period.

For the demonstration that viable organisms exist in rocks isolated from the biosphere for geological periods of time, evaporites have many positive features. Other rocks, carefully examined, may be suitable as well and will sample other geological environments. With that backdrop in mind, let us examine each of the various microbiological points in terms of the work that has already been accomplished and suggest possible improvements.

5.7 MICROBIOLOGICAL CONSIDERATIONS

For microbiologists, the most difficult aspect of sampling underground formations is generally obtaining the samples. Basically, there are only two possible choices. First is remote drilling and coring (Dombrowski, 1963) or the remotely triggered sampling bottle (Huval and Vreeland, 1991). The second choice is to obtain samples directly from a mine area, which is the technique that has been used most often (Denner et al., 1994; Norton et al., 1993; Reiser and Tasch, 1960; Tasch, 1963a, b; Vreeland et al., 1998). Each of these sampling techniques has different strengths and liabilities.

The drilling affords the opportunity to reach regions of formations that have never been exposed at all. Its disadvantages are the costs of drilling and the problem of trying to maintain some type of sterility control. Since the drilling procedures use large amounts of a heavy type of "mud" to keep weight and lubrication on the drill bits, the amount of potential contamination of the sample prior to recovery by the microbiologist is relatively high. Further, no amount of sampling of the excess mud will provide sufficient data to eliminate the chance of introduced contamination.

Obtaining samples directly from a mine environment is less expensive, and certainly easier to control. However, there are other problems involved with sampling in these facilities. The most notable problem in sampling a mine is the formation has been opened to the external environment for a considerable period of time. Due to the confined areas, limited airflow, or potential presence of methane gas pockets, it may be impossible to use sterilizing agents within the mine at the moment of sampling. Consequently, the samples must be taken with only a minimum of handling and must then be protected from further potential contamination during transport. Given these sampling constraints, this section of the chapter focuses on the microbiological considerations of handling the samples once they have been removed from the formation.

5.7.1 Isolation/Sterility Following Field Sampling

For the most part, researchers who have studied underground salt beds have taken pains to at least try to sterilize the surfaces of the crystals being examined. The most common technique used has generally involved the use of ethanol as a sterilizing agent.

The earliest workers in salt formations (Dombrowski, 1963; Reiser and Tasch, 1960; Tasch, 1963a, b) utilized common alcohol flames to provide a measure of sterility. There are several problems with this particular methodology that have never been adequately addressed by microbiologists. First, an alcohol flame is short lived, consequently any type of sterilization must occur very rapidly. Also, a flame like this may work when dealing with hard, non-porous materials such as stainless steel or glass, but its effectiveness on the irregular surfaces of a salt crystal has never been proven. Flame sterilization techniques are based on simply heating the sample; the higher the temperature achieved, the more rapid the sterilization. Unfortunately, the visible alcohol flame is really the combustion of the vapor phase, which has a temperature of only about 78°C (Lide, 1996). Prescott et al. (1995) show that, to kill a mesophilic bacterium at this temperature, the flame would need to burn for at least 10 minutes, which simply doesn't happen. NaCl is an excellent heat conductor, therefore rapid heating at the crystal surface could manifest itself by a rapid heating of the internal brine. This could result in a pasteurization effect on the trapped brine.

The more recent work on salt crystal isolations has been performed with more care, using not only an ethanol soak, but supported by the use of laboratory experiments in which two different halophiles were used as marker organisms. One halophile was incorporated into the crystal, while a second, phenotypically distinguishable, organism was added to the surface of the crystal. The disappearance of the surface organism and the retention of viability of the crystallized strain were taken as evidence for the efficacy of the sterilization procedure (Norton et al., 1993). These data were also cited by Denner et al. (1994), who placed their crystal samples into absolute alcohol immediately after removing them from the mine rib (the wall) and soaked the sample for four or five hours until sampling occurred in the laboratory. This technique is certainly better than simply flaming the crystal. Unfortunately, the phenol coefficient for alcohol is only 0.04 and the most effective ethanol concentration for disinfection is actually 70%, not absolute. Actually, ethanol is a better antiseptic than disinfectant, meaning that while it could eliminate a sensitive test organism, it might only inhibit, not destroy, an organism residing on the surface of a natural crystal (Prescott et al., 1995). Further, the use of a marker organism on the surface of a laboratory-grown test crystal does not provide any quantitative measure of the level of sterility assurance.

Often the people who are critical of claims of the antiquity of a bacterial culture cite a lack of proven sterility of the sample as reason to disbelieve the organism is really old. Certainly, any sample used for this type of work should be surface cleaned or sterilized. It is, in fact, on the issue of sterility that many arguments of the existence of subsurface bacteria have turned (Kaiser and Bollag, 1990; Phelps et al., 1989). Part of the problem with arguments over sterility relate to the probability aspects of sterilization. Succinctly stated, one can never truly guarantee sterility, we can only cite a probability of sterilization. For the most part, microbiologists ignore this aspect of our work, since we normally autoclave media to a sterility assurance level between $1*10^{-3}$ and $1*10^{-6}$. This later value was also the value accepted by NASA for the current and past Mars explorations (Klein, 1992). One of the problems faced by all of the previous researchers is a lack of any type of protocol that can be used to demonstrate a sterility assurance level for the surface sterilization of ancient crystals.

5.7.2 Isolation During Culturing

There is little doubt that the direct isolation of an organism from inside a surface-sterilized crystal would be the single most powerful evidence for the antiquity of the culture. There is also little argument in stating that this something none of this book's authors have accomplished. The original work of Dombrowski (1963), Reiser and Tasch (1960) and Tasch (1963a, b) used intact crystals. Huval and Vreeland (1991) concentrated on artesian brines that moved through ancient salts (these were probably the worst samples of all). Norton et al. (1991), Denner et al. (1994), and Vreeland et al. (1997) all used various descriptions of salt crystals including "efflorescences" (Norton et al., 1991), rock salt and simply crystals (Vreeland et al., 1997). It should be stated at the outset that, prior to 1988, there were virtually no methods available for penetrating and sampling the inside of a salt crystal (Lazar and Holland,

1988). In fact, microbiological sampling of the inside of an ancient structure was not accomplished with any degree of precision prior to 1995 (Cano and Borucki, 1995), when samples were obtained from Dominican amber. This type of work has still not been accomplished with a salt crystal. There are, of course, several unique problems involved with this type of sampling in salt, some of which have already been addressed in this chapter.

Naturally, the key to isolating an organism from a crystal is identifying a crystal that is likely to contain surviving bacterial cells. Since there are currently no non-invasive biological sensors available, it becomes necessary to try to select primary crystals from regions of a formation that have a detectable bacterial population. This is no mean feat. Indeed, in a recently completed survey of an underground formation, Vreeland et al. (1997) found that some crystal samples did produce surprisingly high numbers of bacteria. At the same time however, these researchers also demonstrated that most samples taken from the formation and simply plated onto media produced no detectable colonies. A more-detailed experiment using 10 primary crystals selected from various mine areas in the Salado also illustrates this point (Table 5.1). Extensive washing of these crystals using detergents and distilled water showed that at least six out of 10 crystals appeared sterile when the washes were sampled. Some of these crystals never yielded viable cultures, others (such as # 3, 6, 7 and 9 in Table 5.1) yielded halophiles in different washes, indicating that the organisms were located in areas of the crystal (possibly micro-fractures) other than the brine inclusions. One crystal (#9 in Table 5.1) yielded a viable culture after all of the washings. While this particular bacterium meets criteria that, under ideal conditions, could lend credence to a claim of antiquity, that claim might prove hard to defend because no sterility-assurance levels can be stated for the crystal surfaces. Also, since the entire crystal was allowed to dissolve in the medium, it is possible that the organism was located in a crack or fissure on the surface of the crystal. However, the possibility is remote, since 45% of the original mass of the crystal was removed during the sterile washes.

Another difficulty involved with sampling inside a crystal relates to actually drilling into the system. Other sections of this chapter have described the considerations and requirements for selecting only primary crystals, but, as Table 5.1 shows, simply sampling a bulk crystal that contains primary inclusions may not be enough. The real key is to be able to drill into the crystal using sterilized drill bits, sterilize the drill hole, then penetrate the inclusion and withdraw the brine in a single motion. This type of drilling technique has already been accomplished by a variety of geochemical researchers, but has not yet been used by microbiologists.

The drilling technique itself, first designed by Lazar and Holland (1988), involves construction of a small drill run by a toy motor. The drill bit is made using a piece of tungsten wire fitted into plastic to allow for changing the bit. The entire assembly is fitted to a micromanipulator to allow for slow controlled movement into the crystal. This type of system has been used to extract brines from inclusions in crystals taken from several different formations. The extensive chemical analyses conducted on these brines by Das et al. (1990), Horita (1990), Horita et al. (1991, 1996) and Land et al. (1995) have convincingly proven the utility of these microdrills. These authors

TABLE 5.1
Isolation of Halophilic Bacteria from Individual Natural Salt Crystals

Crystal #	Mine Origin[a]	% Weight Lost	1st SDS wash	1st H$_2$O wash	2nd SDS wash	2nd H$_2$O wash	Growth Tube
1	Rm 3R	62	–	–	–	–	–
2	Rm 3R	93	–	–	–	–	–
3	Rm 3R	87	–	–	–	+	–
4	Rm 3R	80	–	–	–	–	–
5	Rm 3M	73	–	–	–	–	–
6	Rm 7R	64	–	+	–	–	–
7	E140 N780	71	–	–	+	–	+
8	Rm A3R	33	–	–	–	–	–
9	E140 N780	45	–	–	–	–	+
10	W30 S2180	90	–	+	–	–	–

[a] The mine origin of each crystal has been specified using normal mine coordinates and terminology. R = rib (or wall) sample. M = "muck," a generic term for salt that has been removed to produce the mine. RM = a designated waste storage room (i.e., room 3 or room 7). The numerical designations refer to the directions and distance (in feet) away from the construction and salt handling shaft (C&SH), which is also a main route of exit or entrance into the mie, i.e., E140N780 is a point that is 140 feet east and 780 feet north of C&SH.

have inferred that inclusions trapped brine from concentrated ocean water at the time of deposition and have remained isolated since that time. This later point, age dating, has been discussed in another section but is a most critical component of this type of research.

In addition to being able to provide information about the age and purity of the crystal samples being studied, the use of microtechniques offers two other important advantages. First is the potential use of capillary methods for the initial isolation and growth of the cultures. Within a capillary system, an individual bacterial cell can be extracted from the inclusion without coming into contact with air. Further, this capillary can be made to allow microscopic viewing of the extracted brine before the introduction of nutrients. This would also aid in the antiquity argument since the sterility assurance level of the capillary can be determined and the presence of shapes reminiscent of cells can be documented. By omitting the contact with air it is also possible to maintain the supersaturated (with respect to NaCl) condition of the inclusion brine.

Second, while the capillary would help slow down crystallization of the ancient brine, it would also allow the slow introduction of nutrients and slightly lower salt concentrations through normal diffusion processes. This latter point may be important since at least one experiment conducted in Vreeland's laboratory indicated that lower nutrient media resulted in a higher percentage of recovery from laboratory-grown crystals. Further, Vreeland et al. (1998) found that, upon initial isolation, bacterial populations from crystals required up to 45 days to develop into colonies, while after several transfers these same cultures produced colonies in only seven to 14 days.

5.7.3 Isolation Media and Growth Conditions

Perhaps one of the first decisions to be made before obtaining samples from a new environment is the type of medium that will be needed. And, since all microbiologists have their favorite formulation, choosing a growth medium is often something that happens so quickly it is generally given relatively little thought. After all, the range of halophilic media in use today have yielded at least nine (possibly 10) distinct genera of halophilic Archaea (Montalvo-Rodriguez et al., 1998). Table 5.2 lists the components of all of the various media that have been used in sampling underground salt formations. This table includes both the non-halophilic and halophilic media. The table shows that with the exception of the glucose medium used by Vreeland et al. (1998) these media share many similar components, particularly such materials as yeast extract, casamino acids, and proteose peptone #3. Using media with similar components is not necessarily a bad idea, since it allows more-direct comparison of the bacterial populations. At the same time, continually using similar media leads to the isolation of similar metabolic types of bacteria, with the inevitable criticism regarding contamination of the sample.

Certainly it would be possible (and probably desirable) to design unique media for use with these ancient rocks. If we were to assume the bacterium was at a simpler evolutionary stage, it might be advantageous to design a medium with simpler nutritional characteristics. The data produced by Vreeland et al. (1998) showing that the Salado Formation contained more halophilic bacteria growing on a glucose-based medium than on the more common amino acid-based media would certainly seem to support this concept.

While designing a unique medium for use with ancient rocks is a difficult task, it may be possible to set some hypothetical parameters for this type of growth medium. Ostensibly, the medium might be based on our knowledge of conditions existing in the ancient seas. First, during the Permian epoch, all land masses were gathered into a single continent, "Pangea." The region in which the Salado formed was bay or lagoon system. Biologically, the world was dominated largely by marine invertebrates and early reptiles. Mammals weren't present, nor were flowering deciduous plants. The end of the Permian was also a time when species diversity was low, due largely to mass extinctions at that time.

Accepting this, it is probably inappropriate to expect a true Permian bacterium to use media containing large amounts of mammalian-based proteins. The mixtures of amino acids used in many of the media would be acceptable since these have been around for some time. However, it is unlikely that these organisms would have experienced high concentrations of yeast extract or Proteose Peptones.

In fact, most of the commonly used halophile media contain relatively high concentrations of nutrients that may actually be inhibitory to cells that have possibly existed in a state of extended long-term starvation (Morita, 1990). In fact, Salado

TABLE 5.2
Formulations of Media Previously Used to Isolate Bacteria from Ancient Salts

Medium Component	Huval and Vreeland 1991	Norton et al. 1993, Norton and Grant 1988	Dombrowski 1963[a]	Tasch 1963[b]	Vreeland et al. 1997
Beef extract			3.0	1.0	
Casamino acids	7.5	7.5			
Cellulose					4.5
Yeast extract	10	10			
Glucose					0.5
Mannitol				10	
Peptone			5.0		
Proteose peptone #3	5.0			10	
Sodium citrate	3.0	3.0			3.0
KCl	2.0	2.0			4.0
$MgSO_4$	20	20			20
$MnCl_2$		0.000036			
$Fe_2(NH_4)_2SO_4$		0.05			
K_2HPO_4	0.5				0.5
$(NH_4)_2SO_4$					1.0
NaCl	200	200	8	75	200
Agar	20	18	15	15	20
pH	8.0	7.5	7.3	7.4	7.4

[a] This medium was simple nutrient broth.
[b] This medium was mannitol salt agar.
Note: All values listed in grams per liter.

Formation samples that were first dissolved in 25% w/v NaCl brine for at least a couple of hours before being plated onto nutrient media produced detectable bacteria more frequently than did salts that were dissolved and immediately plated onto media. This may also be indicated by the data in Table 5.1, where crystal #9 produced a viable culture from a region of the Salado that consistently appeared sterile in field samples (Vreeland et al., 1998). Ability to provide a brief period of slow rehydration of the cells could be one of the potential advantages arising from the use of capillary-isolation techniques.

It may also be advantageous to use media with relatively low amounts of nutrient materials in the form of fish or even invertebrate extracts rather than the more common modern-day materials. While compounds such as glucose have been used to some success (Vreeland et al., 1998), these metabolically popular nutrients would probably have been fully utilized before reaching the halophiles. Therefore, the bacteria that existed in these terminal environments might be better served by supplying even simpler molecules such as glycerol or acetate (Oren 1995a, b; Oren and Gurevich, 1995).

At this point, it might also be worthwhile to consider the conditions to be used for actually growing the isolates once they are removed from the crystal. All of the experiments and isolations conducted to date have incubated the isolates at common temperatures such as 30 to 37°C. Yet, as we have pointed out, this may not be appropriate. If the geological information is considered, formations such as the Salado formed at temperatures that may have reached 70°C. Consequently, we may find that higher temperature incubations sealed in vessels to prevent moisture loss may be more appropriate.

In summary, there are many possibilities that can be considered in terms of the growth conditions to be used with ancient rocks. By considering the geological history of the site, the age of the rock, conditions of formation, and even fossil information existing during the epoch being studied, it may be possible to tilt the scales in favor of some unique halophiles.

5.8 CONCLUSION

The previous sections of this chapter have focused on things that must be considered to provide at least some substantiation to the idea that a particular bacterial isolate was trapped in the rock when it originally formed in some archaic brine pool, and that this organism survived in this rock for thousands of millennia. This would naturally be no easy feat, and the finding of such an organism would generate at least some excitement. Since most of this chapter has focused on a Permian-age formation, it might be appropriate to finish by mentioning some other formations and some of the things that might be learned from ancient microbes.

A variety of different formations underlie many regions of the world (Zharkov, 1981). Many of those located in North America, amazingly, appear to be have been relatively isolated through geological time. These formations range from Cambrian all the way up to Jurassic in age. Nearly all of the various epochs are represented, providing the possibility of obtaining samples in almost 100 million-year increments. There is even a variety of younger depositional environments starting around 50,000 years and moving toward the present.

Once proven rigorous isolation techniques are available, nearly all of these regions become potential study areas. The time sequence exemplified by these sites provides an amazing opportunity for evolutionary studies. Comparisons of similar molecules from isolates might provide a potential calibrator for a molecular evolutionary clock. Further, studies of the physiological abilities of any isolates would provide a tremendous amount of information about the environment in which these bacteria thrived.

Another obvious study area regards the mechanisms for long-term survival of non-spore-forming bacteria locked into a crystalline matrix. Finally, if we can prove that microbes have survived for millions of years in crystals it might be possible to use these same, or similar, techniques to find life's remnants on planets such as Mars and beyond.

ACKNOWLEDGMENTS

The writing of this manuscript was supported in part by grant # EAR-9714203 to RHV from the U.S. National Science Foundation Life in Extreme Environments program.

REFERENCES

Beauheim, R.L., R.M. Roberts, T.F. Dale, M.D. Fort, and W.A. Stensrud. 1993. *Hydraulic testing of Salado Formation evaporites at the Waste Isolation Pilot Plant site: Second Interpretive Report*: SAND92-0533. Sandia National Laboratories, Albuquerque, NM.

Brookins, D.G. 1980. Use of evaporite minerals for K-Ar and Rb-Sr geochronology: evidence from bedded Permian evaporites, southeastern New Mexico, USA. *Naturwissenschaften* 67:604.

Brookins, D.G., J.K. Register, M.E. Register, and S.J. Lambert. 1980. Long-term stability of evaporite minerals: geochronological evidence, in *Scientific Basis for Nuclear Waste Management*, C.J.M. Northrup, Jr., Ed. Materials Research Society Annual Meeting, Boston, MA, November 27-30, 1979. Plenum, New York. Vol. 2, 479–486.

Cano, R.J. and M.K. Borucki. 1995. Revival and identification of bacterial spores in 25 to 40 million year old Dominican amber. *Science* 268:1060–1064.

Casas, E. and T.K. Lowenstein. 1989. Diagenesis of saline pan halite: comparison of petrographic features of modern, Quaternary and Permian halites. *J. Sed. Petrol.* 59:724–739.

Das, N., J. Horita, and H.D. Holland. 1990. Chemistry of fluid inclusions in halite from the Salina group of the Michigan basin: implications for late Silurian seawater and the origin of sedimentary brines. *Geochim. Cosmochim. Acta* 54:319–327.

Denner, E.B.M., T.J. McGenity, H.J. Busse, W.D. Grant, G. Wanner, and H. Stan-Lotter. 1994. *Halococcus salifodinae* sp. nov. and archaeal isolate from an Austrian salt mine. *Int. J. Syst. Bacteriol.* 44:774–780.

Dombrowski, H.J. 1963. Bacteria from palaeozoic salt deposits. *Ann. N. Y. Acad. Sci.* 108:453–460.

Gard, L.M., Jr. 1968. *Geologic studies, Project Gnome, Eddy County, New Mexico*. US Geological Survey Professional Paper 589, 33 pp.

Gold, M. 1981. Who pulled the plug on Lake Peigneur? *Science-81*:56–63.

Holt, R.M. and D.W. Powers. 1990a. Halite sequences within the late Permian Salado Formation in the vicinity of the Waste Isolation Pilot Plant, in *Geological and Hydrological Studies of Evaporites in the Northern Delaware Basin for the Waste Isolation Pilot Plant (WIPP), New Mexico*, Field Trip #14 Guidebook, D.W. Powers, R.Holt, R.L. Beauheim, and N. Rempe, Eds. Geological Society of America, Dallas Geological Society. 45–78.

Holt, R.M. and D.W. Powers. 1990b. Geologic Mapping of the Air Intake Shaft at the Waste Isolation Pilot Plant: DOE/WIPP 90-051. US Department of Energy, Carlsbad, NM.

Horita, J. 1990. Stable isotope paleoclimatology of brine inclusions in halite: modeling and application to Searles Lake California. *Geochim. Cosmochim. Acta* 54:2059–2073.

Horita, J.T., J. Friedman, B. Lazar, and D.H. Holland. 1991. The composition of Permian seawater. *Geochim. Cosmochim. Acta* 55:417–432.

Horita, J., A. Weinberg, N. Das, and D.H. Holland. 1996. Brine inclusions in halite and the origin of the middle Devonian prairie evaporites of western Canada. *J. Sedim. Res.* 66:956–964.

Huval, J.H. and R.H. Vreeland. 1991. Phenotypic characterization of halophilic bacteria isolated from groundwater sources in the United States, in *General and Applied Aspects of Halophilic Bacteria,* F. Rodriguez-Valera, Ed. Plenum, New York. 53–62.

Jones, C.L. 1972. Permian basin potash deposits, southwestern United States, in *Geology of Saline Deposits,* UNESCO, Earth Science Series 7:191–201.

Kaiser, J.P and J.M. Bollag. 1990. Microbial activity in the terrestrial subsurface. *Experientia* 46:797–806.

Keradjopoulos, D. and A.W. Holldorf. 1977. Thermophilic character of enzymes from extreme halophilic bacteria. *FEMS Microbiol. Lett.* 1:179–182.

Klein, H.P. 1992. *Planetary Protection Issues for the MESUR Mission: Probability of Growth.* NASA Conference Publication 3167.

Land, L.S., R.A. Eustice, L.E. Mack, and J. Horita. 1995. Reactivity of evaporites during burial: an example from the Jurassic of Alabama. *Geochim. Cosmochim. Acta* 59:3765–3778.

Lazar, B. and D.H. Holland. 1988. The analysis of fluid inclusions in brine. *Geochim. Cosmochim. Acta* 52:485–490.

Lide, D.R., Ed. 1996. *CRC Handbook of Chemistry and Physics,* 76th ed. CRC, Boca Raton.

Lowenstein, T.K. 1988. Origin of depositional cycles in a Permian "saline giant": the Salado (McNutt Zone) evaporites of New Mexico and Texas. *Geol. Soc. Am. Bull.* 100:592–608.

Mansure, A. and M. Reiter. 1977. *An Accurate Equilibrium Temperature Log in AEC #8, a Drill Test in the Vicinity of the Carlsbad Disposal Site.* Open-file Report #80, NM Bureau Mines and Mineral Resources, Socorro, NM.

Montalvo-Rodríguez, R., R.H. Vreeland, A. Oren, M. Kessel, C. Betancourt, and J. López-Garriga. 1998. *Halogeometricum borinquense* gen. nov., sp. nov., a novel halophilic archein from Puerto Rico. *Int. J. Syst. Bacteriol.,* in press.

Morita, R.Y. 1990. The starvation survival of microorganisms in nature and its relationship to the bioavailable energy. *Experentia* 46:813–817.

Nehrkorn, A. 1966. Ökologische Untersuchungen über die Keimbesiedlung von Salinen-wässern. *Arch. Hyg. Bakteriol.* 150:232–236.

Nehrkorn, A. and W. Ritzerfield. 1966. Untersuchungen zur Salztoleranz paläozoischer und rezenter Bakerien. *Zeitschr. Allg. Mikrobiol.* 6:189–196.

Norton, C.F. and W.D. Grant. 1988. Survival of halobacteria within fluid inclusions in salt crystals. *J. Gen. Microbiol.* 134:1365–1373.

Norton, C.F., T.J. McGenity, and W.D. Grant. 1993. Archaeal halophiles (halobacteria) from two British salt mines. *J. Gen. Microbiol.* 139:1077–1081.

Onstott, T.C., C. Mueller, K. Mikulki, and D.W. Powers. 1995. $^{40}Ar/^{39}Ar$ laser microprobe dating of polyhalite from bedded, Late Permian evaporites. *Eos–Trans. Am. Geophys. Union* 76: S285.

Oren, A. 1995a. Uptake and turnover of acetate in hypersaline environments. *FEMS Microbiol. Ecol.* 18:75–84.

Oren, A. 1995b. The role of glycerol in the nutrition of halophilic archaeal communities: a study of respiratory electron transport. *FEMS Microbiol. Ecol.* 16:281–290.

Oren, A. and P. Gurevich 1995. Occurrence of the methyl glyoxal bypass in halophilic bacteria. *FEMS Microbiol. Lett.* 125:83–88.

Phelps, T.J., C.B. Fliermans, T.R. Garland, S.M. Pfiffner, and D.C. White. 1989. Methods for recovery of deep terrestrial subsurface sediments for microbial study. *J. Microb. Meth.* 9:267–280.

Powers, D.W. and B.W. Hassinger. 1985. Synsedimentary dissolution pits in halite of the Permian Salado Formation, southeastern New Mexico. *J. Sed. Petrol.* 55:769–773.

Powers, D.W. and R.M. Holt. 1995. *Regional Geological Processes Affecting Rustler Hydrogeology.* Report to Westinghouse Electric Corporation, Carlsbad, NM.

Prescott, L.M., J.P. Harley, and D.A. Klein. 1995. *Microbiology,* 3rd ed. Wm. C. Brown, Dubuque, IA.

Reiser, R. and P. Tasch. 1960. Investigation of the viability of osmophile bacteria of great geological age. *Trans. Kans. Acad. Sci.* 63:31–34.

Roedder, E. 1984. Fluid inclusions, in *Review in Mineralogy,* v. 12, P.H. Ribbe, Ed. Mineralogical Society of America.

Roedder, E. 1990. Fluid inclusion analysis–prologue and epilogue. *Geochim. Cosmochim. Acta* 54:495–507.

Roedder, E. and H.E. Belkin. 1977. *Preliminary Report on Study of Fluid Inclusions in Core Samples from ERDA No. 9 Borehole, Nuclear Waste Site, New Mexico.* US Geological Survey, Reston, VA.

Roedder, E.W., M. d'Angelo, A.F. Dorrzapf, and P.J. Aruscavage. 1987. Composition of fluid chemistry of inclusions in Permian salt beds, Palo Duro Basin, TX. *Chem. Geol.* 61:79–80.

Shearman, D.J. 1970. Recent halite rock, Baja California, Mexico. *Inst. Mining Metallurg. Trans. B* 79:155–162.

Tasch, P. 1963a. Dead and viable fossil bacteria. *Univ. Wichita Bull.* 56:3–7.

Tasch, P. 1963b. Fossil content of salt associated evaporites, in *Symposium on Salt,* A.C. Berkister, K.E. Hoekstra and J.F. Hall, Eds. Northern Ohio Geol. Soc. 96–102.

Tseng, H.-Y., T.C. Onstott, R.C. Burruss, and D.S. Miller. 1996. Constraints on the thermal history of the Taylorsville basin, Virginia, USA, from fluid inclusions and fission track analyses: implications for geomicrobiology experiments. *Chem. Geol.* 127:297–311.

Vreeland, R.H., A.F. Piselli, S. McDonnough, and S.S. Myers. 1998. Distribution and diversity of halophilic bacteria in a subsurface salt formation. *Extremophiles,* in press.

Zharkov, M.A. 1981. Distribution and number of Paleozoic evaporite sequences and basins, in *History of Paleozoic Salt Accumulation,* A.L. Yanshin, Ed. Springer-Verlag, New York.

6 Poikilotrophic Response of Microorganisms to Shifting Alkalinity, Salinity, Temperature, and Water Potential

Anna A. Gorbushina and Wolfgang E. Krumbein

CONTENTS

6.1 Introduction 75
6.2 The Environment of the Poikilotrophic Microorganism 77
6.3 Definition and Description of the Poikilotrophic Microorganism 78
6.4 Conclusion 83
6.5 Summary 83
Acknowledgments 84
References 84

6.1 INTRODUCTION

In 1979, Shilo wrote: "The subject of adaptation to extreme environments has become focal in recent years. Symposia took place and new books have been written on this subject." At that time, a temperature of 91°C was considered extreme (Brock, 1978). Today, the known upper temperature limit for life is nearly 120°C (Karl, 1995), and assumptions go as far as 150°C as a possible upper temperature extreme for metabolic activity. An expansion of an extremity of 30°C on a scale of 0–150 means an expansion of adaptation potential of at least 20% in less than 20 years of research. Actually, it can be stated that the probability is high that within the 120 years since Cohn (1862) studied the Karlsbad hot springs, or the 2,000 and 500 years, respectively, that have passed since Pliny and Dante studied the infernally hot springs of Italy, our concept of the uppermost temperature for life on earth has tripled from 50°C to 150°C. Similarly, our concept of life in salty or bitter water has advanced from biblical times until today (Krumbein and Friedman, 1985; Javor, 1989). Formerly, it was thought that salt or salt water (pickling brine) would sterilize food. Now we know that bacteria and fungi can metabolize even inside large crystals of halite.

Hausmann and Kremer (1994) approached the topic of "extreme environments" in a way that was already partially adopted by Brock (1978). They admit that the idea of a book on extremophiles may be based on a misunderstanding or misinterpretation. They realize that the assignments "extreme" to any relatively stable environment and "extremophile microorganism" to any microbe living in such environments, is extremely anthropocentric and even inadequate. They further conclude that humankind in its self-centered views does not realize that all environments on earth are "normal" for living beings, and will be inhabited by organisms fit for their proper environment in more or less narrow limits. The microorganism will metabolize quite normally and successfully in any appropriate niche. Only humankind behaves differently. All activities of this species tend toward heretofore non-existing and inappropriate environments that create an ideal or "paradise-like" environment that is really "extreme" or "hell" for any other organism adapted to normal terrestrial habitats.

The endothermic (37°C) and catabolic behavior of humankind, as one representative of the homoiotherm group makes it, has created the first environment that could be called extreme in terms of "normal" conditions for life on earth, namely the "living-room." Thus, the terminology is sometimes awkward. Microorganisms hitherto regarded as "extremophiles" such as the thermophilic, psychrophilic or extremely oligophilic and highly endemic inhabitants of very peculiar environments (e.g. the South American *tepuis*), can be regarded as organisms that are highly adapted to very narrow limits of environmental conditions, and thus should be called "normaphilic" because they can only live in a very limited and normative environmental space. On the contrary, the extremotolerant organism will be the real expert for survival and metabolic capacity in very wide limits of environmental conditions. Therefore, these organisms will be the true extremophiles. The environments in which to look for such organisms will not be extremely salty, highly acid or alkaline, hot, cold, dry or wet. They will not be free of oxygen or saturated with hydrogen sulfide, not loaded with methane exhibiting lack of carbon dioxide and also not almost free of inorganic and organic nutrients or saturated with the latter. The environment for this extremophile will be either very "normal" or exhibit major or even extreme shifts between highest and lowest at regular, irregular or episodic time intervals.

The environment of the true extremophile actually is not determined by space, concentration or any other physical/chemical variable. The environment of this organism will be a time-lapse-determined environment. The organism must exhibit characteristics that humankind usually lacks — patience and the equipment for survival under irregularly occurring changes of environmental conditions. In this chapter we will give some examples of such microbiota, their habitats, and conditions under which they live. We also suggest, instead of the terms "true extremophile" or "extremotolerant," the use of a new term that we derived from the nutrient demand of microorganisms, namely the term poikilotroph (Gorbushina et al., 1996b; Krumbein, 1988). This term, however, can also be applied to any of the mentioned environmental demands. The classification of microorganisms into oligotrophic and eutrophic (copiotroph or zymotroph have been used as partial synonyms) comes from the limnological literature describing environments rich in energy sources or nutrients (eutrophy) or poor in energy sources or nutrients (oligotrophy). Therefore, we suggest poikilotrophy as a new term for microorganisms adapted to regularly or irregularly occurring extreme changes of environmental conditions. The microorganisms living and surviving under such conditions then can be designated as poikilotrophic microbes.

6.2 THE ENVIRONMENT OF THE POIKILOTROPHIC MICROORGANISM

The first isolates of such microorganisms were gained from wind- and sun-exposed walls of monuments made of stone in Central Europe (Krumbein, 1966). Following this observation, a new set of microorganisms was found and described for rock-varnished hills of the Negev desert (Friedmann et al., 1967; Krumbein, 1969), and from the calcareous slopes of the foothills near Jerusalem (Krumbein, 1968). Folk and Lynch (1993) for many years tried to attract the attention of microbiologists to nanobacteria from extremely salty, hot or dry environments. They claimed that Krumbein (1983) showed such bacteria, but did not describe them. Staley et al. (1982) hinted at the desert environment as the habitat of a certain class of fungi, which they termed microcolonial. They claimed that these chemoorganotrophic microbiota must depend on external sources of all ingredients for life, namely water, energy sources, and nutrients. Thus, their environment is practically nonexistent. These microorganisms live where they should not live and thrive on accidentally supplied raw materials (Figure 6.1). Such an environment can even be imagined for a meteorite or a planet without life. However, many earthly environments remain in which the search for life is just starting. Eppard et al. (1996) have described the same desert environments for a group of ill-defined actinomycetes and coryneform bacteria, the *Geodermatophilus* group. In a recent program "Life in Extreme Environments" (LEXEN), the National Science Foundation of the USA is supporting research for organisms and life strategies in unusual environments. Among the proposed research places were hot and cold deserts, deeply buried rock environments accessible only with deep drills, deep mines, deep-sea sediments, hot springs, salt ponds, salt mines and others. Museums and collections in botanical, zoological, and mineralogical institutes around the world might provide fertile ground for finding poikilotrophic microorganisms that may await an external trigger to resume metabolic activity after up to 350 years (Gorbushina and Krumbein, in press; Rebrikova and Manturovskaya, 1993). Other sites could be the yet-unexplored tombs of the pharaohs in Egypt.

A very typical site for poikilotrophic microorganisms is the multilayered surface of murals and cave paintings. Some of these environments exhibit drastic changes in temperature, illumination, water activity, salinity, alkalinity and nutrient availability. Cave paintings are known to exceed 30,000 years BP, and some special sites for poikilotrophs hidden in soil can be accurately dated because they were used or treated by humans, e.g. flint tools of sites in Africa and Asia Minor buried in dry soil or caves more than 500,000 years BP (Gorbushina et al., 1996a). Some of these sites have been

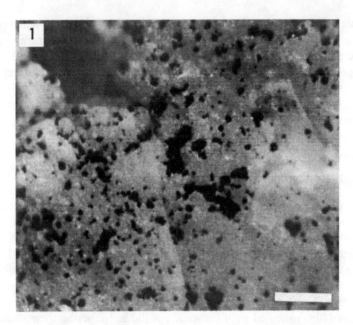

FIGURE 6.1 Compact fungal colonies on the air-exposed surface of marble from the Chersonesus museum complex (Crimea, Ukraine). This typical growth pattern in scattered round colonies is observed also for other rock-inhabiting poikilotrophic microorganisms (e.g. cyanobacteria). Bar = 10 mm.

analyzed by molecular ecology techniques, and have been shown to contain a large variety of signals including halophilic Archaea and some *Frankia*-like actinomycetes (Rölleke et al., 1996). These, however, have not yet been brought into culture.

If we assume that the uppermost 10–50 m of the earth's crust are well searched, another 3–10 km, where the temperatures are below 150°C and may contain subcritical or reactive water, still await exploration. The assumption seems justified that the unexplored portion of the earth, in which poikilotrophic microorganisms might await detection, isolation, and description, could represent as much as 99.9% of the crust. In view of the earth's history, it is tempting to consider that the environments for poikilotrophic microorganisms possibly prevailed during prolonged periods and could even have been the major surface environments in the microbial era, i.e., from 3.5 Ga until 0.5 Ga before present.

6.3 DEFINITION AND DESCRIPTION OF THE POIKILOTROPHIC MICROORGANISM

Examples of isolates from true deserts (Namib, Mojave, Sinai), or from pseudo-deserts such as sun-exposed building surfaces or mural paintings, showed the occurrence of strangely shaped and strangely behaving microorganisms (Krumbein, 1966; Staley et al., 1982). Many of them have been microscopically observed, their metabolic activity ascertained *in situ* and in the laboratory. It was, however, impossible to get sufficient growth to study the physiology and molecular composition. Some cultures transferred from the laboratory of Staley into ours did not attain colonies larger than a small pinhead within 18 months of sub-cultivation. Even less information exists on the microflora of dry salt mines, deeply buried ore deposits under rock cover and other remote places (Hofmann, 1989; de Hoog, personal communication). All these subaerial environments have been shown to exhibit microbial biofilms. The latter, in contrast to subaquatic biofilms, which represent 99.9% microbially stabilized water, may be termed poikilotroph biofilms and are characterized by the opposite: these biofilms represent 99.9% biomass surviving and metabolizing with a minimum of water. Whether nanobacteria biofilms described by Folk and Lynch (1993), for example, are survival stages and strategies, or whether nanobacteria stay nanobacteria when subcultured, is solved only in very few cases. Picoplankton cyanobacteria e.g. remain picoplankton in culture, while some other coccoid cyanobacteria exhibit major changes in morphology when subcultivated (Palinska, 1996).

Rock-inhabiting microorganisms can be classified into three groups according to their survival strategies. When a rock is excavated or exposed by erosion to the atmosphere it is characterized in the beginning by oligotroph (i.e. growing and metabolizing slowly but steadily at low concentrations of energy sources, nutrients and water); or eutroph (i.e. growing and metabolizing fast at high energy-source, nutrient and water levels but being forced into survival and escape strategies like spores and conidia when the conditions on and in the rock change) microbial communities, depending on the initial levels of energy sources, nutrients and water supply. Gromov and Pavlenko (1989) have already realized that, within the juxtaposition of the eutroph and oligotroph principles, there are some points that necessitate perhaps another type of response or a subdivision of the oligotroph principle (see Table 6.1).

TABLE 6.1
Classification of Oligotroph Micro-Organisms According to Growth Features

Growth features	Representatives
On organic medium grow only in first dilution, do not stand any transfer	Bacteria of unusual morphology: hexagonal, toroid, triangular cells
In the first transfer grow only on poor media, later can be cultivated on rich ones as well	*Pseudomonas, Agrobacterium, Photobacterium, Vibrio, Aeromonas, Micrococcus, Arthrobacter. Corynebacterium* etc.
Isolation and cultivation is possible only on poor media	*Hyphomicrobium, Caulobacter, Microcyclus, Leptothrix, Ochrobium, Metallogenium*
Not cultivable in lab conditions. Found in natural water reservoirs during SEM observation	Prosthecate bacteria; some bacteria with gas vesicles

Gromov and Pavlenko (1989) also noted that oligotroph bacteria sometimes change into a copiotrophic growth pattern after several passages on apparently "richer" media. This possibly indicates a heterogeneity of the so-called oligotroph cluster or the existence of a different poikilotroph strategy, which is implied in this chapter.

Under prolonged absence of nutrients and energy sources and periodic or permanent extreme dryness rarely or sporadically interrupted by short periods of wetness, and a supply of organic compounds from the atmosphere, the poikilotrophic type of strategy will be more and more established on and in the rocks. The poikilotrophic microorganism is characterized by generally (or even extremely) slow growth. The colony on the substrate or in laboratory culture is dense, and usually irregular node-like packages of cells of irregular shapes occur. Rarely or never do specialized propagative or survival cells or structures form. On the contrary: each individual cell and cell cluster is equipped and structured to survive as a whole (Figure 6.2). Photosynthetic representatives (e.g. *Chroococcidiopsis, Stigonema*)

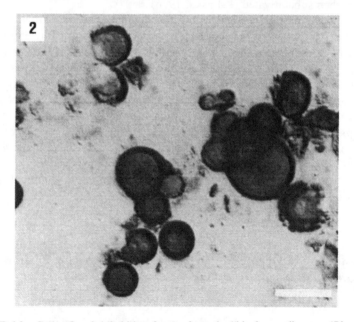

FIGURE 6.2 Cells of rock-inhabiting fungus from the "black yeast" group (*Phaeococcus* sp.) in culture, showing budding cells with expressively thick and melanized cell walls. These cells in culture are forming extremely slow-growing compact colonies similar to those depicted on the original substrate in SEM (Figure 6.4). Bar = 5 μm.

grow at low or extremely high light and water activity. Chemoorganotrophic representatives (e.g. *Geodermatophilus, Monodictys, Acrodictys, Phaeosclera, Trimmatostroma*) grow slowly by budding or similar division types (Figure 6.2 and 6.4) and exhibit typical small, leathery or even stone-like microcolonies (Figure 6.1 and 6.4) that can develop into braid-like structures or macrohyphae-like intertwined strands,

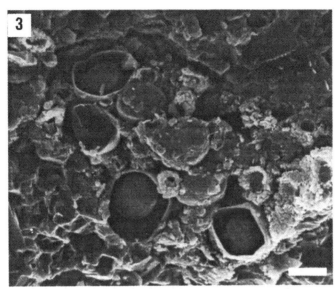

FIGURE 6.3 Algal cells living on an air-exposed rock surface that protected themselves from environmental hazards (e.g. UV light or desiccation) by excreting a thick slime sheath (extracellular polymeric substance). Bar = 5 μm.

not unlike the slime-embedded bundles of another representative of the phototrophic poikilotroph type (*Microcoleus* sp. or *Hydrocoleum* sp. from desert soil and rock). The maintenance of each individual cell formed in one place for prolonged periods of up to 50 years and more is a typical feature.

Microorganisms dwelling on and in poikilotroph environments need to be able to maintain all produced cell material as long as possible. They must be able to survive long periods of dormancy or extremely reduced metabolic rates until conditions are favorable for photosynthesis or chemosynthesis. In many rock surface and sedimentary environments, enough energy sources (including light), nutrients and water are available simultaneously only occasionally. Thus, optimal conditions for life processes are rarely guaranteed in such environments. The poikilotrophic microbe's cells have an optimal volume-to-surface ratio. Figures 6.2 and 6.4 show spherical yeast-like growth in fungi instead of hyphae development. This allows a delay in the net rate of evaporation.

The poikilotrophic microorganism is also characterized by a resistant cell wall with photoprotective pigments (dark to black) (Figures 6.1 and 6.2) and is well equipped for survival of desiccation and drought with special coatings and envelopes (Figure 6.3), usually made of polysaccharides in addition to proteins, fatty acids and some very important macromolecules such as melanins, carotenes, lignin and lignin-related substances such as sporopollenins. Scytonemine is typical for *Scytonema*, *Stigonema*, *Chroococcidiopsis* and other poikilotroph cyanobacteria. The poikilotroph microbe channels energy and nutrients into polysaccharide production for the maintenance and accumulation of the extracellular polymeric substance layers. It produces compounds like trehalose, a non-reducing disaccharide which, coupled

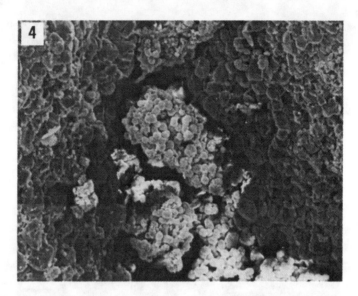

FIGURE 6.4 Fungal colonies on a marble surface of the Dionysos theater in Athens (Greece). Compact colonies with roundish meristematic yeast-like cells (having approximately the same diameter as the marble grains) occupy the space between the rock grains and enlarge the cavities to establish their own habitat near the hostile rock surface. Neither hyphal nor propagative structures are visible. Bar = 30 µm.

with the synthesis of lipids with an elevated level of nonsaturation, keeps phospholipids in a liquid-crystal phase (Potts, 1994; Schrödinger, 1944). Spherical cells of these organisms also form a compact spherical colony, following the same principle of reducing the evaporation rate and other environmental hazards. This holds true even for fungi, whose normal growth pattern is different from spherical cell clusters. Microcolonial fungi first described from desert varnish by Staley et al. (1982), and then discovered on rock surfaces but termed black yeast-like fungi (Gorbushina et al., 1993; Wollenzien et al., 1995), are perfect examples of poikilotrophic microorganisms able to survive extreme changes in environmental conditions (Palmer et al., 1987; Sterflinger and Krumbein, 1995).

Some cyanobacteria of the pleurocapsalean group, the actinomycete *Geodermatophilus*, the yeast-like black fungi *Coniosporium*, *Monodictys*, *Phaeosclera* and *Sarcinomyces* and other genera belong to this new metabolic class detected mainly in the rock biofilm environment (Eppard et al., 1996), mural paintings (Petersen et al. 1993), and salterns (de Hoog, personal communication).

The special environment of mural paintings and rock paintings, as well as the underground environment of mines, finally deserve some more words for historical reasons and for their importance to the cultural history of humankind. All the environments mentioned here have in common that (1) light for phototrophs occurs rarely and sporadically, (2) alkalinity and salinity can have drastic changes from summer to winter, from restoration to restoration and by special technical changes in the environment, and (3) there might be sporadic changes in nutrient and humidity supply. Alexander von Humboldt mentioned the cave, mine or lantern flora in the first of his many publications, written in Latin about the flora of the mine of Freiberg in Saxonia, the oldest mining school in the world.

6.4 CONCLUSION

A relatively large number of bacteria and cyanobacteria, as well as fungi and actinomycetes isolated from such environments, still need further characterization and identification. One of the major obstacles to achieving this is that all of them grow quite slowly. Another obstacle is that some of the poikilotrophic microorganisms have the tendency to be associated with other bacteria in an intimate yet unknown relationship, sometimes called symbiotic, and sometimes proto-lichenic. Some lichens are related by their metabolic and morphological characteristics to the organism type we identified as poikilotroph. Especially the longevity of the individual cells and the way of propagation by vegetative thallus parts suggests the relatedness to poikilotroph life strategy. However, the well-established symbiotic relationship between the photo- and mycobiont in lichens brings about morphologically much more complex propagative structures with the same functions. The morphological complexity and the difficulty in laboratory cultivation of these organisms are restricting the work to the original material only, and separate lichens into a special group of poikilotrophic organisms.

As in the case of other organisms specially adapted to very narrow and unusual habitat conditions (Archaea, hind-gut flora, etc.), this new group of specially adapted microorganisms poses many questions. It is not surprising, however, that we could identify representatives from many different groups of microorganisms. This can easily be explained by the fact that this group is not characterized by extremely narrow limits of growth — albeit deviating extremely from our environmental demands — but by wide limits of growth with very short periods of time during which metabolic activities are possible.

In conclusion, the strategy of the poikilotrophic organism can be regarded as an example of an interesting adaptational pattern. It is an organism that is suited to surviving considerable spans of time without active metabolism. Further, it usually metabolizes slowly even under optimal growth conditions, as if it does not dare to believe in good times. Microbial dormancy or survival of vegetative cells and structures was another approach to this phenomenon discussed widely in microbiology. We think that this discussion of the poikilotroph principle in the frame of a symposium on hypersaline environments, like the organism itself, has a potential for survival, even if not permanently, but merely sporadically, nourished.

6.5 SUMMARY

Extreme environments are sometimes misinterpreted as extremely hot, cold, salty, or dry (compared with the physiology and ecology of humankind). The environments in which any organism (including microorganisms) really has extreme problems of adaptation exhibit special phenomena and processes on the side of the environment and the organism living within. Such environments are characterized by sudden or oscillating changes from very high to very low temperatures, from very high to very low salinity, alkalinity, etc. The organisms faced with such extremes usually develop special strategies. A response pattern of several groups of microorganisms to desert or desert-like conditions on and in rocks and other hard substrates is described. Examples stem from true deserts (e.g. the Negev, Sinai and Namib) or from pseudo-deserts such as building surfaces and mural paintings. The microorganisms living routinely and consistently under such conditions are termed poikilotrophs. Often lacking nutrients, energy sources and moisture, only rarely or sporadically interrupted by short wet periods and a supply of organic energy sources from the atmosphere, the poikilotrophic type of strategy will become more and more established. The poikilotrophic microorganism is characterized by generally slow growth of dense and usually irregular node-like packages of cells of irregular shape. Specialized propagative or survival cells or structures are rarely or never formed. On the contrary, each individual cell and cell cluster is equipped and structured to survive as a whole. In addition, irradiation-resistant pigmentation (dark to black), high water storage potential through thick cell walls, and extracellular polymeric substances are frequently found. Each individual cell formed in one place at a certain moment may be maintained for prolonged periods of up to 50 years or more. The major strategy of the poikilotroph, thus, is self-protection against extreme changes in environmental conditions.

ACKNOWLEDGMENTS

A great deal of our knowledge on poikilotrophs stems from work on grant Nr. Kr 333/25 of the DFG and grant Nr. 02895 of the DBU. We acknowledge discussions of the topic with B.V. Gromov.

REFERENCES

Brock, T.D. 1978. *Thermophilic Microorganisms and Life at High Temperatures.* Springer-Verlag, New York, 465 pp.

Cohn, F. 1862. Über die Algen des Karlsbader Sprudels, mit Rücksicht auf die Bildung des Sprudelsinters. *Abhandlungen der Schlesischen Gesellschaft für vaterländische Kultur, Abt. Naturwiss. Med.* II:35–55.

Eppard, M., W.E. Krumbein, C. Koch, E. Rhiel, J.T. Staley, and E. Stackebrandt. 1996. Morphological, physiological and molecular biological investigations on new isolates similar to the genus *Geodermatophilus* (Actinomycetes). *Arch. Microbiol.* 166:12–22.

Folk, R.L. and F.L. Lynch. 1997. The possible role of nanobacteria (dwarf bacteria) in clay-mineral diagenesis and the importance of careful sample preparation in high-magnification SEM study. *J. Sed. Res.* 67:583–589.

Friedmann, E.I., Y. Lipkin, and R. Ocampo-Paus. 1967. Desert algae of the Negev. *Phycologia* 6:185–200.

Gorbushina, A.A. and W.E. Krumbein. Sub-aerial microbial mats and their effects on soil and rock. in *Microbialites*, R. Riding, Ed. Springer-Verlag, Berlin, in press.

Gorbushina, A.A., W.E. Krumbein, C.-H. Hamann, L. Panina, S. Soukharjevski, and U. Wollenzien. 1993. On the role of black fungi in colour change and biodeterioration of antique marbles. *Geomicrobiol. J.* 11:205–221.

Gorbushina, A.A., W.E. Krumbein, A. Rosenfeld, and N. Goren-Inbar. 1996a. On the microbiology of flint tools and silica skins, in *Abstracts of the International Union of Microbiological Societies–8th Int. Congress of Bacteriology and Mycology Division of IUMS, Jerusalem.* p. 103.

Gorbushina, A.A., W.E. Krumbein, and D.Y. Vlasov. 1996b. Biocarst cycles on monument surfaces, in *Preservation and Restoration of Cultural Heritage*, R. Pancella, Ed. Proceedings of the 1995 LPC Congress, EPFL, Lausanne. 319–332.

Gromov, B.V. and G.V. Pavlenko. 1989. *Ecology of Bacteria* (in Russian). Publishing House of LSU, Leningrad. 248 pp.

Hausmann, K. and B.P. Kremer. 1994. *Extremophile. Mikroorganismen in ausgefallenen Lebensräumen.* VCH, Weinheim, 419 pp.

Hofmann, B. 1989. *Genese, Alteration und rezentes Fließsystem des Uranerzlagerstätte Krunkelbach (Menzenschwand, Südschwarzwald).* Ph.D. thesis, Bern, 195 pp.

Javor, B. 1989. *Hypersaline Environments. Microbiology and Biogeochemistry.* Springer-Verlag, New York. 419 pp.

Karl, D.M., Ed. 1995. *Deep-sea Hydrothermal Vents.* CRC, Boca Raton, 299 pp.

Krumbein, W.E. 1966. *Zur Frage der Gesteinsverwitterung (Über geochemische und mikrobiologische Bereiche der exogenen Dynamik).* Ph.D. thesis, Würzburg, 130 pp. and appendices.

Krumbein, W.E. 1968. Geomicrobiology and geochemistry of the "Nari-Lime-Crust," in *Recent Developments in Carbonate Sedimentology in Central Europe,* G. Müller and G.M. Friedman, Eds. Springer-Verlag, Berlin. 138–147.

Krumbein, W.E. 1969. Über den Einfluß der Mikroflora auf die exogene Dynamik (Verwitterung und Krustenbildung). *Geol. Rundsch.* 58:333–363.

Krumbein, W.E., Ed. 1983. *Microbial Geochemistry.* Blackwell, Oxford, 330 pp.

Krumbein, W.E. 1988. Biotransformation in monuments–a sociobiological study. *Durability of Building Materials* 5:359–382.

Krumbein, W.E. and G.M. Friedman. 1985. *Hypersaline Ecosystems–The Gavish Sabkha.* Springer-Verlag, Berlin. 484 pp.

Palinska, K.A. 1996. *Ecophysiological and Taxonomical Studies of Unicellular Cyanobacteria Isolated from Microbial Mats of the German Wadden Sea.* Ph.D. thesis, University of Oldenburg, 122 pp.

Palmer, F.E., D.R. Emery, J. Stemmler, and J.T. Staley. 1987. Survival and growth of microcolonial rock fungi as affected by temperature and humidity. *New Phytol.* 107:155–162.

Petersen, K., W.E. Krumbein, N. Häfner, E. Lux, and A. Mieth. 1993. Aspects of biocide application on wall paintings–report on EUROCARE project EU 489. Biodecay, in *Conservation of Stone and Other Materials. Vol. 2. Prevention and Treatments,* M.J. Thiel and F.N. Spon, Eds. London, 597–604.

Potts, M. 1994. Desiccation tolerance in prokaryotes. *Microbiol. Rev.* 58:755–805.

Rebrikova, N.L. and N.V. Manturovskaya. 1993. Study of factors facilitating the viability of microscopic fungi in library and museum collections, in *ICOM Congress Proceedings, Vol. II.* Dresden. 887–880.

Rölleke, S., G. Muyzer, C. Wawer, G. Wanner, and W. Lubitz. 1996. Identification of bacteria in a biodegraded wall painting by denaturing gradient gel electrophoresis of PCR-amplified gene fragments coding for 16S rRNA. *Appl. Environ. Microbiol.* 62:2059–2065.

Schrödinger, E. 1944. *What is Life?* Cambridge University Press, Cambridge, UK.

Shilo, M., Ed. 1979. *Strategies of Life in Extreme Environments*. Verlag Chemie, Weinheim. 519 pp.

Staley, J.T., F. Palmer and J.B. Adams. 1982. Microcolonial fungi: common inhabitants on desert rocks? *Science* 215:1093–1095.

Sterflinger, K. and Krumbein, W.E. 1995. Multiple stress factors affecting growth of rock-inhabiting black fungi. *Botanica Acta* 108:490–496.

Wollenzien, U., G.S. de Hoog, W.E. Krumbein, and C. Urzi. 1995. On the isolation of microcolonial fungi occurring on and in marble and other calcareous rocks. *Science of the Total Environment.* 167:287–294.

Section II

Biogeochemistry

7 Hypersaline Depositional Environments and Their Relation to Oil Generation

Zeev Aizenshtat, Irena Miloslavski, Dorit Ashengrau, and Aharon Oren

CONTENTS

7.1 Introduction 89
7.2 Models of Oil Generation Based on Processes in the Open Sea 90
7.3 Hypersaline Depositional Environments as a Possible Source for
    Oil Generation — a Literature Survey 90
7.4 Prolific Production of Biomass in Hypersaline Environments 91
7.5 Recent Hypersaline Benthic Microbial Communities and Possible
    Resemblance to Precambrian Stromatolites 92
7.6 Case Studies 93
    7.6.1 The Monterey Formation (USA) 96
    7.6.2 The Jordan Valley Rift (Dead Sea Area): Bituminous Rocks,
          Asphalts and Oil Shales 98
    7.6.3 Tataria and Perm Basins (Volgo-Ural, Russia) 99
7.7 Geochemical Tracers for the Identification of Carbonate-Rich,
    Organic-Rich Sediments Deposited in Hypersaline, Highly
    Reducing, Restricted Environments 100
7.8 Conclusion 102
Acknowledgments 104
References 105

7.1 INTRODUCTION

In spite of the consensus that oil deposits originated from the accumulation of organic material in marine environments, little is known of the conditions that led to the kind of amassing of the amounts of organic matter sufficient to explain the size of the oil deposits worldwide. Scenarios proposed include the sedimentation of marine plankton on the sea bottom, with or without the formation of anaerobic conditions that supposedly favor the preservation of organic matter. In our opinion, the explanations brought forward in these models are unsatisfactory. We propose that hypersaline depositional environments in shallow basins in coastal marine environments could fulfill all the requirements for the collection of organic matter, which would lead to the formation of at least certain types of oil deposits.

In this chapter, we present evidence to support the theory that the existence of hypersaline conditions might have been a prerequisite for the formation of source rocks capable of generation of petroleum, especially type IIS kerogens.

7.2 MODELS OF OIL GENERATION BASED ON PROCESSES IN THE OPEN SEA

Before describing our model involving hypersaline depositional environments, we must discuss other potential sources of organic matter contribution to marine sediments, such as the sinking of plankton from the upper part of the water body to the sea bottom. Plankton proliferates mostly in the photic zone, with an optimum located generally between 10 to 75 m (Hite and Anders, 1991; Williams, 1984). Plankton production, even under optimal conditions, will not lead to massive collection of organic matter on the bottom, as bacterial degradation makes survival of the organic matter very limited. Hence, the amounts reaching the bottom are small. On the sea bottom, the organic matter remains subject to microbial degradation, either aerobically in the oxygenated water, or in the anaerobic sediment by dissimilatory sulfate reduction.

The Black Sea has been presented as a possible model for the formation of black oil shales, a model based on the accumulation of organic matter transported from rivers. However, the deep water body, even if anoxic, still supports high rates of organic matter degradation, using sulfate as the terminal electron acceptor (Pedersen and Calvert, 1990), and thereby making a long-term accumulation on the bottom improbable.

It has been claimed that upwelling effects such as occur off the Peruvian continental shelf, might also lead to mass accumulation of organic material on the sea bottom (Reimers, 1981). These sediments show dense communities of sulfide-oxidizing *Beggiatoaceae*, which are only in form similar to the cyanobacterial mats. *Beggiatoa* does not contain chlorophylls or carotenoids, derivatives of which are typically found in the types of oils and source rocks encountered. Thus, there is no chemical evidence that *Beggiatoa* mats are the source of the oil (Williams, 1984).

7.3 HYPERSALINE DEPOSITIONAL ENVIRONMENTS AS A POSSIBLE SOURCE FOR OIL GENERATION — A LITERATURE SURVEY

The importance of carbonaceous organic-rich sediments as potential source rocks for oil generation was only hesitantly mentioned before 1984. The 1984 symposium on "Petroleum Geochemistry and Source-Rock Potential of Carbonate Rocks" (Palacas, 1984) contributed much to change our views in this respect. The papers published in the proceedings of that meeting encouraged the inclusion of carbonate sediments as potential source rocks in petroleum exploration (Oehler et al., 1984; Palacas, 1984; Palacas et al., 1984).

The most recent comprehensive review of the relation between petroleum and evaporites was published by Hite and Anders (1991). Illustrated by a number of case studies, this review covers different types of depositional environments from carbonate through anhydrite to sediments in which halite and potash are the major evaporite constituents, at salinities increasing from 35‰ (marine) to more than 350‰ (hypersaline).

In most cases, the formation of hypersaline conditions by evaporation creates a combination of the development of a benthic microbial community in the shallow waters and accumulation of organic material in a deeper anoxic water body or sediment. The relation between anoxia and productivity in connection with the formation of organic-rich sediments is still debated. Dismissing the prevailing explanation that anoxic conditions alone suffice to invoke the formation of organic-rich deposits, Pedersen and Calvert (1990) claim that primary productivity provides first-order control for the accumulation of organic-rich rocks. It is obvious that if both a high productivity and anoxia with reducing conditions exist simultaneously or in sequence, the conditions for sedimentation of an organic-rich facies are fulfilled. Stromatolites — layered deposits, rich in organic matter — are often formed under these conditions, and they present one of the most intriguing formations of evaporites.

The reconstruction of spatio-temporal relationships of petroleum formation and accumulation of evaporites, based on the data collected by Parparova et al. (1981) and Zharkov (1981) as plotted by Hite and Anders (1991), show a reasonable coincidence of peak production for both. One should take into account that not in all cases the age of the source rocks is known, and that for most oils only the reservoir stratigraphy is recorded. It is also known that fine-grained carbonates can form good cap-rocks, and that salt domes create good traps for petroleum. Hence, this correlation can only be employed in general terms. In spite of all this, ample evidence exists for the presence of abundant amounts of organic matter in proven carbonaceous source rocks, and examples will be discussed below.

Large basins with rich type-IIS kerogens generally exhibit carbonate as major evaporitic constituent. In some cases, a siliceous member is present as a remnant of a diatomaceous contribution, with very little terrestrial-derived smectites. Such cases mentioned in the literature are referred to as black shales, marls or bituminous rocks, such as the Smackover Formation (Mississippi, Alabama and Florida, dating from the Jurassic period) (Oehler et al., 1984), and the Paradox basin (Utah). Hite and Anders (1991) also include some of the Green River Formation sections in this category. However, we prefer not to adopt this idea because of the freshwater origin of this unusual inland-lake depositional environment. The lack of sulfate in the water body of this lake and the absence of significant bacterial sulfate reduction led to the formation of the unique type I kerogens.

7.4 PROLIFIC PRODUCTION OF BIOMASS IN HYPERSALINE ENVIRONMENTS

The most prolific conditions for the accumulation of biomass are present in benthic microbial communities known as cyanobacterial mats. Primary production in these mats can be exceedingly high, and the salinity of the ecosystem, often in excess of 200‰, protects the organic matter formed by excluding burrowing and grazing animals. It may be difficult to assess to what extent planktonic productivity contributes to the biomass reaching the sediment. This question was addressed earlier, and no direct molecular biomarkers could be found to differentiate between benthic and planktonic contributions (Aizenshtat, 1989a). In addition, the "chemical postmortem" of the benthic microbial communities is also enigmatic, but some of the lipids of the major bacterial types have been identified (Cohen and Rosenberg, 1989; Cohen et al., 1984; Hite and Anders, 1991). Although the lipid fractions are very informative as potential biomarkers (see below), the bulk of the biomass that dominates the formation of the sedimentary organic matter is becoming rapidly insoluble, forming a protokerogen rich in bonded sulfur in the case of benthic microbial communities, or forming humic matter in the case of planktonic and terrestrial-derived matter, which is transformed into kerogen upon aging and as a result of thermal processes.

We have reviewed the interrelationships of carbon and sulfur cycles in depositional environments (Aizenshtat, 1989b; Aizenshtat et al., 1995). Here we want to mention that both carbon- and sulfur-stable isotope ratios can be used as indicators for the existence of restrictive depositional conditions. The rate at which carbon is fixed into biomass strongly influences the isotope selectivity. If productivity is very high, CO_2 limitation leads to a microbial biomass with a heavy carbon signature (Aizenshtat et al., 1984). The supply of sulfate to the hypersaline waters determines the degree of isotopic discrimination by the sulfate-reducing bacteria. In closed systems, the pool of polysulfides will gradually become isotopically similar to the sulfate supplied. However, when the supply of sulfate is unlimited, the enrichment of the organic matter with sulfur originating from dissimilatory sulfate reduction results in a light sulfur ($\delta^{34}S$ around –40‰) (Aizenshtat et al., 1984; Dinur et al., 1980). It is important to note that the availability of iron in the sedimentary environment is a prime factor determining the competition between pyrite formation and the secondary enrichment of the organic matter by the reduced sulfur. Hence, floods from terrestrial sources or deltas of big rivers have a twofold influence on possible sabkhas: they reduce the salinity, with formation of lacustrine conditions, and they import iron-rich smectites. Under these conditions the nature of the biomass formed will be different, and so will the post-depositional chemical conditions, leading to formation of kerogen of type II or the coaly type III.

7.5 RECENT HYPERSALINE BENTHIC MICROBIAL COMMUNITIES AND POSSIBLE RESEMBLANCE TO PRECAMBRIAN STROMATOLITES

Hypersaline cyanobacteria-dominated mats are presently scattered, mostly in places where the oceans meet arid climates. Some of these mats are formed in the tidal zones and some are more lagoonal. Such mats were described from the Guerrero Negro (Baja California, Mexico) (D'Amelio et al., 1989) and from Solar Lake (Sinai, Egypt) (Krumbein, 1983; Krumbein and Cohen, 1974). The models presented in Figure 7.1 show various possibilities for the development of anoxia and basin formation. The shallow water areas, where the high benthic productivity can produce large amounts of biomass that can be sedimented in protected environments, are thus of great interest. We and others have initiated in-depth studies of such environments, which could be responsible for the formation of laminated carbonate with layers rich in organic matter. Our studies concentrated on the Solar Lake (Sinai) and other sabkhas in the area, while some lagoons and salt ponds were also included. The results of these interdisciplinary studies were presented in three books (Cohen and Rosenberg, 1989; Cohen et al., 1984; Friedman and Krumbein, 1985). The mats at these sites consist of layers rich in organic material, interlaminated with carbonate and some biogenic silicates deposited by diatoms, which grow at the water-sediment interface. The interstitial water is highly alkaline, and very rich in dissolved polysulfides formed by the intense activity of sulfate-reducing bacteria (Aizenshtat et al., 1984).

Despite the considerable geographical separation, the mats from the abovementioned locations are very similar. In structure, some Australian mats show similar benthic microbial communities (Bauld, 1984), however, no in-depth comparison of the lipids extracted from such mats has yet been performed. Attempts to artificially mature the mats yielded a highly sulfur-enriched kerogen. Some of this sulfur is still thermally unstable, and its functionality differs from that found in petroleum (Aizenshtat et al., 1995; Krein and Aizenshtat, 1995). Sediments a few million years older, such as the Senonian bituminous rocks of Israel, the Monterey Formation (Miocene), and the oldest suspected Domanic (Devonian) of the Volgo-Ural area, carry in their bulk parameters and, to some extent, in their biomarkers, the memory of the depositional conditions. Hite and Anders (1991) use other examples to establish the importance of various degrees of hypersalinity without pointing to the possibility of the involvement of benthic microbial communities. Below we will describe the depositional models in an attempt to reconstruct the prevailing chemical conditions, to prove that at least type IIS kerogen could be formed from organic matter deposited by benthic microbial communities.

7.6 CASE STUDIES

To obtain a better understanding of the relationship between potential source rocks capable of generation of oil and depositional environments, we have made an extensive search for geochemical parameters relevant to petroleum generation. In the present overview we concentrate on the reconstruction of specific hypersaline restricted environments leading to production of sediments rich in type IIS kerogen. These types of kerogens yield sulfur-rich oils with a characteristic constitution. We will discuss the possible link between the origin of these source rocks and the presence of a restricted biomass of halophilic microorganisms, and the changes these caused in the interstitial waters of the sediments. Three case studies will be discussed below: the Monterey Formation (USA), the bituminous rocks, asphalts, and oil shales of the Senonian Formation deposits in the Dead Sea area, Israel, and the Tataria and Perm basins (Russia). These case studies were not selected because they are unique; on the contrary, they coincide with global maxima of evaporite production, and large basins with rich type IIS kerogens are more and more often encountered. The

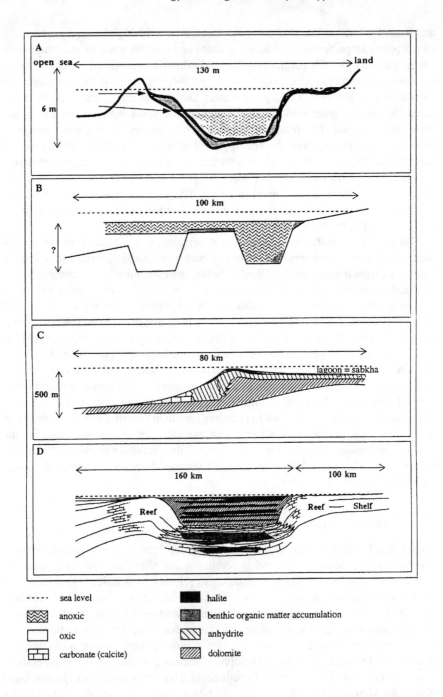

mid–late Devonian bituminous shales of Horn River, Canada, present an additional case, as the geochemical characterization of the potential source rocks suggested that the depositional environment was rich in carbonate and hypersaline (Feinstein et al., 1991).

Less relevant in the framework of the present discussion are the restricted hypersaline environments that occur in deep marine basins such as the Tyro Basin (eastern Mediterranean), a brine-filled anoxic environment (ten Haven et al., 1987). In such environments, organic material can accumulate, as the high salinity prevents effective microbial breakdown. The organic matter found in these sediments is sapropelic in nature, young, and of reworked planktonic origin. However, such environments are, and probably were, very limited in extent throughout geological history.

Some of the points to be discussed are: (1) What types of sedimentary organic matter are considered to originate from hypersaline depositional environments? (2) Is there a direct connection between the formation of an anoxic environment and the formation of certain organic type IIS kerogens? (3) What is the connection between the recent benthic formation of microbial mats in sabkhas and more-mature bituminous rocks, such as those found in the Israeli deposits? (4) What is the reason for the existence of the gross similarities between deposits of rich organic sediments also rich in organically bound sulfur? Such potential source rocks produce petroleum not only rich in organic sulfur but also in asphaltenes, and show a high VO/Ni ratio. In the past, we have investigated the correlation between the geochemical environment and the factors controlling the metallation of porphyrins (Lipiner et al., 1988). These studies show that the complexation of tetrapyrroles by the first transition metal ions can be controlled at the diagenetic stage by phase transfer catalysts or by carrier molecules such as fatty acids, long-chain thiols and hydroxyaromatic compounds. Moreover, Ni^{2+} and VO^{2+} complexes are the result of stability-controlled sequences

---

FIGURE 7.1 Schematic models for the formation of hypersaline conditions leading to the formation of either benthic microbial mats or anoxic basinal bottoms. A: The conditions for the Solar Lake (Sinai, Egypt), with benthic microbial community formation and seasonal stratification of the water column. Conditions are not representative for normal lagoons because of the development of a high temperature bellow (Cohen et al., 1984). The arrows indicate seepage of sea water into the hypersaline lake. B: Proposed two-lagoon model for the depositional environments in the Monterey Formation (Miocene, California), based on Pisciotto and Garrison (1981). The shaded area is anoxic. C: A typical sabkha tidal zone, alternately connected and disconnected to the open sea. The mode of accumulation of organic benthic microbial community matter is described by D'Amelio et al. (1989) for the Guerrero Negro (Baja California Sur, Mexico) lagoon. D: Scheme for the possible formation of a salt (halite) plug sealing off a reef. This may explain the limestone potential source rocks in the Delaware basin (West Texas), in which halite is present as both cap-rock and trap.

rather than the kinetically controlled mechanisms that will preferentially incorporate other metals. The domination of the VO complex is favored under the same conditions suggested for the secondary sulfur enrichment of sedimentary organic matter (Lipiner et al., 1988). (5) Is it coincidental that a low availability of iron and absence of floods or terrestrial input (evaporitic conditions in desert-bordering lagoons or sabkhas) led to little pyrite formation and accumulation of large amounts of reduced sulfur, resulting in secondary sulfur enrichment of the organic matter? (6) It has been suggested that the molecular biomarkers such as the C_{19} and C_{24} triterpanes are in most cases dominated by the C_{23} triterpane, derived from an unknown source of microorganisms typical to the above-mentioned conditions (Feinstein et al., 1991). Some investigators attributed these C_{19} and C_{24} triterpanes to the presence of tasmanite, shales in which microscopically recognizable *Tasmanales* algae have been identified by their characteristic fluorescence (Aquino Neto et al., 1989; Feinstein et al., 1991). However, the true nature of the tasmanite is still elusive to the geochemists.

7.6.1 THE MONTEREY FORMATION (USA)

Because the Monterey Formation can be found at various depths along the coast of California, it is presumptuous to offer a single model for the development of this depositional environment. Since the early years of this century, it has been suggested that the Monterey Formation is the principal source for California oil, and the formation was identified as the reservoir for these oils (Isaacs, 1984). Therefore, it is surprising that, despite the interest in and importance of this formation, comparatively little has been published on the Monterey as a source rock or reservoir. Published information on the depositional character of the Monterey organic matter and its variability according to the various basins is scant.

Some of the early thoughts about the origin of the various basins composing the Monterey Formation (also designated Monterey shale) were expressed by Reed (1933) and by Kleinpell (1938). The deposition of these "shales" was related to the middle and late Miocene period, when the ocean covered the basins that trapped voluminous quantities of biogenic silica. The occurrence of rhythmic bedded diatomaceous rocks was also explained by this model (Bramlette, 1946). However, in addition to the siliceous members, the Monterey "shale" contains carbonaceous sequences, sometimes rich in phosphate. Some of the laminations observed in the Monterey Formation were suggested to mimic microbial mats (Williams, 1984). Whether they really represent ancient microbial mats remains an open question. Involvement of algal mats is not contradicted by the presence of fossils of fish and other respiring animals. Their occurrence required some modification of the model proposed for the depositional conditions. A new model, suggested by Pisciotto and Garrison (1981) (see Figure 7.1B), assumes an anoxic bottom below an aerobic water body to explain the finding of planktonic fossils and fossilized fish. Lewan (1980) examined outcrop samples of various basins in California, and reported total organic carbon contents from below 2.5% for many of the "black shales" to more than 5% in a few samples. The diversity of the mineral matrix was discussed later by Isaacs (1984), and the kerogen was found to be especially rich in organic-bound sulfur. Moreover, elemental sulfur was found in relatively large amounts in the extracts (bitumen). Lewan's analysis of five bitumens for V/V+Ni ratio showed an excess of nickel in most of them, and only one sample showed a slight domination of vanadium. These results contradict previous studies that correlate a high sulfur content with dominance of VO-porphyrins (Dunning et al., 1953; Hodgson, 1971). The many analyses of Monterey oils also showed a relatively high content of VO-porphyrins (Orr, 1986).

The controlling factors relating sulfur enrichment and vanadium in porphyrins indicate different genetic stages in which this metallation can occur (Aizenshtat, 1989b; Aizenshtat and Sundararaman, 1989; Aizenshtat et al., 1984). The domination of V over Ni correlates well with the presence of highly reducing polysulfides in the depositional environment. Some chlorines still exist in the Monterey, and since the nature of the early metallation of chlorines is still not resolved, it is possible that some kinetic effects may lead to the formation of more Ni-chlorines. Heavy oils produced from Monterey type IIS kerogens produce oils of low API, high in S, and with a high V/Ni ratio.

The model for the Monterey deposition as based on shallow-water benthic prolific microbial mats with active sulfate-reducing bacterial activity contradicts the prior interpretations to some extent. Most investigators of the Monterey depositional sites have emphasized the importance of basin-floor deposition, but paleogeographic reconstructions (Isaacs, 1984) indicate extensive slope and shelf deposition with laminated, anoxic facies. Moreover, hints of extensive benthic foraminiferal assemblages were found, dominated by bolivinids (Williams, 1984). It is, of course, difficult to reconstruct the relative quantities of the various contributions to the organic matter by the identification of fossils. The $\delta^{13}C$ values recorded for most sedimentary organic matter in the Monterey and Monterey-derived oils are relatively heavy. As for the kerogens and some asphaltic shows, they are even slightly heavier than the more matured oils (Orr, 1986). This trend is expected if, for example, we correlate kerogen of +8‰ with oil of an average +6‰. The Monterey organic matter is in most cases enriched by 8‰ in ^{13}C in comparison with other marine-derived organic matter. Consequently, the Monterey oils show a similar trend. A similar isotope-distribution difference, even much more drastic, was observed in the Solar Lake (Sinai) and in solar saltern evaporation ponds cyanobacterial mats, an effect explained by benthic CO_2 (HCO_3^-)-limited cyanobacterial growth (Gazit-Yaari et al., this volume,. Schidlowski et al., 1984, 1985). If we take into consideration the fact that some organic matter could be introduced by floods from a terrestrial source, then the effect is even more pronounced, as terrestrial organic matter is lighter (^{12}C-enriched) than the planktonic marine-derived organic matter (Aizenshtat, 1989a).

Most geochemists who studied the sulfur-enriched sedimentary organic matter and oils derived from these potential source rocks agree that the enrichment of sulfur is secondary. This means that this reduced sulfur must be derived from the activity of dissimilatory sulfate-reducing bacteria. These show an isotopic fractionation effect for the $SO_4^{2-} \rightarrow HS^-$ reaction of as much as −40‰ (Goldhaber and Kaplan, 1974). However, values of $\delta^{34}S$ pyritic and organic matter in the Monterey range from $^-10$ to +24‰. As in other cases, the pyrite is always isotopically lighter (^{32}S-enriched) than the sulfur in bitumen, kerogen, or the oils produced. However, the sulfur in most sedimentary organic matter samples from the Monterey was 6–10‰ heavier than the pyrite. In bituminous marls of Israel, this value was about 20‰. The difference in δ ^{34}S between the sulfur-rich organic matter and pyrite encountered worldwide ranges from –32 to –4‰ (Aizenshtat et al., unpublished results).

The small difference in sulfur isotopic distribution between the pyrite and the sulfur in the sedimentary organic matter can be attributed to the slow release of iron available to form pyrite, and the presence of a large excess of reduced sulfur with relatively active organic matter. Another possibility is that some of the pyrite was formed during the early stages of deposition and the remainder was secondary. This secondary pyrite could be formed by equilibration with the labile sulfur released during later stages of diagenesis, or even in the early catagenesis stage. Microscopic examination of the Monterey pyrite shows that only a small portion is framboidal. Analysis of several Monterey sediments and oil samples revealed that the pyrite is always isotopically lightest, the kerogens somewhat heavier, and all the oils are enriched in ^{34}S. At the time we first reported this phenomenon (Aizenshtat, 1989b), the idea was that there is a possibility that sulfur isotope discrimination could occur during maturation. However, Orr (1986) claimed that no isotope changes do occur between kerogens/asphalt and petroleum of the Monterey. We believe that we now have compelling evidence that our interpretation was correct, but this is beyond the scope of the present discussion. We have offered an explanation in which the H_2S released from the rich IIS kerogen is enriched in ^{32}S, while the sulfur stabilized in C-S-C bonds is enriched in ^{34}S. This was shown in thermal (pyrolysis) experiments and now recorded by the Institut Français de Pétrole (Everlien et al., 1997). The rather heavy δ^{34}S values recorded for the Monterey oils (+16 to +24‰) presented a puzzle, which drove some investigators to offer the chemical reduction of sulfate to organically bound sulfur as a possible explanation. We do not believe that the high temperatures required for such a mechanism did prevail in any of the basins studied. On the contrary, the production of heavy oils from the Monterey Formation potential source rocks is suggested to have occurred at relatively low temperatures. Hence, we suggest that the benthic sulfate reducing bacteria in the mats were operating under conditions alternating between closed, i.e., a restricted supply of sulfate throughout the water body, or a limited supply of sulfate to the interstitial water, and open, i.e., with an unlimited supply of sulfate. The fact that no or very little sulfate-containing minerals were found in the Monterey assemblage indicates that most of the sulfate was removed.

7.6.2 THE JORDAN VALLEY RIFT (DEAD SEA AREA): BITUMINOUS ROCKS, ASPHALTS AND OIL SHALES

Our group has extensively studied the organically rich Senonian bituminous rocks, sometimes referred to as the "oil shales" of the Dead Sea area. We have compared the asphalts and some oil shows taken from various drill holes with the Senonian bituminous rock as a potential source rock and the asphalts as early maturation products. The vast amount of geochemical information gathered, describing the various sedimentary organic shows of the Jordan Valley Rift, can be found in the publications by Spiro (1980), Tannenbaum (1983), and Krein (1993). Additional studies of the east members of the bituminous Senonian in Jordan were performed, but no comparison with the Israeli bituminous rocks has yet been published (see Kohnen et al., 1990).

We can summarize the information relevant for the present discussion as follows: (1) Although diversity is found in some specific geochemical indices, most organic matter has unique characteristics, showing that it has been deposited under similar conditions. (2) Sediments are mostly carbonaceous (marls). (3) The organic matter is very rich in sulfur (8–12%). (4) The VO/Ni ratio, mostly in the asphaltene fraction, is very high. (5) Very little terrestrial input, if any, can be demonstrated in both the inorganic and the organic matrix; vitrinite is absent. (6) Some cases of lamination and interbedding with fine-grained carbonates occur. (7) Sulfur isotopic compositions are uniform for each sub-basin, but differ for different basins, possibly indicating alternations between closed and open systems.

The oils found are very different in comparison with the oils found in the coastal plain of Israel. It is also important to note that, in some cases, the total organic carbon of the bituminous rocks reaches 30%, whereas in general, the total organic carbon content is 9–14%.

The reconstruction of the conditions prevailing during deposition indicates that benthic microbial communities, formed under slightly hypersaline conditions, may have been involved. A comparison of the geochemical markers in the Jordan Valley samples and in the other sites studied (Monterey Formation, Tataria and Perm basins) follows.

7.6.3 TATARIA AND PERM BASINS (VOLGO-URAL, RUSSIA)

The Tataria and Perm basins are two prolific oil-producing basins in the Volgo-Ural region, among the greatest Paleozoic oil basins in the world. Paleozoic crustal epeirogenesis (continent formation) and transgression-regression cycles resulted in a regional stratigraphic pattern of clastic carbonate rock alternations, whereas contemporaneous extensive faulting and block rotation during sedimentation resulted in relatively sharp lithofacies and variations (Gutman and Osipova, 1982; Kozlov et al., 1995; Kropotkin, 1971). This combination of regional transgression-regression cycles with block faulting and lithofacies variation form optimal conditions for development of both source and reservoir rocks, as well as trapping conditions. Pay zones in the Tatarian and Permian basins comprise a wide diversity of clastic and carbonate Paleozoic reservoir rocks, ranging in age from Devonian to Permian, and distributed over an immense terrain. We analyzed 45 oil samples from 30 fields across the Tatarian and Perm basins. These samples represent a vast area and a wide diversity of Devonian, Carboniferous and Permian reservoir rocks of variable lithologies. A detailed organic and inorganic chemical characterization was performed to enable a comprehensive oil-oil correlation, a classification of the oils into genetic families, a definition of their thermal maturity, and a differentiation of their possible secondary modifications.

Based on oil analyses and the scant literature on the possible source rock (the Devonian Domanic formation), we have attempted a reconstruction of the depositional environment as possible source rock for the oils. The oil samples are all rich in vanadium and nickel, with a distinctive dominance of vanadium (VO/Ni > 4). The high VO/V (> 0.5) suggests that the majority of the vanadium in the oils is complexed as metallo-porphyrins. This, along with the relatively high abundance of S (> 3% for most samples) and the low pristane/phytane ratio (< 0.6) suggests that the oils studied originated from a source rock deposited in an environment with a relatively low redox potential and probably under somewhat hypersaline-restricted marine conditions (Aizenshtat and Sundararaman, 1989; Didyk et al., 1978; Hughes, 1984; Lewan, 1984; Lipiner et al., 1988; Moldowan et al., 1985; Palacas, 1984). Some of the molecular geochemical markers have been described by Aizenshtat et al. (1997).

7.7 GEOCHEMICAL TRACERS FOR THE IDENTIFICATION OF CARBONATE-RICH, ORGANIC-RICH SEDIMENTS DEPOSITED IN HYPERSALINE, HIGHLY REDUCING, RESTRICTED ENVIRONMENTS

The extensive formation of reduced sulfur mostly as polysulfides was suggested as the possible source for the secondary enrichment of the organic matter in sulfur (Aizenshtat, 1989b). This process can be followed using the stable isotopes of sulfur as tracers. The isotopic composition of the sulfur is sensitive to the type of restrictive environment — open environments leading to much larger deviations in the isotopic composition of the sulfur than of the carbon, as compared with closed environments. The carbon isotope signature of the CO_2 deprivation is of less selectivity, as isotopic fractionation of carbon can be caused by both biological and physical effects (Gazit-Yaari et al., Chapter 8). The major enigma is the nature of the chemically controlled transformations during the geo-maturation, and its complex mechanisms can only be guessed. In general terms, the more mature kerogens, and hence the petroleum formed, gradually become enriched in the light carbon isotope by release of heavy CO_2, up to the formation of isotopically light thermal gas. For sulfur the picture is much more complex. Orr (1986) claimed that no maturation changes occur in sulfur isotopes during the conversion from kerogen to oils. We oppose this hypothesis, based on experiments showing that oils preferentially stabilize the heavier sulfur (see section 7.6.1).

Oils originating from sedimentary organic material will be relatively rich in sulfur and very rich in vanadium, mostly in the porphyrins, leading to a high VO/Ni ratio. Also other transition metals are enriched under hypersaline conditions (Lyons and Gaudette, 1985; Renfro, 1974). Examples of the occurrence of early sulfur enrichment are shown in Figure 7.2. In the two cases studied (Saanich Inlet and Solar Lake), the bottom is anoxic, but the origin of their biomass is very different. The Saanich Inlet is rich in organic matter of terrestrial origin, whereas the Solar Lake has a benthic microbial community. Sulfur enrichment was also found for the Peruvian organic matter in the upwelling zones off the continental shelf, but there the total organic carbon content is much lower than that found in the examples discussed above (Figure 7.2).

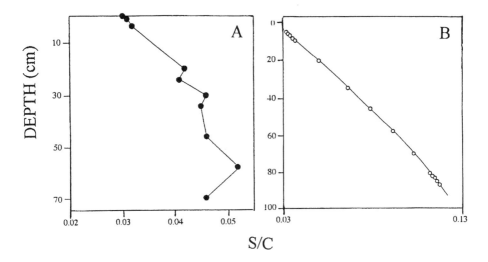

FIGURE 7.2 Two examples for organic sedimentary matter enrichment in sulfur, given as S/C vs. depth: Saanich Inlet (fjord/British Columbia, Canada) (A) and Solar Lake (Sinai, Egypt) (B). For further discussion, see Aizenshtat et al. (1995).

It is impossible to prove a perfect correlation between the bulk parameters of the potential source rock and the properties of the oils generated. However, the chemically stable skeletons of certain molecular structures used as biomarkers may retain the resemblance during the thermal restructuring of the kerogens to yield petroleum, and much information can be obtained by studying these compounds. The saturated hydrocarbon fraction in the extractable organic material from sediments (bitumen) forms one such group of biomarkers useful in the reconstruction of the highly reducing hypersaline carbonate environment (for examples, see Figures 7.3–7.5): (1) In total saturated hydrocarbons with chain lengths exceeding 20 carbons, no preference for odd over even carbon numbers is found; in some cases, this ratio slightly favors the even-saturated hydrocarbons. (2) The pristane/phytane ratio is much lower than unity, in some cases even less than 0.5. (3) The nC_{17}/pristane and the nC_{18}/phytane ratios are generally much lower than for clay-rich sediments. (4) In general, a high abundance of isoprenoid-saturated hydrocarbons is found, including large amounts of cyclic biomarkers such as steranes and hopanes. (5) The hopanes exceed steranes in their relative concentration. (6) In many cases, the tricyclic compounds, usually the C_{23}, are abundant to dominant.

An additional example that correlates very well with the total sulfur abundance is the content of thiophenes in the aromatic fraction, in bitumens and related oils. It is also accepted that the type-IIS kerogens start to generate heavy petroleum at lower temperatures (low API value). The examples discussed here complement the case studies detailed in the review by Hite and Anders (1991).

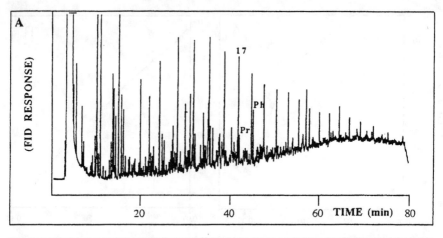

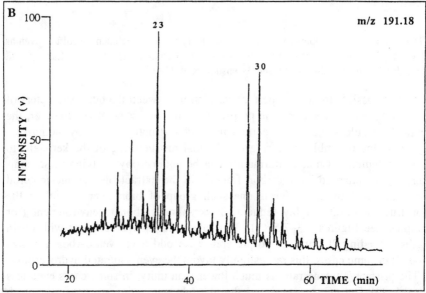

FIGURE 7.3 Gas chromatogram of Dead Sea asphalt. A: hydrocarbon fraction, showing phytane (Ph) much higher than pristane (Pr); B, 191.18 m/z massfragmentogram for terpanes (same asphalt sample as in A), two maxima in distribution centered around C_{23} tricyclics and C_{30} "normal" hopanes (chromatography conditions were not identical for A and B).

7.8 CONCLUSION

In the overview given above we documented the connection between evaporites and petroleum generation. We have introduced hypersalinity as a critical factor leading to deposition of organic matter, and having a profound influence on the biogeochemical cycles of both carbon and sulfur and the chemical "post mortem."

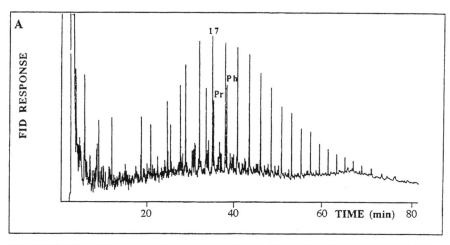

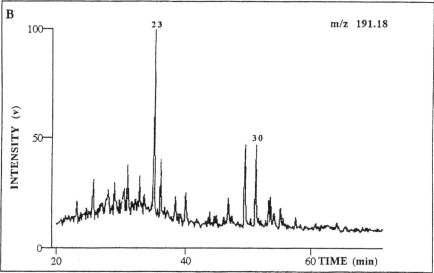

FIGURE 7.4 Gas chromatogram for the hydrocarbon fraction of a Dead Sea crude oil (A), and GC/MS 191.18 m/z massfragmentogram (B). Conditions were as in Figure 3.

Many problems and unanswered questions still remain. For example, we can study very young and old sediments, but we lack a continuous time and temperature series to refer to, so much of our ideas on still older oil-generating rocks are based on extrapolations. Also microbiologists have yet to discover the connection between hypersaline conditions and the formation of a special alginite referred to as tasmanite with its characteristic abundance of the C_{19}–C_{24} tricyclic distribution. In any case, attempts to relate large deposits of sedimentary organic-rich formations with environments with seawater salinity, whether or not involving upwelling phenomena, fall short in comparison to the model of the involve-

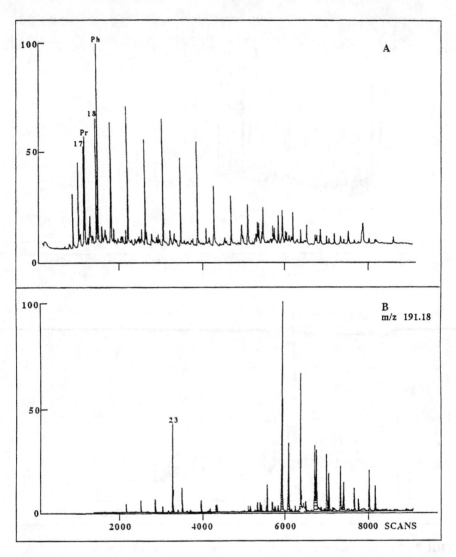

FIGURE 7.5 Total ion current GC/MS of the saturated hydrocarbons fraction for a Monterey oil sample, showing the C_{15+} section (A). Note that phytane > pristane, and phytane > nC_{18}. (B), 191.18 m/z massfragmentogram, showing a tricyclic C_{23} and C_{28} hopane-dominated distribution (chromatography conditions were identical for A and B).

ment of benthic microbial communities as very prolific systems protected by the high salinity environment.

ACKNOWLEDGMENTS

We thank the Moshe Shilo Minerva Center for Marine Biogeochemistry for the support of our group in some of the studies leading to this review, Dr. I. Podolsky for part of the literature survey, and the organic geochemistry group at the Casali Institute of Applied Chemistry for technical assistance and help with the preparation of the figures.

REFERENCES

Aizenshtat, Z. 1989a. Chemical postmortem, in *Microbial Mats. Physiological Ecology of Benthic Microbial Communities,* Y. Cohen and E. Rosenberg, Eds. American Society for Microbiology, Washington, DC, 431–437.

Aizenshtat, Z. 1989b. Proposed mechanism for sulfur enrichment of sedimentary organic matter and its fate during maturation, in *Abstracts of the 197th ACS Meeting on Geochemistry of Sulfur in Fossil Fuels*, Dallas, TX, Section A. 4 pp.

Aizenshtat, Z. and P. Sundararaman. 1989. Maturation trend in oils and asphalts of the Jordan Rift: utilization of detailed vanadyl-porphyrin analysis. *Geochim. Cosmochim. Acta* 53:3185–3188.

Aizenshtat, Z., G. Lipiner, and Y. Cohen. 1984. Biogeochemistry of carbon and sulfur cycle in the microbial mats of the Solar Lake (Sinai), in *Microbial Mats: Stromatolites,* Y. Cohen, R.W. Castenholz, and H.O. Halvorson, Eds. Alan R. Liss, New York. 281–312.

Aizenshtat, Z., E.B. Krein, M.A. Vairavamurthy, and T.P. Goldstein. 1995. Role of sulfur in the transformations of sedimentary organic matter: a mechanistic overview, in *Geochemical Transformations of Sedimentary Sulfur,* M.A. Vairavamurthy and A.A. Schoonen, Eds. American Chemical Society, Washington, DC. 16–38.

Aizenshtat, Z., S. Feinstein, I. Miloslavski, Z. Yakubson, and C.I. Yakubson. 1997. Oil–oil correlation and potential source rocks for oils in Paleozoic reservoir rocks in the Tatarian and Permian basins, Russia, in *Abstracts of the 18th International Meeting on Organic Geochemistry,* Maastricht. 817–819.

Aquino Neto, R.F., J. Triguis, D.A. Azevedo, R. Rodrigues, and B.R.T. Simoneit. 1989. Organic geochemistry of geographically unrelated tasmanites, in *Abstracts of the 14th International Meeting on Organic Geochemistry,* Paris. Abstract no. 189.

Bauld, J. 1984. Microbial mats in marginal marine environments: Shark Bay, Western Australia and Spencer Gulf, South Australia, in *Microbial Mats: Stromatolites,* Y. Cohen, R.W. Castenholz, and H.O. Halvorson, Eds. 1984. Alan R. Liss, New York. 39–58.

Bramlette, M.N. 1946. The Monterey formation of California and the origin of its siliceous rocks. *United States Geological Survey Professional Paper 212.* 57 pp.

Cohen, Y. and E. Rosenberg, Eds. 1989. *Microbial Mats. Physiological Ecology of Benthic Microbial Communities.* American Society for Microbiology, Washington, DC.

Cohen, Y., R.W. Castenholz, and H.O. Halvorson, Eds. 1984. *Microbial Mats: Stromatolites.* Alan R. Liss, New York.

D'Amelio, E., Y. Cohen, and D.J. Des Marais. 1989. Comparative functional ultrastructure of two hypersaline submerged cyanobacterial mats: Guerrero Negro, Baja California Sur, Mexico, and Solar Lake, Sinai, Egypt, in *Microbial Mats. Physiological Ecology of Benthic Microbial Communities,* Y. Cohen and E. Rosenberg, Eds., American Society for Microbiology, Washington, DC. 97–113.

Didyk, B.M., B.R.T. Simoneit, S.C. Brassel, and G. Eglinton. 1978. Organic geochemical indicators of paleoenvironmental conditions of sedimentation. *Nature* 272:216–222.

Dinur, D., B. Spiro, and Z. Aizenshtat. 1980. The distribution and isotopic composition of sulfur in organic-rich sedimentary rocks. *Chem. Geol.* 31:37–51.

Dunning, T.V., J.W. Moore, and M.O. Denekas. 1953. Interfacial activities and porphyrin contents of petroleum extracts, *Ind. Eng. Chem.* 45:1759–1765.

Everlien, G., C. Leblond, and A. Prinzhofer. 1997. Isotopic characterization of thermogenic hydrogen sulphide from sulphur-rich organic matter, in *Abstracts of the 18th International Meeting on Organic Geochemistry,* Maastricht. 803–804.

Feinstein, S., G.K. Williams, L.R. Snowdon, P.W. Brooks, M.G. Fowler, F. Goodazri, and T. Gentzis. 1991. Organic geochemical characterization and hydrocarbon generation potential of mid-late Devonian Horn River bituminous shales, southern Northwest Territories. *Bull. Can. Petrol. Geol.* 39:192–202.

Friedman, G.M. and W.E. Krumbein, Eds. *Hypersaline Ecosystems — The Gavish Sabkha.* Springer-Verlag, Berlin.

Goldhaber, M.B. and I.R. Kaplan. 1974. The sulfur cycle, in *The Sea,* E.D. Goldberg, Ed. John Wiley & Sons, New York. 569–655.

Gutman, C.I. and G.Z. Osipova. 1982. The nature of the deposition of reservoir formations in erosional environments of terrigenous and carbonate rocks of the early carboniferous formations in the NW Tataria. *Geology of Oil and Gas* 6:6–12 (in Russian).

Hite, R.J. and D.E. Anders. 1991. Petroleum and evaporites, in *Evaporites, Petroleum and Mineral Resources,* J.L. Melvin, Ed. Elsevier, Amsterdam. 349–411.

Hodgson, G.W. 1971. Origin of petroleum chemical constraints in origin and refining of petroleum. *Am. Chem. Soc. Adv. Chem. Ser.* 103:1–29.

Hughes, W.B. 1984. Use of thiophenic organosulfur compounds in characterizing crude oils derived from carbonate versus siliciclastic sources. *Am. Assoc. Petrol. Geol. Stud. in Geol.* 18:181–196.

Isaacs, C. M. 1984. Monterey — key to offshore California boom. *Oil & Gas J.* 82:75–81.

Kleinpell, R. M. 1938. *Miocene Stratigraphy of California.* Am. Assoc. Petrol. Geol., Tulsa. 450 pp.

Kohnen, M.E.L., J.S. Sinninghe Damsté, W.I.C. Rijpstra, and J.W. de Leeuw. 1990. Alkylthiophenes as sensitive indicators of paleoenvironmental changes: a study of a cretaceous oil shale from Jordan, in *Geochemistry of Sulfur in Fossil Fuels,* W.L. Orr and C.M. White, Eds. American Chemical Society, Washington, DC. 444–483.

Kozlov, V.I., Z.A. Sinitsyna, E.I. Kulagina, V.N. Pazukhin, V.N. Puchkhov, N.M. Kochetkova, A.N. Abramova, T.V. Klimenko, and N.D. Sergeeva. 1995. *Guidebook of Excursion for the Paleozoic and Upper Precambrian Sections of the Western Slope of the Southern Ural and Preuralian Regions.* Russian Academy of Sciences, Ufa Scientific Center. 165 pp.

Krein, E.B. 1993. Organic sulfur in the geosphere: analysis, structure and chemical processes, in *The Chemistry of the Sulfur-Containing Functional Groups,* S. Patai and Z. Rappoport, Eds. John Wiley & Sons, Chichester. 975–1032.

Krein, E.B. and Z. Aizenshtat. 1995. Proposed thermal pathways for sulfur transformations in organic macromolecules: laboratory simulation experiments, in *Geochemical Transformations of Sedimentary Sulfur,* M.A. Vairavamurthy and A.A. Schoonen, Eds. American Chemical Society, Washington, DC. 110–137.

Kropotkin, P.N. 1971. Deep structure and deformation of ancient platforms (in connection with problems concerning oil and gas presence in platform cover), in *Deep Tectonics of Ancient Platforms of the Northern Hemisphere,* P.N. Kropotkin, B.M. Valyadov, R.A. Gafarov, I.A. Solovieva, and Y.A. Trapeznikov, Eds. Nauka, Moscow. 321–364 (in Russian).

Krumbein, W.E. 1983. Stromatolites — the challenge of a term in space and time. *Precamb. Res.* 20:493–531.

Krumbein, W.E. and Y. Cohen. 1974. Biogene, klastische und evaporitische Sedimentation in einem mesothermen monomiktischen ufernahen See (Golf von Aqaba). *Geol. Rundschau* 63:1035–1065.
Lewan, M.D. 1980. *Geochemistry of Vanadium and Nickel in Organic Matter of Sedimentary Rocks.* Ph.D. thesis, University of Cincinnati.
Lewan, M.D. 1984. Factors controlling the proportionality of vanadium to nickel in crude oils. *Geochim. Cosmochim. Acta* 48:2231–2238.
Lipiner, G., I. Willner, and Z. Aizenshtat. 1988. Correlation between geochemical environments and controlling factors in the metallation of porphyrins, in *Advances in Organic Chemistry and Geochemistry,* L. Mattaveli and L. Novelli, Eds. Vol. 13. Pergamon Press, Oxford. 747–756.
Lyons, W.B. and H.E. Gaudette. 1985. Trace metal concentrations in sediments from the Gavish Sabkha, in *Hypersaline Ecosystems — The Gavish Sabkha,* G.M. Friedman and W.E. Krumbein, Eds. Springer-Verlag, Berlin. 346–349.
Moldowan, J.M., P. Sundararaman, and M. Schoell. 1985. Sensitivity of biomarker properties to depositional environment and/or source input in the Lower Toarcian of SW Germany. *Adv. Org. Geochem.* 10:915–926.
Oehler, J.H. 1984. Carbonate source rocks in the Jurassic Smackover trend of Mississippi, Alabama, and Florida, in *Petroleum Geochemistry and Source-Rock Potential of Carbonate Rocks,* J.G. Palacas, Ed. *Am. Assoc. Petrol. Geol. Stud. in Geol* 18:71–96.
Orr, W.L. 1986. Kerogen/asphaltene/sulfur relationships in sulfur-rich Monterey oils. *Org. Geochem.* 10:499–516.
Palacas, J.G. 1984. *Petroleum Geochemistry and Source-Rock Potential of Carbonate Rocks. Am. Assoc. Petrol. Geol. Stud. in Geol.* 18. 208 pp.
Palacas, J.G., D.E. Andres, and J.D. King. 1984. Southern Florida basin, a prime example of carbonate source rocks of petroleum, in *Petroleum Geochemistry and Source-Rock Potential of Carbonate Rocks,* J.G. Palacas, Ed. *Am. Assoc. Petrol. Geol. Stud. in Geol.* 18:63–69.
Parparova, G.M., S.G. Neruchev, A.V. Zhukor, and P.A. Trushkov. 1981. Catagenesis and oil and gas productivity, in *Kalgenez e Neftegazonosnost,* S.G. Neruchev, Ed., Nedra. Leningrad. 240 pp.
Pedersen, T.F. and S.E. Calvert. 1990. Anoxia vs. productivity: what controls the formation of organic-carbon-rich sediments and sedimentary rocks? 1. *Am. Assoc. Petrol. Geol. Bull.* 74:454–466.
Pisciotto, K.A. and R.E. Garrison. 1981. Lithofacies and depositional environments of the Monterey Formation, California, in: *The Monterey Formation and Related Siliceous Rocks of California.* SEPM Pacific Section Symposium Volume. 97–122.
Reed, R.D. 1933. *Geology of California.* Amer. Assoc. Petrol. Geol., Tulsa. 355 pp.
Reimers, C.E. 1981. *Sedimentary Organic Matter: Distribution and Alternation Processes in the Coastal Upwelling Region off Peru.* Ph.D. thesis, Oregon State University, Corvallis.
Renfro, A.R. 1974. Genesis and evaporite associated stratiform metalliferous deposits — a sabkha process. *Econ. Geol.* 69:33–45.
Schidlowski, M., U. Matzigkeit, and W.E. Krumbein. 1984. Superheavy organic carbon from hypersaline microbial mats. *Naturwissenschaften* 71:303–308.
Schidlowski, M., U. Matzigkeit, E.G. Mook, and W.E. Krumbein. 1985. Carbon isotope geochemistry and ^{14}C ages of microbial mats from the Gavish Sabkha and the Solar Lake, in *Hypersaline Ecosystems — The Gavish Sabkha,* G.M. Friedman and W.E. Krumbein, Eds. Springer-Verlag, Berlin. 381–401.

Spiro, B. 1980. *Geochemistry and Mineralogy of Bituminous Rocks in Israel.* Ph.D. thesis, The Hebrew University of Jerusalem (in Hebrew).
Tannenbaum, E. 1983. *Researches in the Geochemistry of Oils and Asphalts in the Dead Sea Area.* Ph.D. thesis, The Hebrew University of Jerusalem (in Hebrew).
ten Haven, H.L., J.W. de Leeuw, J. Rullkötter, and J.S. Sinninghe Damsté. 1987. Restricted utility of the pristane/phytane ratio as a paleoenvironmental indicator. *Nature* 330:641–643.
Williams, L.A. 1984. Subtidal stromatolites in Monterey Formation and other organic rich rocks as suggested source contributors to petroleum formation. *Am. Assoc. Petrol. Geol.* 68:1879–1893.
Zharkov, M.A. 1981. *History of Paleozoic Salt Accumulation.* Springer-Verlag, New York.

8 Field Evidence for ^{13}C Depletion Due to Atmospheric CO_2 Invasion in Hypersaline Microbial Mats

Naama Gazit-Yaari, Boaz Lazar, and Jonathan Erez

CONTENTS

8.1 Introduction 109
8.2 Materials and Methods 112
    8.2.1 Setting of the New Evaporation System 112
    8.2.2 Field Methods 112
8.3 Results and Discussion 112
8.4 Conclusion 114
References 117

8.1 INTRODUCTION

The invasion of atmospheric CO_2 into alkaline solutions produces dissolved inorganic carbon (DIC) highly depleted in ^{13}C (Baertschi, 1952; Craig, 1953, 1954). This depletion is caused by the direct reaction of the invading CO_2 with OH^- to form bicarbonate: $CO_2 + OH^- \rightarrow HCO_3^-$ (Siegenthaler and Munnich, 1981; Usdowski and Hoefs, 1986; Wanninkhof, 1985). This isotopic phenomenon was first observed under natural conditions in a soft-water lake during DIC depletion caused by a phytoplankton bloom (Herczeg and Fairbanks, 1987). Lazar and Erez (1990, 1992) observed such extreme isotopic depletion in brines from the solar evaporation pans of the Israel Salt Company, Eilat, and termed it the Baertschi effect. The Eilat solar salt plant concentrates Gulf of Aqaba (northern Red Sea) seawater in a series of large shallow evaporation reservoirs to precipitate halite (Figure 8.1). The degree of evaporation is defined as the concentration ratio of Br^- in the brine to the Br^- in mean ocean water and is termed here as E_{Br} (a more detailed explanation is provided

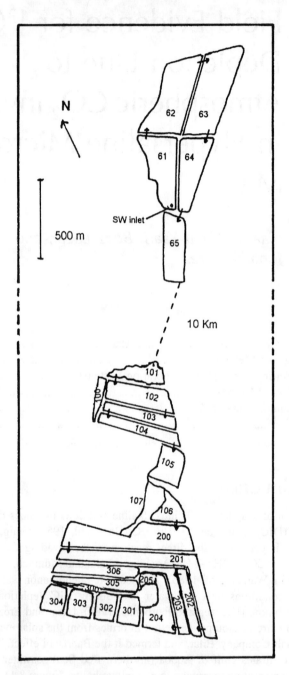

FIGURE 8.1 Map of the evaporation pan system of the Israel Salt Company. The two groups of pans (Ein Evrona pans in the north and Eilat pans in the south) are connected by a 10-km-long pipeline, represented by a dashed line.

Field Evidence for ^{13}C Depletion Due to Atmospheric CO_2 Invasion

in section 8.2.2). As the brine concentrates, the behavior of $\delta^{13}C$ of the DIC ($\delta^{13}C_{DIC}$) can be described as: an initial increase, a sharp decrease (ca. 11â), and a gradual increase during the final stages of the evaporation process (line 26.8.86 in Figure 8.2). The initial increase was explained by removal of ^{12}C due to photosynthesis, and the increase toward the end of the evaporation process is caused by reequilibration with atmospheric CO_2. The significant decrease in $\delta^{13}C_{DIC}$ occurred in pans whose bottoms were extensively covered by microbial mat communities dominated by photosynthetic microorganisms commonly found in such hypersaline environments (Cohen and Rosenberg, 1989). The depletion of 11‰ in $\delta^{13}C_{DIC}$ is opposed to an expected ^{13}C enrichment because of the intense photosynthetic activity of the microbial mats. A similar decrease in $\delta^{13}C_{DIC}$ has been observed by others (Des Marais et al., 1989; Pierre et al., 1987; Stiller et al., 1985), and has been attributed to oxidative decomposition of ^{13}C-depleted organic matter and to precipitation of $CaCO_3$. Lazar and Erez (1990, 1992) rejected this explanation because the DIC concentrations in these pans decreased, rather than increased, as would be expected if the dominant process were oxidation of organic matter. They suggested that the ^{13}C depletion was driven by the Baertschi effect, as explained above. While this seems a logical explanation, the occurrence of the Baertschi effect in the brines was not proven directly, and the explanation was based mainly on elimination of other possible scenarios.

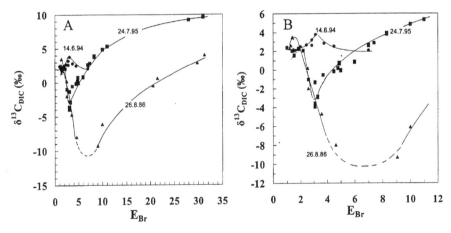

FIGURE 8.2 $\delta^{13}C_{DIC}$ as a function of the degree of evaporation (E_{Br}) in August 1986 (Lazar and Erez, 1990, 1992), June 1994, and July 1995. (A) The entire evaporation path ($0 \leq E_{Br} \leq 35$); (B) enlarged representation of the range $0 \leq E_{Br} \leq 12$.

Since 1992, the evaporation area of the Israel Salt Company has doubled (Figure 8.1). The addition of the new ponds disrupted the salinity regime of the old pan system, and the microbial mat communities were redistributed according to the new regime. The present study exploits this perturbation, in which the system was driven out of steady state, to test the hypothesis that the development of a microbial mat community can drive a Baertschi effect type of fractionation.

8.2 MATERIALS AND METHODS

8.2.1 Setting of the New Evaporation System

The study area is shown in Figure 8.1. As described above, the evaporation area was expanded to include the northern Ein Evrona section. The new system consists of a pipeline introducing seawater to pan 61, which then flows by gravitation to the consecutive pans. The addition of the new system of pans at Ein Evrona disrupted the salinity regime of the old pans, and the microbial mat communities on the bottom of the pans were redistributed. For example, pan 101, to which seawater was pumped originally (Figure 8.1) and had an E_{Br} value of 1.2, has now developed a massive growth of microbial mats, and has reached an E_{Br} of 3. All the pans that follow pan 101 showed massive microbial mat growth up to an E_{Br} of about 4. Similarly, pans 200 and 202 now developed a thick gypsum layer on top of the previously well-developed microbial mat.

8.2.2 Field Methods

The pans were sampled four times within the 16 months following the expansion. Temperature, density, dissolved oxygen, DIC and $\delta^{13}C_{DIC}$ were measured using techniques described earlier (Lazar and Erez, 1992). Briefly, the samples for DIC and $\delta^{13}C_{DIC}$ were pressure filtered on-site, using Whatman filters (GF/C) and a 100-ml glass syringe. Duplicate samples were collected in 100-ml glass bottles, carefully avoiding contact with the atmosphere, and poisoned with 1 ml saturated $HgCl_2$ solution.

The CO_2 for $\delta^{13}C_{DIC}$ analyses was extracted on a vacuum line, and DIC was measured manometrically. The isotopic composition ($\delta^{13}C_{DIC}$) was determined on a VG-602 mass spectrometer, and the results are reported on a PDB scale. The analytical precision for DIC was 0.07 µmol kg^{-1} and for $\delta^{13}C_{DIC}$ was 0.1‰. Dissolved oxygen was determined using the Winkler titration (Grasshoff, 1990), with a precision of ±5 µmol l^{-1}.

The density of the brines (specific gravity) was transformed to E_{Br}, using polynomes based on our previous data on evaporated seawater (Lazar and Erez, 1990, 1992). E_{Br}, representing the degree of brine evaporation, is the molal concentration of Br$^-$ in the brine divided by its concentration in "mean" ocean water (salinity of 35 g kg^{-1}). Bromide is used because it is a conservative ion that does not form a mineral phase until a very high degree of evaporation. Its application is useful even in the halite precipitation range, because its coprecipitation with halite has a very low distribution coefficient. All other measured parameters are therefore presented as a function of E_{Br}.

8.3 RESULTS AND DISCUSSION

Figures 8.2 and 8.3 show plots of $\delta^{13}C_{DIC}$ and DIC vs. E_{Br}. For the sake of clarity, we present only the first and last sampling periods. The first data set was three months after the expansion of the pan area (June 1994), and the last was sampled 16 months later (July 1995). In addition, we plotted the data from the summer of

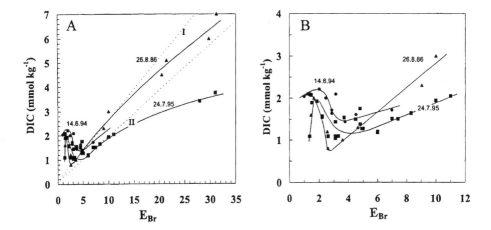

FIGURE 8.3 DIC as a function of the degree of evaporation (E_{Br}) in August 1986 (Lazar and Erez, 1990, 1992), June 1994, and July 1995 (conservative lines I and II are referred to in the text). (A) The entire evaporation path ($0 \leq E_{Br} \leq 35$); (B) enlarged representation of the range $0 \leq E_{Br} \leq 12$.

1986 (Lazar and Erez, 1990, 1992), which represent the former steady-state conditions achieved 12 years after the original construction of the evaporation pans. The latter curve showes an initial increase of 3‰ in $\delta^{13}C_{DIC}$ until E_{Br} reached 1.8 (Figure 8.2B). This increase is due to photosynthesis of macroalgae. This was followed by a sharp decrease of 11‰ (until E_{Br} reached about 7) in brines overlying pans with massive microbial mat growth. This decrease has been interpreted to represent a natural Baertschi effect type of fractionation. The gradual increase of 14‰ from E_{Br} of 9 to the end of the evaporation process was interpreted as reequilibration with atmospheric CO_2. The DIC curve at steady state showed an initial conservative increase until the E_{Br} reached 1.8. This is followed by a sharp decrease of more than 1 mmol kg^{-1} to E_{Br} of 4.5 (Figure 8.3B), coinciding with the sharp isotopic depletion shown in Figure 8.2. The decrease in DIC was caused mainly by photosynthesis of the microbial mat communities, which provided the driving force for atmospheric CO_2 invasion into the brine. This flux caused the large (chemically enhanced) negative carbon isotopic fractionation described earlier as the Baertschi effect.

The flooding of the new evaporation system perturbed the steady-state conditions, and thus provided the environment to perform a field experiment to test the validity of the Baertschi effect hypothesis. Pans that in the past supported growth of microbial mat communities were now precipitating gypsum over the older mats. New microbial mats developed in pans that in the past had lower salinity, and were characterized by macroalgal growth and isotopically heavy DIC. Three months after the new system was put into use (June 1994) a slight (2‰) decrease in $\delta^{13}C_{DIC}$ was already observed in the pans with E_{Br} values of 3–6, although the overall $\delta^{13}C_{DIC}$ values were still high (Figure 8.2B). During this sampling period, these pans were covered with a thin veneer of newly formed microbial mat, which might account for the slight decrease in $\delta^{13}C_{DIC}$. DIC concentrations were higher than under steady-state conditions, and showed a decrease from 2.2 to 1.7 mmol kg^{-1} between E_{Br} of 2 to 4 (Figure 8.3B). By July 1995 the $\delta^{13}C_{DIC}$ had already decreased by 7‰ to reach a value of –4‰ at E_{Br} of 3, followed by a continuous increase in $\delta^{13}C_{DIC}$ until the end of the sampled evaporation path (Figure 8.2B). The DIC concentration increased at the beginning, followed by a decrease from 2 to 1 mmol kg^{-1}, i.e. almost to the level of the 1986 steady state (Figure 8.3B). In July 1995, we observed lower DIC values relative to the 1986 steady state in ponds with E_{Br} values between 5 and 30. This could be explained by continued photosynthetic activity of the older microbial mats that were covered by the newly precipitated gypsum. Indeed, diurnal measurements in these pans showed high photosynthetic activities, with oxygen concentrations exceeding saturation at peak hours.

A positive correlation between development of microbial communities and ^{13}C depletion is shown in the time-dependent plots of $\delta^{13}C_{DIC}$ and DIC (Figures 8.4, 8.5). No significant change in $\delta^{13}C_{DIC}$ of pans that developed microbial mats was observed during the first 12 months after flooding with brines of higher salinity (Figure 8.4A). However, during the next three months, $\delta^{13}C_{DIC}$ decreased by 7‰. A similar trend can be seen in the DIC concentration, decreasing from 1.7 mmol kg^{-1} just after the salinity change to 1.1 mmol kg^{-1} 16 months later (Figure 8.4B). As for $\delta^{13}C_{DIC}$, the sharp decrease in DIC also occurred during the last three months of sampling, when a thick and continuous microbial mat cover had developed. A comparison with the $\delta^{13}C_{DIC}$ values in pans that lack growth of profuse microbial mats (pans with too high or too low E_{Br} values) is provided in Figure 8.5; no changes in $\delta^{13}C_{DIC}$ were observed, and the DIC concentrations show only a very slight decrease.

The direct correlation between the development of microbial mat communities, the decrease in DIC, and the ^{13}C depletion provide solid evidence that the Baertschi effect is the dominating mechanism controlling the carbon isotopic depletion in the Israel Salt Company evaporation system. The new accumulation of microbial mats over the earlier silty bottoms indicates net accumulation of organic carbon in these pans. Therefore, the possibility that the isotopic depletion is caused by oxidation of particular and/or dissolved organic matter such as demonstrated in the Guerrero Negro salt ponds (Canfield and Des Marais 1993; Des Marais et al., 1989) is highly unlikely in the Israel Salt Company system. The Baertschi effect can occur in other natural systems as well, from the ecosystem level to the scale of single organisms in situations where chemical reactions of CO_2 are enhanced.

8.4 CONCLUSION

Dissolved inorganic carbon highly depleted in ^{13}C is commonly found in marine-derived brines. Our earlier observations in the solar salt evaporation pans of the Israel Salt Company in Eilat suggested that the isotopic depletion was caused by atmospheric CO_2 invasion into the brine and its reaction with OH$^-$ to form bicarbonate. The DIC deficit was caused by intense photosynthetic activity of the microbial mat communities in the pans. We termed this isotopic depletion Baertschi effect, after Baertschi's experiments on CO_2 invasion into akaline solutions performed in the early 1950s. The presence of the Baertschi effect in marine brines was not proven

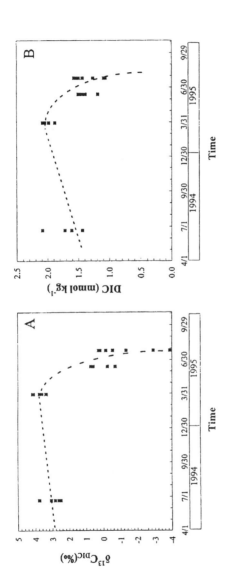

FIGURE 8.4 $\delta^{13}C_{DIC}$ (A) and DIC (B) as a function of time in pans that developed active benthic microbial mat communities (pans 101, 102, 103, 104, 105, 106, see Figure 8.1). The origin indicates the date of flooding of the new pan system.

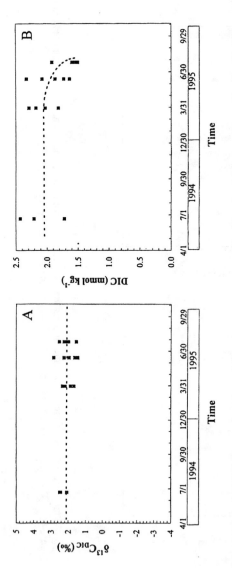

FIGURE 8.5 δ $^{13}C_{DIC}$ (A) and DIC (A) as a function of time in pans not covered with microbial mats (pans 61, 62, 63, 201, see Figure 8.1). The origin indicates the date of flooding of the new pan system.

directly, but could be inferred by elimination of other possible scenarios. The present study used a perturbation (addition of a new area to the evaporation pans, which drove the system out of steady state) to test the hypothesis that the development of a new microbial mat community cover will drive a Baertschi effect type of fractionation. Brine sampling and analyses were conducted four times over a 16-month period at the new evaporation pan system. A positive correlation between ^{13}C depletion, microbial mat development, and DIC depletion was observed. With time, when the microbial mat cover was fully developed, the ^{13}C depletion increased dramatically. In pans that lacked microbial mat growth, such ^{13}C and DIC depletions were not observed. These data support the Baertschi effect mechanism, as opposed to the alternative explanation that the isotopic depletion was caused by oxidation of organic matter.

REFERENCES

Baertschi, P. 1952. Die Fraktionierung der Kohlenstoffisotopen bei der Absorption von Kohlendioxyd. *Helvetica Chimica Acta* 53:1030–1036.
Canfield, D.E. and D.J. Des Marais. 1993. Biogeochemical cycles of carbon, sulfur, and free oxygen in a microbial mat. *Geochim. Cosmochim. Acta* 57:3971–3984.
Cohen, Y. and E. Rosenberg, Eds. 1989. *Microbial Mats. Physiological Ecology of Benthic Microbial Communities*. American Society for Microbiology, Washington, DC.
Craig, H. 1953. The geochemistry of stable carbon isotopes. *Geochim. Cosmochim. Acta* 3:53–92.
Craig, H. 1954. Carbon 13 in plants and the relationships between carbon 13 and carbon 14 variations in nature. *J. Geol.* 62:115–149.
Des Marais, D.J., Y. Cohen, H. Nguyen, T. Cheatham, and E. Munoz. 1989. Carbon isotopic trends in hypersaline ponds and microbial mats at Guerrero Negro, Baja, California Sur, Mexico: implication for pre-Cambrian stromatolites, in *Microbial Mats. Physiological Ecology of Benthic Microbial Communities*, Y. Cohen and E. Rosenberg, Eds. American Society for Microbiology, Washington, DC. 191–203.
Grasshoff, K. 1983. Determination of oxygen, in *Methods of Seawater Analysis*, 2nd ed., K. Grasshoff, M. Ehrhardt, and K. Kremling, Eds. Verlag Chemie, Weinheim. 61–72.
Herczeg, A.L. and R.G. Fairbanks. 1987. Anomalous carbon isotope fractionation between atmospheric CO_2 and dissolved inorganic carbon induced by intense photosynthesis. *Geochim. Cosmochim. Acta* 51:895–899.
Lazar, B. and J. Erez. 1990. Extreme ^{13}C depletions in seawater-derived brines and their implications for their past geochemical carbon cycle. *Geology* 18:1191–1194.
Lazar, B. and J. Erez. 1992. Carbon geochemistry of marine derived brines: I. ^{13}C depletions due to intense photosynthesis. *Geochim. Cosmochim. Acta* 56:335–345.
Pierre, C., D. Thouron, J.J. Pueyo, R. Utrilla, and M. Ingles. 1987. Carbon isotope behaviour of CO_2 during sea water evaporation in salt pans. *Terra Cognita* 7:413–414.
Siegenthaler, U. and K.O. Munnich. 1981. C-13/C-12 Fractionation during CO_2 transfer from air to sea, in *Carbon Cycle Modeling*, B. Bolin, Ed. *SCOPE 1*. John Wiley & Sons, Geneva. 249–257.
Stiller, M., J.S. Rounick, and S. Shasha. 1985. Extreme carbon isotope enrichments in evaporating brines. *Nature* 316:434–435.

Usdowski, E. and J. Hoefs. 1986. $^{13}C/^{12}C$ partitioning and kinetics of CO_2 absorption by hydroxide buffer solutions. *Earth Planet. Sci. Lett.* 80:130–134.

Wanninkhof, R. 1985. Kinetic fractionation of the carbon isotopes ^{13}C and ^{12}C during transfer of CO_2 from air to seawater. *Tellus* 37B:128–135.

Section III

The Dead Sea

9 The Dead Sea–A Terminal Lake in the Dead Sea Rift: A Short Overview

Ittai Gavrieli, Michael Beyth, and Yoseph Yechieli

CONTENTS

9.1 Introduction 121
9.2 The Recent Evolution of the Dead Sea 122
9.3 Future Evolution of the Dead Sea 124
9.4 The Proposed Dead Sea–Red Sea Canal 125
References 126

9.1 INTRODUCTION

The Dead Sea is a terminal lake (Figure 9.1) located in one of the rhomb-shaped grabens that developed along the sinistral transform fault of the Dead Sea Rift (Freund, 1965; Garfunkel, 1981). The size of its drainage basin is about 40,000 km², and includes five countries from both sides of the rift. The Dead Sea water level (411 m below mean seawater level in 1997) is the lowest exposed surface on earth, while its bottom (730 m below mean seawater level) forms the lowest continental surface. As a lake, it has a unique Ca-chloride composition [i.e., $Ca^{2+} > (SO_4^{2-} + HCO_3^-)$] forming one of the world's saltiest lakes with a total dissolved salt concentration of 340 g l^{-1} and a density of about 1.235 g ml^{-1} (Gavrieli, 1997). Similar, but more diluted Ca-chloride brines are common as groundwater in the Rift Valley, and have been proposed to have developed from seawater following evaporation, mineral deposition, and water-rock interaction (Starinsky, 1974).

The Dead Sea is the last in a series of lakes that occupied the Dead Sea Rift since its formation in the Neogene. Its precursor, the Lisan Lake, existed between 70,000 and 15,000 years ago and occupied a much larger area, from the Sea of Galilee (Lake Kinneret) in the north to about 25 km south of the southern basin of the Dead Sea (Begin et al., 1974). Typical for terminal lakes, the Dead Sea water level represents the balance between evaporation and inflow of water. Therefore, until recent massive artificial intervention in its water balance, the Dead Sea level reflected the climatic changes in the area (Klein, 1985). However, over the last several decades, due to large irrigation projects on both sides of the rift, there has

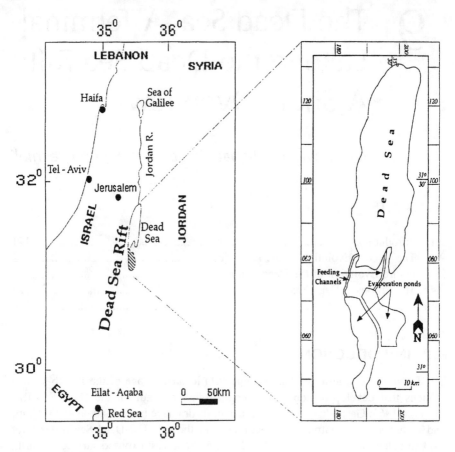

FIGURE 9.1 Dead Sea location map.

been a continuous decline in the inflow of floodwater to the lake, mainly from the Jordan River. Whereas the Jordan River input is estimated to have been 1,100 million m^3 annually (Ben-Zvi, 1982) at the turn of the century, its present inflow is about 250 million m^3 (Klein, 1990).

Recently, there has been a growing interest in the Dead Sea as part of the peace treaty between Israel and Jordan. A major project currently being considered is a Dead Sea–Red Sea Canal, which will have two major goals: (1) supply of desalinized drinking water to this arid area with its growing population and (2) stabilization of the Dead Sea level. The following sections present an overview of the recent and future evolution of the Dead Sea, and some historical aspects of the proposed Dead Sea–Red Sea Canal.

9.2 THE RECENT EVOLUTION OF THE DEAD SEA

Starting at the end of the 1920s, a chemical industry developed along the Dead Sea, exploiting the rich and unusual mineral composition of the brine. The current production of minerals by the Dead Sea Works Ltd. and the Arab Potash Co. is the core for the development of the area. The major products of these industries include potash, bromine and magnesium. The processes involved in the production of these chemicals include pumping of the brines to shallow evaporation ponds in which halite (NaCl) first precipitates, followed by carnallite ($KMgCl_3 \cdot 6H_2O$). This process increases the volume of water that evaporates naturally from the Dead Sea by 10 to 20%. The sodium- and potassium-depleted brines (end brines) from these processes are returned to the Dead Sea.

The negative water balance of the Dead Sea since the beginning of the century has resulted in a decrease in its water level by about 20 m (Figure 9.2). This has been accompanied by an increase in the salinity of the brine and by various changes in the chemical and physical characteristics of the lake. In 1976, following a decline in the water level, the shallow southern basin was disconnected and dried out from the main water body. In 1979, the Dead Sea water column overturned, thereby ending several hundred years of meromictic (stratified) conditions, during which the lower water body was anoxic (Beyth, 1980; Neev and Emery, 1967).

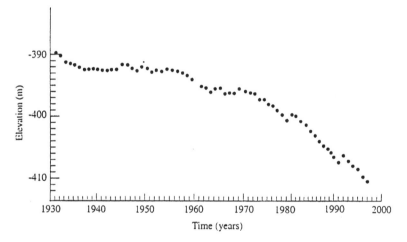

FIGURE 9.2 The Dead Sea water level since 1930.

Since 1983, with a short break between 1992 and 1995, holomictic conditions have prevailed in the Dead Sea. Its hydrography is characterized by an annual cycle during which stratification develops in winter. Despite evaporation and the formation of a destabilizing halocline in summer, the stratification is maintained by a stabilizing thermocline. However, as the surface layer cools in autumn, its salinity increases, and eventually an overturn takes place in November (Anati and Stiller, 1991; Anati et al., 1987).

The above annual cycles were interrupted following the exceptionally rainy winter of 1991/92 when approximately 1,500 million m³ water flowed into the Dead Sea, leading to a level rise of nearly 2 m and to the development of a relatively diluted upper water layer (Beyth et al., 1993). The lowered salinity of this water mass led to the development of a dense community of unicellular green algae and red Archaea. Initially, the diluted layer was only a few meters thick, but over the next three years it deepened to more than 40 m, while its salinity increased. The red Archaea remained above the pycnocline and became a sensitive tracer for the physical state of the water column (Anati et al., 1995). In November 1995, after more than three years of meromictic conditions, the density of the upper and lower water masses of the Dead Sea equalized, and the Dead Sea once again overturned.

In 1983, as a result of the overall negative water balance of the lake and its increased salinity, massive halite precipitation began and, apart from a short period, has continued till today (Gavrieli, 1997; Gavrieli et al., 1989). The continuous halite precipitation from the brine and the industrial activities of the potash industries south of the Dead Sea (Figure 9.1) have resulted in a change in the composition of the brine. This has been exhibited by decreasing Na/Cl and increasing Mg/K ratios over the last four decades (Figure 9.3). The annual weight of halite that precipitated in the lake since 1983 was estimated by various calculation methods to be >$100*10^6$ tons (Gavrieli, 1997). This does not include the halite that precipitates in the industrial evaporation ponds, estimated at about $35*10^6$ ton year^{-1}.

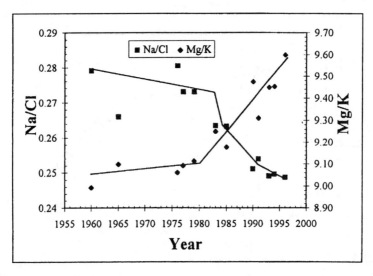

FIGURE 9.3 Na/Cl and Mg/Cl ratios as a function of time in the Dead Sea since 1960. (From Gavrieli and Yechieli, 1997. With permission.)

9.3 FUTURE EVOLUTION OF THE DEAD SEA

The expected future chemical evolution of the Dead Sea brine due to evaporation and salt precipitation was studied in a series of evaporation experiments. The changes in the density and the Na/Cl ratio as a function of the remaining volume of the original brine are depicted in Figure 9.4. Halite precipitated throughout the experiment, leading to a continuous decrease in the Na/Cl ratio of the brine, while carnallite began to precipitate when the volume of the brine was reduced to about half of its initial value.

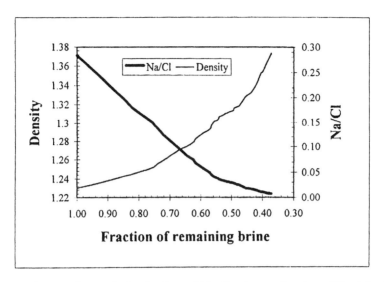

FIGURE 9.4 The changes in the density and Na/Cl ratio as a function of Dead Sea brine volume left in evaporation experiments. (From Gavrieli and Yechieli, 1997. With permission.)

Will the Dead Sea brine evolve in a similar manner to that observed in the experiment, and will the Dead Sea volume continue to shrink until complete dryness? A model based on the above results indicates that under present conditions (similar water input, similar climatic conditions over the lake) the Dead Sea will not die. In the model, the rate of evaporation, which will gradually decrease as the salinity of the brine increases, was estimated on the basis of a simplified relationship between density and evaporation rate (Mero and Simon, 1985). The preliminary results suggest that a new equilibrium between evaporation and input will be reached within 400 years, at a water level of about 500 m below mean seawater level, some 90 m below the present level (Gavrieli and Yechieli, 1997). At the new equilibrium level, the Dead Sea volume will have decreased to two-thirds of its present volume, while its salinity and density will be about 380 g l^{-1} and 1.27 g ml^{-1}, respectively. This will be accompanied by a continuous halite precipitation and a decrease in the Na/Cl ratio from 0.25 to about 0.1. However, the lake will not attain saturation with respect to carnallite, which is the mineral precipitating and exploited in the industrial evaporation ponds.

9.4 THE PROPOSED DEAD SEA–RED SEA CANAL

The 400 m difference in elevation between mean seawater level and the Dead Sea surface has triggered the imagination of travelers in the area from the end of the last century. Their idea was to use this hydrostatic potential to produce hydroelectric energy (Herzl, 1902). This idea was seriously considered by the government of Israel following the 1973 energy crisis. In the mid 1980s, the Israeli Mediterranean–Dead Sea Company Ltd. submitted its recommendation following a thorough study and suggested an alignment for the canal from the southern Mediterranean coast to the southwestern margins of the Dead Sea. The idea was to install an 800-MW-capacity power plant adjacent to the Dead Sea, to produce electricity during peak hours. This recommendation was later rejected by the government of Israel because of political and economical considerations. However, the idea of using the hydroelectric potential as local pumped storage at the margins of the Dead Sea was adopted by the Israel Electrical Company, and is still considered. During the late 1980s and early 1990s, a pre-feasibility study was initiated to examine the hydrostatic potential to desalinize seawater by reverse osmosis to produce drinking water (Ministry of Energy and Infrastructure, 1995).

On October 26th, 1994 a peace treaty between Israel and the Hashemite Kingdom of Jordan was signed. Part of the treaty includes a pre-feasibility study that revises the idea of desalinization of seawater by constructing a Red Sea–Dead Sea canal along the Dead Sea Rift to produce 800 million m^3 of desalinized seawater annually. This pre-feasibility study has since been carried out by the Harza Group, financed by the World Bank.

REFERENCES

Anati, D.A. and M. Stiller. 1991. The post-1979 thermohaline structure of the Dead Sea and the role of double-diffusive mixing. *Limnol. Oceanogr.* 36:342–354.
Anati, D.A., M. Stiller, S. Shasha, and J.R. Gat. 1987. Changes in the thermohaline structure of the Dead Sea: 1979-1984. *Earth Planet. Sci. Lett.* 84:109–121.
Anati, D.A., I. Gavrieli, and A. Oren. 1995. The residual effect of the 1991–1993 rainy winters on the Dead Sea stratification. *Israel J. Earth Sci.* 44:63–70.
Begin, Z.B., A. Ehrlich, and Y. Nathan. 1974. Lake Lisan, the late Pleistocene precursor of the Dead Sea. *Israel Geol. Surv. Bull.* 63. 30 pp.
Ben-Zvi, A. 1982. Forecast of Dead Sea level. *Israel Nat. Council Res. Dev.* 10/82. 1-4 (in Hebrew).
Beyth, M. 1980. Recent evolution and present stage of the Dead Sea brines, in *Hypersaline Brines and Evaporitic Environments,* A. Nissenbaum, Ed. Elsevier, Amsterdam. 155–166.
Beyth, M., I. Gavrieli, D.A. Anati, and O. Katz. 1993. Effects of the December 1991–May 1992 floods on the Dead Sea vertical structure. *Israel J. Earth Sci.* 42:45–48.
Freund, R. 1965. A model of the structure development of Israel and adjacent areas since Upper Cretaceous times. *Geol. Mag.* 102:189–205.
Garfunkel, Z. 1981. Internal structure of the Dead Sea leaky transform (rift) in relation to plate kinematics. *Tectonophysics* 80:81–108.
Gavrieli, I. 1997. Halite deposition in the Dead Sea: 1960-1993, in *The Dead Sea — The Lake and its Setting,* T. M. Niemi, Z. Ben-Avraham, and J.R. Gat, Eds. Oxford University, 162–171.
Gavrieli, I. and Y. Yechieli. 1997. The future evolution of the Dead Sea in light of its present changes, in *Terra Nostra, GIF 13 Meeting–The Dead Sea Rift as a Unique Global Site,* M. Gavish and M. Horesh, Eds. Alfred-Wegner-Stiftung, Köln.
Gavrieli, I., A. Starinsky, and A. Bein. 1989. The solubility of halite as a function of temperature in the highly saline Dead Sea brine system. *Limnol. Oceanogr.* 34:1224–1234.
Herzl, T. 1902. *Altneuland.* Wein Verlag, Berlin. 330 pp.
Klein, C. 1985. Climatological fluctuations in Israel and Dead Sea level changes during historical times, *Meteorol. Israel* 85/3 (in Hebrew).

Klein, Z. 1990. Dead Sea level changes. *Israel J. Earth Sci.* 39:49–51.
Mero, F. and E. Simon. 1985. A daily simulation for evaluation of future Dead Sea level, in *Scientific Basis for Water Management.* Int. Assoc. Hydrol. Sci. Publ. 153:265–276.
Ministry of Energy and Infrastructure. 1995. *Economic Reassessment of the Dead Sea Hydro Project,* 8 pp.
Neev, D. and K.O. Emery. 1967. *The Dead Sea: Depositional Processes and Environments of Evaporites.* Israel Geological Survey Bulletin 41. 147 pp.
Starinsky, A. 1974. *Relationship between Ca-chloride Brines and Sedimentary Rocks in Israel.* Ph.D. thesis, The Hebrew University of Jerusalem. 104 pp. (in Hebrew).

10 The Rise and Decline of a Bloom of Halophilic Algae and Archaea in the Dead Sea: 1992–1995

Aharon Oren

CONTENTS

10.1 Introduction 129
10.2 A Remote Sensing Study of the Development of an
 Algal Bloom in the Dead Sea 130
10.3 Characterization of the Dominant Halophilic Archaea in the
 Dead Sea by Polar Lipid Analysis 132
10.4 Occurrence of Virus-Like Particles in the Dead Sea 133
10.5 Biological and Physical Evidence for the Termination of the
 Dead Sea 1991–1995 Stratification 135
10.6 Conclusion 136
Acknowledgments 137
References 137

10.1 INTRODUCTION

More than 60 years have passed since Benjamin Elazari-Volcani (Wilkansky) first showed that the Dead Sea is inhabited by an indigenous community of microorganisms adapted to the extremely high salinity and the unusual ionic composition of its waters.

Systematic quantitative studies on the biology of the Dead Sea have been performed only since 1980. It is now well established that the Dead Sea biota are dominated by the unicellular green alga *Dunaliella parva* and by red halophilic Archaea, belonging to the family *Halobacteriaceae*. Blooms of red Archaea occur in the Dead Sea whenever conditions become suitable for development of *Dunaliella*, which supplies the carbon sources for the heterotrophic bacteria. Algal and archaeal blooms in the Dead Sea are relatively rare events, whose occurrence is closely correlated with the physical structure of the water column.

Since the overturn of the lake in the beginning of 1979 that led to the formation of a mixed water column, the Dead Sea has known periods of stratification, with meromixis lasting for several years (1980–1982, 1992–1995), and a period of holomixis (1983–1991) (Anati and Stiller, 1991; Beyth et al., 1993). During the holomictic period, the salinity of the surface water was so great (total dissolved salt concentration about 340 g l^{-1}) that it prevented the development of mass blooms of algae and bacteria in the lake. Consequently, bacterial densities were low, and algae were absent altogether from the water column (Oren, 1988, 1993).

When, in the beginning of 1980, the upper water layers of the Dead Sea became sufficiently diluted as a result of the inflow of massive amounts of fresh water from the Jordan river and rain floods from the catchment area, a dense *Dunaliella* bloom developed (up to 8,800 cells ml^{-1}), followed by a mass development of halophilic Archaea (up to $2.2 \cdot 10^7$ cells ml^{-1}), which imparted a red coloration to the water (Oren, 1988; Oren and Shilo, 1982). The algal bloom remained present for a few months only; the archaeal community declined to about half of its peak density soon after the end of the *Dunaliella* bloom, followed by a long period of stability of the bacterial community size.

A new meromictic episode in the Dead Sea began at the end of 1991. The meromixis was a result of a cold-air temperature anomaly in the Middle East during the winter of 1991–1992 that resulted in massive amounts of rainwater being transported to the Dead Sea, which caused an increase in level of almost 2 m (Beyth et al., 1993). A dilution of the upper 5 meters to a salinity as low as 245-250 g l^{-1} triggered a short-lived mass development of *Dunaliella* (April–May 1992, more than 15,000 cells ml^{-1}) (Oren, 1993; Oren et al., 1995a) and a prolonged bloom of red Archaea (from May 1992 onward, up to $3.5 \cdot 10^7$ cells ml^{-1}) (Oren, 1993; Oren and Anati, 1996; Oren and Gurevich, 1995) (Figure 10.1, upper and middle panel). This new period of meromixis gave us the opportunity for a renewed study of biological phenomena in the Dead Sea. In this chapter different aspects of these studies are outlined, relating to the origins of the *Dunaliella* bloom, the nature of the dominant Archaea in the bacterial community, and the causes of its decline.

10.2 A REMOTE SENSING STUDY OF THE DEVELOPMENT OF AN ALGAL BLOOM IN THE DEAD SEA

Since no *Dunaliella* cells were observed in the Dead Sea from 1982 till 1991, it can be asked what the inoculum source was that enabled the rapid development of the 1992 *Dunaliella* bloom at the moment conditions in the lake became suitable for algal growth. Two hypotheses have been brought forward in the past as to the source of the inoculum: (1) from less saline springs on the shore of the lake (Oren, 1989; Oren and Shilo, 1982), or (2) from resting stages (cysts) in the bottom sediments, originating from cells that developed during a previous bloom.

The Rise and Decline of a Bloom of Halophilic Algae and Archaea

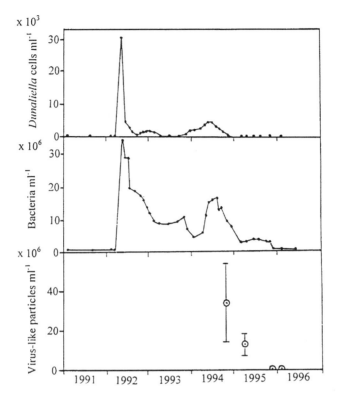

FIGURE 10.1 Numbers of *Dunaliella* cells (upper panel), bacteria (middle panel), and virus-like particles (lower panel) in the Dead Sea surface water, 1991–1996. (Modified from Oren et al., *Extremophiles*, 1, 143, 1997, with permission.)

Multispectral LANDSAT images of the Dead Sea area were analyzed to obtain spatial and temporal information on the development of the 1992 algal bloom (Oren and Ben-Yosef, 1997). Two images were obtained: the first, on April 15 at the onset of the *Dunaliella* bloom, and the second, on June 22, when the algal bloom had declined and the Dead Sea was colored red by a dense community of halophilic Archaea. For comparison, a LANDSAT image was also obtained on May 5, 1991, at a time when algae were absent and bacterial numbers were negligible. The search for algal and bacterial signatures on the LANDSAT images is based on the effect of the microorganisms on the volume reflectance of the water body. The spectral distribution of the reflected solar radiation is modified by pigments present in the microorganisms. The main interaction commonly measured in such studies is the absorption of sunlight by the red absorption band of chlorophyll contained in the algae, and the ratio of reflectance in the 0.75–0.9 μm band to the 0.63–0.69 μm band (commonly known as "algal index" or "chlorophyll index") is most useful. Due to the broad-band scattering effects of the algae, the radiance in the long-wavelength band will tend to increase with increasing algal concentration, while due to the specific absorption, the radiance in the 0.63–0.69 μm band will tend to decrease. The use of the ratio neutralizes most of the interfering atmospheric effects.

While the archaeal carotenoids did not produce a recognizable signal, the presence of chlorophyll-containing algae in high densities in April 1992 was easily detected. The average band ratios in May 1991, and April and June 1992 were 0.59, 0.66, and 0.42, respectively. As expected, the highest values were observed at the onset of the *Dunaliella* bloom, with low values in June, when *Dunaliella* had virtually disappeared from the lake. The intermediate value calculated for May 1991 was unexpected, as at this time the lake was almost devoid of biota. However, radiance values in the 0.525–0.605 μm band and the 0.63-0.69 μm band were higher in June 1992 than in May 1991, possibly indicating light scattering as a result of the massive presence of bacteria (Oren and Ben-Yosef, 1997).

The image obtained at the time of the onset of the bloom suggested that the algal bloom originated at the shallow areas near the shore. The most probable explanation is that we witnessed here the onset of the *Dunaliella* bloom, and that this bloom originated from resting stages present in the sediment. We have observed the formation of such resting cells or cysts during the decline of the 1992 bloom (Oren et al., 1995a), and it can be expected that similar resting cells were deposited at the end of the previous bloom in 1980. The presence of chlorophyll in the bottom sediments of the Dead Sea has been documented earlier (Nissenbaum et al., 1972). The finding of a similarly high chlorophyll index near the southern shores of the lake, where freshwater springs are scarce or absent, supports the hypothesis that cells present in the surface sediment, and not cells derived from less saline springs in the area, formed the inoculum from which the bloom developed. These results also show the first indications of the existence of patchiness in the spatial distribution of the algae, already suggested by Nissenbaum (1975). Thus, analysis of LANDSAT images has enabled us to obtain information on the spatial and temporal distribution of the biota in the Dead Sea — information that is difficult to obtain otherwise.

10.3 CHARACTERIZATION OF THE DOMINANT HALOPHILIC ARCHAEA IN THE DEAD SEA BY POLAR LIPID ANALYSIS

Species of halophilic Archaea, belonging to different genera, have been isolated in the past from the Dead Sea: *Halorubrum sodomense* (Oren, 1983a), *Haloarcula marismortui* (Oren et al., 1990), and *Haloferax volcanii* (Mullakhanbhai and Larsen, 1975). They were isolated following enrichment in liquid media, and no quantitative estimations have been made on the contribution of these (and possible other) species to the microbial community of the Dead Sea. Attempts to obtain information on the nature of the dominant Archaea in the 1992 bloom by isolation and characterization of bacteria developing on agar plates proved unsatisfactory in view of the low plating efficiencies obtained. This is not a specific problem for the Dead Sea: little quantitative information exists on the contribution of these and other species to the archaeal communities in hypersaline lakes (Oren, 1994).

The genera of the *Halobacteriaceae* differ in the nature of the polar lipids in their membranes, and especially in their glycolipids. These differences have formed the basis for the classification of the halophilic Archaea into the presently recognized genera (Torreblanca et al., 1986; Tindall, 1992).

We attempted to exploit the differences in polar lipid composition — notably of the glycolipid component — between the halophilic archaeal genera to obtain information on the community structure of the archaeal bloom that developed in the Dead Sea in 1992. The method does not involve cultivation, and thus no bias is introduced in favor of those types that are easy to isolate. Cells collected by centrifugation of 5–20 l of brine were extracted, and lipids were separated by one- and two-dimensional thin-layer chromatography on silica gel plates, followed by detection of total lipids, phospholipids and glycolipids, using specific stains. Three polar lipids were detected in extracts of the biomass collected from the Dead Sea during the 1992 bloom (Oren and Gurevich, 1993): the diether derivatives of phosphatidylglycerol, phosphatidylglycerol phosphate, and a single glycolipid, co-chromatographing with the sulfated diglycosyl diether lipid of the genus *Haloferax*. No significant amounts of those glycolipids that would indicate the presence of other Dead Sea Archaea, such as *Halorubrum sodomense* or *Haloarcula marismortui*, were found. The diether derivative of phosphatidylglycerosulfate, found thus far in all genera with the exception of *Haloferax*, was not detected. On the basis of these results, it was first postulated that the dominant Archaeon that developed in the Dead Sea during 1992 belonged to the genus *Haloferax*. However, it is possible that this conclusion requires revision, because a novel type of halophilic Archaea was subsequently isolated from the lake with a similar polar lipid composition. On the basis of 16S rRNA sequence data, the isolate was classified in a new genus as *Halobaculum gomorrense* (Oren et al., 1995b).

The finding of a high content of bacteriorhodopsin toward the end of the 1980–1981 archaeal bloom in the Dead Sea (Oren and Shilo, 1981) might signify that the community then differed from the one that developed in 1992. The purple color of bacteriorhodopsin was not observed in the biomass collected in 1992. Bacteriorhodopsin was never reported in any representative of the genus *Haloferax*, nor did we find it in the new genus *Halobaculum* (Oren et al., 1995b), but it is produced by the Dead Sea isolate *Halorubrum sodomense* (Oren, 1983a).

10.4 OCCURRENCE OF VIRUS-LIKE PARTICLES IN THE DEAD SEA

When attempting to explain the causes for the initial sharp decrease in bacterial community size following the peak of the bloom, it should be remembered that the Dead Sea is an environment lacking protozoa and higher trophic levels. Amoeboid or ciliate protozoa have been isolated from the Dead Sea in the past (Elazari-Volcani, 1943, 1944), but such organisms were not observed in the lake in recent years, and their contribution to the decline in bacterial communities can be considered negligible or altogether absent.

The existence of bacteriophages attacking halophilic Archaea has been well documented, and the possibility that bacteriophages might play a role in the decrease in bacterial densities in the Dead Sea has been suggested before (Oren, 1983b).

Different double-stranded DNA phages, showing head and contractile tail morphology, have been isolated from lysing *Halobacterium salinarum* strains (Torsvik and Dundas, 1974; Wais et al., 1975). At least some of these phages can enter a lysogenic state. Other phages, such as HF1, isolated from an Australian saltern pond, have a broad host range (Nuttall and Dyall-Smith, 1993). Little is known on the distribution and species specificity of halophilic bacteriophages and their role in nature, but electron-microscopic examination has shown the presence of different types of bacteriophages within cells of the square Archaea occurring in saltern ponds (Guixa-Boixareu et al., 1996), suggesting they may be important in regulating community densities of halophilic Archaea.

We harvested viruses, bacteria, and other particles directly on electron microscope grids by ultracentrifugation and enumerated them in the electron microscope, using established techniques (Bergh et al., 1989; Børsheim et al., 1990). Virus-like particles (size range 50–100 nm) were found in high numbers, the most abundant ones being spindle-shaped (Oren et al., 1997). In addition, hexagonal virus-like particles and tailed bacteriophages were seen. Occasionally, aggregates of the particles were observed, resembling a recent burst event of a bacterium, releasing mature bacteriophages. Other types of particles were frequently found, such as unidentified algal scales, and virus-sized, star-shaped particles.

It is to be regretted that no data on the abundance of virus-like particles in the Dead Sea are available for the periods October–December 1980 and June–July 1992, when rapid declines in bacterial community size were observed (Oren, 1983b; Oren and Gurevich, 1995). The first enumerations of virus-like particles in the Dead Sea were performed in October 1994 (between $0.9–7.3*10^7$ ml^{-1} in the upper 20 m of the water column), at a time in which a dense archaeal community (about 10^7 cells ml^{-1} in the upper 18 m of the water column) was still present. In this period some *Dunaliella* cells were still found (around 50 cells ml^{-1}), and the bacterial numbers were declining. The ratio between virus-like particles and bacteria found (4.4 on the average) is similar to the values reported for freshwater, estuarine, and marine systems, suggesting that viruses may play a quantitatively similar role in extremely hypersaline ecosystems as in conventional aquatic habitats. Samples collected during 1995 contained low numbers of both bacteria and virus-like particles ($1.9–2.6*10^6$ and $0.8–4.6*10^7$ ml^{-1} in April 1995), with viral numbers sharply declining afterward (less than 10^4 ml^{-1} in November 1995 – January 1996) (Oren et al., 1997) (Figure 10.1, lower panel).

Plaque count assays to enumerate virulent bacteriophages in the Dead Sea have never been performed, but their possible contribution to the understanding of the importance of phages in the ecosystem is expected to be low. One of the reasons is that the nature of the dominant Archaeon in the lake, to be used as host in such assays, is still unknown (Oren and Gurevich, 1993, see above). Moreover, most phages in aquatic environments are probably temperate rather than virulent, and the discovery of lysogenic phages in halophilic Archaea suggests that the same may be true for hypersaline environments. If the bacteria used host lysogenic bacteriophages, they will be immune to reinfection, and they will thus not be useful as host bacteria for counting of plaque-forming units of the phage they carry.

Lysis of Archaea by halophilic bacteriophages may have a profound impact on the availability of organic carbon and nitrogen to the remaining bacterial community. Moreover, the occurrence of bacteriophages in the dense archaeal communities that develop in certain periods in the Dead Sea may have important implications for possible gene transfer between different strains of halophilic Archaea.

10.5 BIOLOGICAL AND PHYSICAL EVIDENCE FOR THE TERMINATION OF THE DEAD SEA 1991–1995 STRATIFICATION

The distribution of bacterial and algal communities in the upper water layers of the Dead Sea closely follows the depth of the pycnocline and/or thermocline that separates the upper water from the concentrated brine below. Thus, the vertical distribution of the biota becomes a sensitive indicator of the physical structure of the water column.

Following the rainy winters of 1991–1992 and 1992–1993, the Dead Sea entered a meromictic phase (Beyth et al., 1993). The stability of the stratification decreased in subsequent years as a result of the relatively dry winters. The stability of the water column stratification in the autumn of 1994 was found to be so weak that an overturn was considered imminent. Occurrence of an overturn is expected to lead to an even distribution of the remaining biological communities and their activities at all depths. Such an event was witnessed during late 1982– early 1983, when the mixing of the water column was accompanied by a sharp decrease in particulate protein and bacterial numbers per unit of volume. Mixing caused the redistribution of the bacterial community (a remainder of the bloom that had developed in 1980), formerly restricted to the upper 20–30 m above the halo/thermocline (Oren, 1988). No overturn took place in the autumn/winter of 1994, a conclusion based (in addition to additional lines of evidence) on the finding of a sharp vertical decrease in bacterial numbers and other parameters associated with the bacterial community below 30–40 m depth (Anati et al., 1995).

The presence of a sufficiently dense archaeal community ($2-4*10^6$ cells ml^{-1}) in the upper 30 m of the Dead Sea water column in the autumn of 1995 provided us with a unique opportunity to follow the extent and the rate of the mixing of the water column at the end of 1995. The paucity of salinity and temperature profiles collected during the period when the overturn was expected provided little information on its exact timing, indicating only that an overturn did occur at some time between August 7, 1995 and January 23, 1996. To obtain a more precise estimate of when and at what rate the overturn event took place, we monitored bacterial numbers and bacterioruberin content in water samples collected during cruises and in water samples collected at Ein Gedi beach, assuming that bacterial numbers in samples collected there are similar to surface waters at the center of the lake. This assumption seems justified based on comparison of bacterial numbers in samples collected on the same day at Ein Gedi beach and in the lake's center in 1995. All microbiological parameters showed a similar trend: after a period of at least six months (April–September 1995), during which bacterial numbers and bacterioruberin content of surface water samples did not change much, a rapid five- to sixfold decline took place within the first three weeks of November 1995. The decrease in bacterial numbers and activities reached its completion by the time of the November 26 cruise, and all parameters stabilized at a new low level. When the overturn causes a passive redistribution of the bacterial community over the whole of the water column, and the bacteria become spread evenly from the upper 40 meters to all depths down to 320 m, an approximately fivefold decrease of all bacterially related parameters in the upper water layers (an estimate based on the hypsometric curve of the Dead Sea) can be expected. This is very similar to the observed values.

The relatively short time (a few weeks only) that elapsed from the onset of the decrease in biological parameters in the upper water layers till complete mixing strongly suggests that double-diffusive mixing may have enhanced the approach to turnover. Double-diffusive mixing is a very effective form of vertical mixing that occurs when the temperature step across the thermocline is stabilizing and the salinity step across the halocline is destabilizing. Double diffusion occurs when evaporation brings about an increase in salinity only within the upper layer, and its salt concentration eventually reaches that of the lower layer. The stability of the pycnocline is then maintained by temperature alone. With continuing evaporation, the salinity step becomes more destabilizing, and, in the upper layer, salt fingers start to form that rapidly penetrate the lower layers.

10.6 CONCLUSION

The microbial bloom that developed in the Dead Sea in the spring of 1992 and remained till the end of 1995 provided us with a unique opportunity for a renewed study of the biology of the lake. The close correlation between biological processes and the chemistry and physics of the lake enabled the use of interdisciplinary approaches involving the monitoring of physical properties (light reflectance and absorption) to obtain information on the distribution of the biological communities, use of data on the distribution with depth and time of the Archaea to draw conclusions on the stratification or mixing of the water column. Use of lipids as biomarkers in a chemotaxonomic approach enabled us to obtain relevant information on the nature of the archaeal bloom.

The overturn of the Dead Sea water column in November 1995 started a new holomictic period. A new algal and bacterial bloom can only be expected to occur following renewed massive winter rain floods, which dilute the upper layers to a sufficiently low salinity to support the rapid growth of *Dunaliella*, which in turn will support a bacterial bloom (Oren, 1988, 1993). Only when a sufficiently diluted epilimnion is formed — an event whose occurrence cannot be predicted — will the Dead Sea water column again become a challenging environment for the study of microbial community dynamics and interactions.

ACKNOWLEDGMENTS

I thank David A. Anati, Nisim Ben-Yosef (the Hebrew University of Jerusalem) and Mikal Heldal and Gunnar Bratbak (University of Bergen, Norway) for their contributions to the results presented in this chapter. The work was supported by grants from the Israel Science Foundation administered by the Israel Academy of Sciences and Humanities, and from the Israeli Ministry for Energy and Infrastructure.

REFERENCES

Anati, D.A. and M. Stiller. 1991. The post-1979 thermohaline structure of the Dead Sea and the role of double-diffusivity. *Limnol. Oceanogr.* 36:342–354.
Anati, D.A., I. Gavrieli, and A. Oren. 1995. The residual effect of the 1991–93 rainy winters on stratification of the Dead Sea. *Israel J. Earth Sci.* 44:63–70.
Bergh, O., K.Y. Børsheim, G. Bratbak, and M. Heldal. 1989. High abundance of viruses found in aquatic environments. *Nature* 340:467–468.
Beyth, M., I. Gavrieli, D.A. Anati, and O. Katz. 1993. Effects of the December 1991–May 1992 floods on the Dead Sea vertical structure. *Israel J. Earth Sci.* 42:45–47.
Børsheim, Y., G. Bratbak, and M. Heldal. 1990. Enumeration and biomass estimation of planktonic bacteria and viruses by transmission electron microscopy. *Appl. Environ. Microbiol.* 56:352–356.
Elazari-Volcani, B. 1943. A dimastigamoeba in the bed of the Dead Sea. *Nature* 152:301–302.
Elazari-Volcani, B. 1944. A ciliate from the Dead Sea. *Nature* 154:335.
Guixa-Boixareu. N., J.I. Caldéron-Paz, M. Heldal, G. Bratbak, and C. Pedrós-Alió. 1996. Viral lysis and bacterivory as prokaryotic loss factors along a salinity gradient. *Aquat. Microb. Ecol.* 11:215–227.
Mullakhanbhai, M.F. and H. Larsen. 1975. *Halobacterium volcanii* spec. nov., a Dead Sea halobacterium with a moderate salt requirement. *Arch. Microbiol.* 104:207–214.
Nissenbaum, A. 1975. The microbiology and biogeochemistry of the Dead Sea. *Microb. Ecol.* 2:139–161.
Nissenbaum. A., M.J. Baedecker, and I.R. Kaplan. 1972. Organic geochemistry of Dead Sea sediments. *Geochim. Cosmochim. Acta* 36:709–727.
Nuttal, S.D. and M.L. Dyall-Smith. 1993. HF1 and HF2: novel bacteriophages of halophilic archaea. *Virology* 197:678–684.
Oren, A. 1983a. *Halobacterium sodomense* sp. nov., a Dead Sea halobacterium with extremely high magnesium requirement and tolerance. *Int. J. Syst. Bacteriol.* 33:381–386.
Oren, A. 1983b. Population dynamics of halobacteria in the Dead Sea water column. *Limnol. Oceanogr.* 28:1094–1103.
Oren, A. 1988. The microbial ecology of the Dead Sea, in *Advances in Microbial Ecology*, Vol. 10, K. C. Marshall, Ed. Plenum, New York. 193–229.
Oren, A. 1989. Photosynthetic and heterotrophic benthic bacterial communities of a hypersaline sulfur spring on the shore of the Dead Sea (Hamei Mazor), in *Microbial Mats. Physiological Ecology of Benthic Microbial Communities*, Y. Cohen and E. Rosenberg, Eds. American Society for Microbiology, Washington. 64–76.
Oren, A. 1993. The Dead Sea–alive again. *Experientia* 49:518–522.

Oren, A. 1994. Ecology of extremely halophilic archaea. *FEMS Microbiol. Rev.* 13:415–440.
Oren, A. and D.A. Anati. 1996. Termination of the Dead Sea 1991–1995 stratification: biological and physical evidence. *Israel J. Earth Sci.* 45:81–88.
Oren, A. and N. Ben-Yosef. 1997. Development and spatial distribution of an algal bloom in the Dead Sea: A remote sensing study. *Aquat. Microb. Ecol.* 13:219–223.
Oren, A. and P. Gurevich. 1993. Characterization of the dominant halophilic archaea in a bacterial bloom in the Dead Sea. *FEMS Microbiol. Ecol.* 12:249–256.
Oren, A. and P. Gurevich. 1995. Dynamics of a bloom of halophilic archaea in the Dead Sea. *Hydrobiologia* 315:149–158.
Oren, A. and M. Shilo. 1981. Bacteriorhodopsin in a bloom of halobacteria in the Dead Sea. *Arch. Microbiol.* 130:185–187.
Oren, A. and M. Shilo. 1982. Population dynamics of *Dunaliella parva* in the Dead Sea. *Limnol. Oceanogr.* 27:201–211.
Oren, A., M. Ginzburg, B.Z. Ginzburg, L.I. Hochstein, and B.E. Volcani. 1990. *Haloarcula marismortui* sp. nov., nom. rev., an extremely halophilic bacterium from the Dead Sea. *Int. J. Syst. Bacteriol.* 40:209–210.
Oren, A., P. Gurevich, D.A. Anati, E. Barkan, and B. Luz. 1995a. A bloom of *Dunaliella parva* in the Dead Sea in 1992: biological and biogeochemical aspects. *Hydrobiologia* 279:173–185.
Oren, A., P. Gurevich, R.T. Gemmell, and A. Teske. 1995b. *Halobaculum gomorrense* gen. nov., sp. nov., a novel extremely halophilic archaeon from the Dead Sea. *Int. J. Syst. Bacteriol.* 45:747–754.
Oren, A., G. Bratbak, and M. Heldal. 1997. Occurrence of virus-like particles in the Dead Sea. *Extremophiles* 1:143–149.
Tindall, B.J. 1992. The family Halobacteriaceae, in *The Prokaryotes. A Handbook on the Biology of Bacteria: Ecophysiology, Isolation, Identification, Applications.* 2nd ed., A. Balows, H.G. Trüper, M. Dworkin, W. Harder, and K.-H. Schleifer, Eds. Springer-Verlag, New York. 768–808.
Torreblanca, M., F. Rodriguez-Valera, G. Juez, A. Ventosa, M. Kamekura, and M. Kates. 1986. Classification of non-alkaliphilic halobacteria based on numerical taxonomy and polar lipid composition, and description of *Haloarcula* gen. nov. and *Haloferax* gen. nov. *Syst. Appl. Microbiol.* 8:89–99.
Torsvik. T. and I.D. Dundas. 1974. Bacteriophage of *Halobacterium salinarium*. *Nature* 248:680–681.
Wais, A.C., M. Kon, R.E. MacDonald, and B.D. Stollar. 1975. Salt-dependent bacteriophage infecting *Halobacterium cutirubrum* and *H. halobium*. *Nature* 256:314–315.

11 Studies on the Microbiota of the Dead Sea–50 Years Later

Antonio Ventosa, David R. Arahal, and Benjamin E. Volcani

CONTENTS

11.1 Introduction 139
11.2 The Dead Sea 140
11.3 Microbiological Studies by B.E. Volcani 140
11.4 Later Microbiological Studies 142
11.5 Recent Studies on Volcani's Enrichments 143
    11.5.1 Halobacteria 143
    11.5.2 Moderate Halophiles 144
Acknowledgments 146
References 146

11.1 INTRODUCTION

Throughout history, the Dead Sea has been considered a hostile environment, devoid of any kind of living forms. The salinity of its water was considered too high to be compatible with life and therefore unable to support growth or survival of fish and other organisms. Thus, the 6th-century Madaba map shows that fish entering the Dead Sea from the Jordan River try to avoid its salty brines (Nissenbaum, 1975).

The same opinion prevailed among the scientific community during the 19th century and into the beginning of the 20th, when several chemical and geological studies were conducted in the lake. Gay-Lussac in 1819, Ehrenberg in 1848, and Barrois in the mid 1880s all stated that there was no sign of life in the Dead Sea. Its apparent sterility motivated Lortet to look for some useful applications of this water as an aseptic liquid, and he therefore inoculated in appropriate media some of the mud collected by Barrois. To his surprise, fast growth occurred of pathogenic anaerobic rods of the genus *Clostridium*, which cause gas gangrene and tetanus (Lortet, 1892). These bacteria can be considered contaminants, since they are unable to grow in this hypersaline environment and merely survive as dormant endospores that can be revived in suitable media. Therefore, in spite of these findings, the Dead Sea was still considered sterile.

When the extreme halophiles were discovered in the beginning of the 20th century, it became obvious that the high salt concentration of the Dead Sea alone could not explain its lifelessness, and it was postulated that the real cause was its high Ca^{2+} ion content (Baas Becking, 1934). By that time, B.E. Volcani (Wilkansky) was about to publish the first of his reports (Wilkansky, 1936) on the isolation and characterization of microorganisms from the Dead Sea, thus ending the misconception that no life occurs in this lake.

11.2 THE DEAD SEA

The Dead Sea is a terminal lake in the Syrian-African rift valley that has always attracted the interest of the civilizations living in its surroundings. Although it was considered sterile and therefore unsuitable for fishery, it has been used as an economic resource for more than 10,000 years for obtaining salt, asphalt, magnesium, potash, bromine, therapeutics and cosmetics, etc. (Nissenbaum, 1993), and most of these resources are still being exploited today.

In the last decades, the lake has undergone far-reaching physical and chemical changes as a result of human activity in the lake itself and its surroundings. The use of water from the Jordan River catchment area for irrigation, initiated in the 1950s, caused a reduction in the inflow of water to the lake and resulted in a steady drop in the water level (Beyth, 1980).

The actual level of the lake is 410 m below sea level; it is the lowest point on the earth's surface. In 1930, the water level was −392 m. At that time, the lake was stratified and consisted of two basins: a deep one in the north and a shallow one in the south. In 1976, the southern basin was detached and today the area is occupied by a series of evaporation ponds that are filled by pumping water from the northern basin. The negative water balance of the Dead Sea also led to an increase in the concentration of salts in the upper layer, resulting in a complete overturn in February 1979 (Steinhorn and Gat, 1983).

The salt composition of the lake is much different from that of thalassohaline brines. The Dead Sea is relatively poor in Na^+ ions, while it contains more Mg^{2+}, Ca^{2+} and Br^- ions than hypersaline environments that originated by evaporation of seawater (Oren, 1993a).

11.3 MICROBIOLOGICAL STUDIES BY B.E. VOLCANI

Benjamin Elazari-Volcani (also known as B. Wilkansky and B. Elazari-Volcani) is often referred to as the pioneer of microbiological studies in the Dead Sea. A short article entitled "Life in the Dead Sea" was published in *Nature* in 1936 (Wilkansky, 1936), opening the way for a series of microbiological studies that continue today. The first microorganisms observed included a phytoflagellate of the genus *Chlamydomonas* or *Dunaliella*, a Gram-positive long filamentous organism and two Gram-negative microorganisms, one of them being a yeast that produced an orange pigment, and the second a small rod-shaped bacterium. Volcani's research on this topic was summarized in his Ph.D. thesis "Studies on the Microflora of the Dead Sea" (Elazari-Volcani, 1940a).

When Volcani's studies were conducted, the lake had a larger surface than it has now, the southern basin was still connected with the northern basin through the Lisan straits, and the northern basin was stratified. Figure 11.1 shows a map of the Dead Sea at that time, indicating Volcani's sampling sites. The salinity of the water, lower than today, was 263.3 g l^{-1} total dissolved salts in surface water as compared with the present level of close to 340 g l^{-1} (Table 11.1).

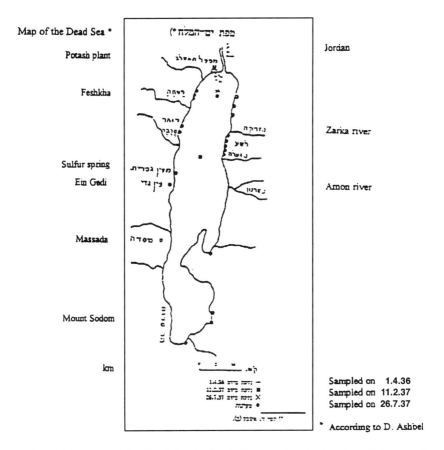

FIGURE 11.1 Map of the Dead Sea in the 1930s, showing the location of the sampling sites and some additional geographic features (Elazari-Volcani, 1940a).

Samples were inoculated in media designed for different physiological groups, and growth was obtained in media for denitrifying bacteria, aerobic cellulose decomposing bacteria, fibrinolytic bacteria, as well as in a peptone brine medium (Elazari-Volcani, 1940a). Volcani described the following categories of microorganisms:

TABLE 11.1
Chemical Composition (% wt/vol) of the
Dead Sea in 1935 and in 1991

	Values in 1935[a]		Value in 1991[b]
	Surface	50 m depth	
NaCl	7.14	9.42	9.37
KCl	1.18	1.47	1.51
$MgCl_2$	14.24	16.37	17.41
$CaCl_2$	3.80	4.75	4.87
$CaSO_4$	0.13	0.06	N.D.
$MgBr_2$	0.39	0.59	0.59
Total	26.88	32.65	33.75

[a] Volcani (1940a).
[b] Calculated on the basis of data provided by Oren (personal communication).
Note: N.D. – not determined.

I. Halo-obligatory microorganisms producing red colonies. He described the presence of the following microorganisms: *Flavobacterium (Halobacterium) marismortui, Flavobacterium (Halobacterium) trapanicum, Micrococcus morrhuae,* and *Sarcina morrhuae.*

II. Halo-tolerant organisms. On the basis of their phenotypic features, three new species were described and named: *Chromobacterium maris-mortui, Pseudomonas halestorgus,* and *Flavobacterium halmephilum.*

III. Halo-resistant organisms, originating from freshwater areas in the surroundings of the Dead Sea. Ten sporulating Gram-positive cultures were isolated.

According to the classification of halophilic microorganisms proposed by Kushner (1985), these three categories would correspond with extreme halophiles, moderate halophiles, and halotolerant microorganisms, respectively. Unfortunately, with the exception of some moderate halophiles, most of the cultures isolated at that time were lost, and they are not currently maintained and available from culture collections.

The phytoflagellate *Dunaliella viridis* was found in the water samples. Volcani was also able to isolate cyanobacteria and several algae (diatoms and green algae) from the bed of the lake (Elazari-Volcani, 1940a, 1940b). In further studies, other organisms such as a dimastigamoeba, a ciliate, and other organisms were isolated and described from water or mud samples (Elazari-Volcani, 1940b, 1943a, 1943b, 1944; Volcani, 1944).

11.4 LATER MICROBIOLOGICAL STUDIES

For many years, no new research on the microbiology of the Dead Sea was performed. In the 1960s and '70s only a few new studies were initiated (Kaplan and

Friedmann, 1970; Nissenbaum, 1975), and in the 1980s and onward a large number of studies dealing with the ecology, physiology and taxonomy of the Dead Sea microorganisms have been conducted, mainly by Oren and co-workers (Oren, 1988, 1993a, 1993b; Oren and Gurevich, 1994).

A number of new bacterial species have been isolated from The Dead Sea and its sediments. Thus, the following new species have been described within the family *Halobacteriaceae*: *Haloferax* (formerly *Halobacterium*) *volcanii* (Mullakhanbhai and Larsen, 1975), *Haloarcula marismortui* (Oren et al., 1990), *Halorubrum* (formerly *Halobacterium*) *sodomense* (Oren, 1983a; McGenity and Grant, 1995) and, more recently, *Halobaculum gomorrense* (Oren et al., 1995). In addition, some studies have been carried out on moderately halophilic bacteria: the species *Chromobacterium marismortui* (Elazari-Volcani, 1940a) was compared with some strains isolated from Mediterranean salterns, and was reclassified in a new genus as *Chromohalobacter marismortui* (Ventosa et al., 1989). Dobson and co-workers (1990) presented an emended description of *Halomonas halmophila*, and strain Ba_1, originally isolated from the Dead Sea and widely used in physiological studies, was described as *Halomonas israelensis* (Huval et al., 1995). Some new, strictly anaerobic and moderately halophilic, species from the Dead Sea are *Halobacteroides halobius* (Oren et al., 1984), *Sporohalobacter lortetii* (Oren, 1983b; Oren et al., 1987), *Orenia marismortui* (Oren et al., 1987; Rainey et al., 1995), and *Ectothiorhodospira marismortui* (isolated from a hypersaline sulfur spring on the shore of the lake) (Oren et al., 1989).

11.5 RECENT STUDIES ON VOLCANI'S ENRICHMENTS

Recently, we have begun to characterize the microorganisms present in several enrichments that were set up by Volcani when he was studying the microbiota of the Dead Sea in the 1930s. Fifteen of these enrichments, prepared with Dead Sea water plus 1.0% (wt/vol) peptone and inoculated with water samples collected in 1936 at surface level in the northern basin of the lake, were kept in 500-ml bottles under aseptic conditions in the dark in a dry place at 18–20°C. These enrichments remained unopened over the entire storage period (57 years).

Using different media for moderate and extreme halophiles and taking inocula from these enrichments, and by repeated streaking onto agar plates, a large number of isolates were obtained in pure culture. The composition of the media and the isolation conditions have been previously described (Arahal et al., 1996; Ventosa et al., 1982). Altogether, 261 strains were isolated. On the basis of their salt range and optimum salinity for growth, they were divided into two physiological groups: 158 strains were extreme halophiles, and 103 strains were moderate halophiles.

11.5.1 HALOBACTERIA

On solid media, all extreme halophiles produced colonies with pink-to-red pigmentation and grew with a minimum of 15% salts and optimally at 20–25% salts; on the basis of these and other phenotypic features they were considered to be members of the family *Halobacteriaceae* (Grant and Larsen, 1989). Twenty-two randomly selected halobacterial isolates were taxonomically characterized in more detail (Arahal et al., 1996). All strains were Gram-negative, motile, catalase and oxidase positive, and they were all able to hydrolyze DNA. Most strains produced acid from glucose and other sugars; variable reactions were obtained for other tests (hydrolysis of gelatin, casein, starch, etc.). On the basis of phenotypic traits, they were divided into five groups: Groups I (five strains), II (five strains), and V (two strains) were identified as members of the genera *Haloferax*, *Haloarcula* and *Halobacterium*, respectively. Group III (six strains) and IV (four strains) were not assigned to any genus, since they showed phenotypic characteristics that did not clearly conform with any of the currently described genera within the family *Halobacteriaceae*.

To determine their phylogenetic position, we determined the 16S rRNA gene sequence of a representative isolate from each of the five phylogenetic groups by DNA extraction, followed by PCR amplification of the 16S rDNA fragment. The purified DNA amplicons were sequenced using a TAQuence Cycle Sequencing Kit, the 16S rRNA sequences were compared with the known sequences of other halobacteria, and a phylogenetic tree was constructed (Figure 11.2). The 16S rRNA sequences of strains E1 (group I) and E8 (group IV) were very similar to each other and to the sequence of *Haloferax volcanii* ATCC 29605 (more than 99.5% similarity in each case). Strain E12 (group V) was very similar to *Halobacterium salinarum* DSM 671. These results are in accordance with the phenotypic characteristics of the strains studied. Strains E2 (group II) and E11 (group III) both belong to the genus *Haloarcula*, as inferred from their phenotypic features (Arahal et al., 1996) and their phylogenetic relationships. The closest relative of strains E2 and E11 is *Haloarcula hispanica* ATCC 33960 (98.9% and 98.8% sequence similarity, respectively). They could represent two new species, but other methods, such as DNA–DNA hybridization should be used to verify that separate species designations are warranted (Oren et al., 1997).

11.5.2 Moderate Halophiles

Among the 103 strains of moderate halophilic bacteria isolated, 91 were Gram-positive endospore-forming rods and 12 were Gram-negative aerobic rods. According to their phenotypic features, the Gram-positive rods could represent a new species within the genus *Bacillus*, growing best at 10% salts. Using a salt-free standard agar medium, Volcani isolated 10 sporulating Gram-positive bacteria (Elazari-Volcani, 1940a). The maximum salt concentration enabling growth was 6% to 12%, and one strain was even able to grow with 24% salts. They were considered contaminants that can resist the high salt concentration by forming endospores. Volcani suggested that it is quite possible that such organisms are able to adapt gradually to relatively high salt concentrations, thus becoming halo-tolerant (moderately halophilic) or even halo-obligatory (extremely halophilic) (Volcani, 1944). These isolates were not studied in detail, and therefore it is not possible to ensure how similar they are to our recent isolates. Nevertheless, there is at least one important difference, since our strains are moderate halophiles and not halotolerant organisms. One reason for the difference observed may be that in our study the concentration of the media used for the isolation was 10% (Ventosa et al., 1982), while Volcani used salt-free media

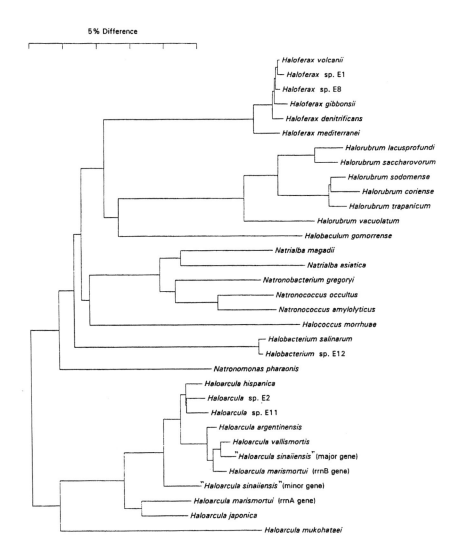

FIGURE 11.2 Phylogenetic tree based on 16S rRNA sequence similarity data for the five isolates included in our study and other species of the family *Halobacteriaceae*. The scale bar represents a 5% nucleotide sequence difference as determined by measuring the lengths of horizontal lines connecting any two species.

that can be selective for non-halophilic organisms. The possibility can also be considered that during the period of storage of the samples these *Bacillus* strains adapted to become moderate halophiles, as suggested before (Volcani, 1944).

The Gram-negative strains are related to the *Halomonas-Deleya* complex according to their phenotypic characteristics. A more in-depth study is required to ascertain the phylogenetic relationship between these isolates and *Halomonas halmophila* described by Volcani (Elazari-Volcani, 1940a).

ACKNOWLEDGMENTS

We are grateful to Drs. B.J. Paster and F.E. Dewhirst for their help with the phylogenetic study and valuable discussions. D.R. Arahal was supported by a fellowship from the Spanish Ministerio de Educación y Ciencia. This work was supported by grants from the European Commission (Generic Projects "Biotechnology of Extremophiles" BIO 2-CT93-0274, and "Extremophiles as Cell Factories" BIO 4-CT96-0488), Spanish Ministerio de Educación y Ciencia (PB93-0920), and Junta de Andalucia.

REFERENCES

Arahal, D.R., F.E. Dewhirst, B.J. Paster, B.E. Volcani, and A. Ventosa. 1996. Phylogenetic analyses of some extremely halophilic archaea isolated from Dead Sea water, determined on the basis of their 16S rRNA sequences. *Appl. Environ. Microbiol.* 62:3779–3786.

Baas Becking, L.G.M. 1934. *Geobiologie of Inleiding tot de Milieukunde*. W.P.V. Stockum & Zn., The Hague.

Beyth, M. 1980. Recent evolution and present stage of Dead Sea brines, in *Hypersaline Brines and Evaporitic Environments*, A. Nissenbaum, Ed. Elsevier, Amsterdam. 155–165.

Dobson, S.J., S.R. James, P.D. Franzmann, and T.A. McMeekin. 1990. Emended description of *Halomonas halmophila* (NCMB 1971^T). *Int. J. Syst. Bacteriol.* 40:462–463.

Elazari-Volcani, B. 1940a. *Studies on the Microflora of the Dead Sea*. Ph.D. thesis, The Hebrew University of Jerusalem (in Hebrew).

Elazari-Volcani, B. 1940b. Algae in the bed of the Dead Sea. *Nature* 145:975.

Elazari-Volcani, B. 1943a. Bacteria in the bottom sediments of the Dead Sea. *Nature* 152:274–275.

Elazari-Volcani, B. 1943b. A dimastigamoeba in the bed of the Dead Sea. *Nature* 152:301–302.

Elazari-Volcani, B. 1944. A ciliate from the Dead Sea. *Nature* 154:335–336.

Grant, W.D. and H. Larsen. 1989. Extremely halophilic archaeobacteria. Order Halobacteriales ord. nov., in *Bergey's Manual of Systematic Bacteriology*, J.T. Staley, M.P. Bryant, N. Pfennig, and J.G. Holt, Eds. Vol. 3. The Williams & Wilkins Co., Baltimore. 2216–2233.

Huval, J.H., R. Latta, R. Wallace, D.J. Kushner, and R.H. Vreeland. 1995. Description of two new species of *Halomonas*: *Halomonas israelensis* sp. nov. and *Halomonas canadensis* sp. nov. *Can. J. Microbiol.* 41:1124–1131.

Kaplan, I.R. and A. Friedmann. 1970. Biological productivity in the Dead Sea. Part I. Microorganisms in the water column. *Israel J. Chem.* 8:513–528.

Kushner, D.J. 1985. The *Halobacteriaceae*, in *The Bacteria: a Treatise on Structure and Function*, C.R. Woese and R. Wolfe, Eds. Vol. 8. Academic, London. 171–214.

Lortet, M.L. 1892. Researches on the pathogenic microbes of the mud of the Dead Sea. *Palestine Expl. Fund.* 1892:48–50.

McGenity, T.J. and W.D. Grant. 1995. Transfer of *Halobacterium saccharovorum*, *Halobacterium sodomense*, *Halobacterium trapanicum* NRC 34021 and *Halobacterium lacusprofundi* to the genus *Halorubrum* gen. nov., as *Halorubrum saccharovorum* comb. nov., *Halorubrum sodomense* comb. nov., *Halorubrum trapanicum* comb. nov., and *Halorubrum lacusprofundi* comb. nov. *Syst. Appl. Microbiol.* 18:237–243.

Mullakhanbhai, M.F. and H. Larsen. 1975. *Halobacterium volcanii* spec. nov., a Dead Sea halobacterium with a moderate salt requirement. *Arch. Microbiol.* 104:207–214.

Nissenbaum, A. 1975. The microbiology and biogeochemistry of the Dead Sea. *Microb. Ecol.* 2:139–161.
Nissenbaum, A. 1993. The Dead Sea–an economic resource for 10000 years. *Hydrobiologia* 267:127–141.
Oren, A. 1983a. *Halobacterium sodomense* sp. nov., a Dead Sea halobacterium with an extremely high magnesium requirement. *Int. J. Syst. Bacteriol.* 33:381–386.
Oren, A. 1983b. *Clostridium lortetii* sp. nov., a halophilic obligatory anaerobic bacterium producing endospores with attached gas vacuoles. *Arch. Microbiol.* 136:42–48.
Oren, A. 1988. The microbial ecology of the Dead Sea, in *Advances in Microbial Ecology*, Vol. 10, K.C. Marshall, Ed. Plenum, New York. 193–229.
Oren, A. 1993a. Ecology of extremely halophilic microorganisms, in *The Biology of Halophilic Bacteria,* R.H. Vreeland and L.I. Hochstein, Eds. CRC, Boca Raton. 25–53.
Oren, A. 1993b. The Dead Sea–alive again. *Experientia* 49:518–522.
Oren, A. and P. Gurevich. 1994. Production of D-lactate, acetate, and pyruvate from glycerol in communities of halophilic archaea in the Dead Sea and in saltern crystallizer ponds. *FEMS Microbiol. Ecol.* 14:147–156.
Oren, A., W.G. Weisburg, M. Kessel, and C.R. Woese. 1984. *Halobacteroides halobius* gen. nov., sp. nov., a moderately halophilic anaerobic bacterium from the bottom sediments of the Dead Sea. *Syst. Appl. Microbiol.* 5:58–70.
Oren, A., H. Pohla, and E. Stackebrandt. 1987. Transfer of *Clostridium lortetii* to a new genus *Sporohalobacter* gen. nov. as *Sporohalobacter lortetii* comb. nov., and description of *Sporohalobacter marismortui* sp. nov. *Syst. Appl. Microbiol.* 9:239–246.
Oren, A., M. Kessel, and E. Stackebrandt. 1989. *Ectothiorhodospira marismortui* sp. nov., an obligately anaerobic, moderately halophilic purple sulfur bacterium from a hypersaline sulfur spring on the shore of the Dead Sea. *Arch. Microbiol.* 151:524–529.
Oren, A., M. Ginzburg, B.Z. Ginzburg, L.I. Hochstein, and B.E. Volcani. 1990. *Haloarcula marismortui* (Volcani) sp. nov., nom. rev., an extremely halophilic bacterium from the Dead Sea. *Int. J. Syst. Bacteriol.* 40:209–210.
Oren, A., P. Gurevich, R.T. Gemmell, and A. Teske. 1995. *Halobaculum gomorrense* gen. nov., sp. nov., a novel extremely halophilic archaeon from the Dead Sea. *Int. J. Syst. Bacteriol.* 45:747–754.
Oren, A., A. Ventosa, and W.D. Grant. 1997. Proposed minimal standards for description of new taxa in the order *Halobacteriales*. *Int. J. Syst. Bacteriol.* 47:233–238.
Rainey, F.A., T.N. Zhilina, E.S. Boulygina, E. Stackebrandt, T.P. Tourova, and A. Zavarzin. 1995. The taxonomic status of the fermentative halophilic anaerobic bacteria: description of Haloanaerobiales ord. nov., *Halobacteroidaceae* fam. nov., *Orenia* gen. nov. and further taxonomic rearrangements at the genus and species level. *Anaerobe* 1:185–199.
Steinhorn, I. and J.R. Gat. 1983. The Dead Sea. *Sci. Am.* 249(4) 102–109.
Ventosa, A., E. Quesada, F. Rodríguez-Valera, F. Ruiz-Berraquero, and A. Ramos-Cormenzana. 1982. Numerical taxonomy of moderately halophilic Gram-negative rods. *J. Gen. Microbiol.* 128:1959–1969.
Ventosa, A., M.C. Gutierrez, M.T. Garcia, and F. Ruiz-Berraquero. 1989. Classification of "*Chromobacterium marismortui*" in a new genus *Chromohalobacter* gen. nov., as *Chromohalobacter marismortui* comb. nov., nom. rev. *Int. J. Syst. Bacteriol.* 4:382–386.
Volcani, B.E. 1944. The microorganisms of the Dead Sea, in *Papers Collected to Commemorate the 70th Anniversary of Dr. Chaim Weizmann.* Daniel Sieff Institute, Rehovoth. 71–85.
Wilkansky, B. 1936. Life in the Dead Sea. *Nature* 138:467.

12 Radiotracer Studies of Bacterial Methanogenesis in Sediments from the Dead Sea and Solar Lake (Sinai)

Mark Marvin DiPasquale, Aharon Oren, Yehuda Cohen, and Ronald S. Oremland

CONTENTS

12.1 Introduction..149
12.2 Methanogenesis in Dead Sea Sediment Samples 151
    12.2.1 February 1996 Samples..152
    12.2.2 August 1996 Samples...153
12.3 Methanogenesis in the Solar Lake 154
12.4 Conclusion 157
Acknowledgments..158
References..158

12.1 INTRODUCTION

Methanogenesis has been studied extensively in freshwater and marine ecosystems, but much less is known about its occurrence in hypersaline environments (Oremland, 1988; Oremland and King, 1988). Of these, most research has been done in moderately hypersaline systems (e.g., salinity range = ~ 50–150 g l^{-1}), such as evaporative bacterial mats and the sediments of alkaline lakes (King, 1988; Ollivier et al., 1994; Oremland et al., 1982). Moderately hypersaline environments harbor methanogenic populations that produce methane by fermentation of precursors like methanol, methylated amines, and dimethyl sulfide, but show little affinity for acetate fermentation or the reduction of carbon dioxide with hydrogen (Giani et al., 1984, 1989; Kiene et al., 1986; King, 1988; Oremland et al., 1982). The reason for this constraint rests with the effectiveness of sulfate-reducing bacteria in out-competing methanogens for hydrogen and acetate, and the fact that hypersaline brines contain an abundance of sulfate several times greater than seawater. However, there have been very few radiotracer experiments with ^{14}C-methanogenic precursors conducted in hypersaline environments to confirm whether these substances are robustly or weakly metabolized by the resident methanogens (e.g., Oremland et al., 1993). In addition, sulfate reduction in the profoundal sediments of highly productive soda lakes occurs to such an extent that a complete depletion of porewater sulfate can be observed with depth in cores (Oremland et al., 1987). Under these conditions, methanogenesis via reduction of carbon dioxide is theoretically possible, but it proceeds at slow rates and it is problematic to detect because of the large dissolved carbonate + bicarbonate pool diluting any added $^{14}HCO_3^-$ (Oremland and Miller, 1993; Oremland et al., 1993, 1994).

Biogeochemical investigations of extremely hypersaline systems (i.e., salinity ≥ 300 g l^{-1}) have focused on three large water bodies: the Orca Basin in the Gulf of Mexico, the Dead Sea, and the Great Salt Lake. Detailed studies on the ecology of methanogenesis are lacking for all three, but aspects of related research in these locations suggest the possibility of active, but constrained, methane cycles. Thus, the anoxic Orca Basin (salinity = ~ 300 g l^{-1}) has supersaturated levels of dissolved methane that are strongly depleted in ^{13}C, thereby indicating a bacterial origin (Wiesenburg et al., 1985). Although simple incubations of Orca Basin sediment slurries failed to demonstrate discernible methanogenic activity (Oremland and King, 1988), nonetheless, its waters have appreciable levels of archaebacterial lipids, which suggests the presence of active methanogens (Dickins and van Vleet, 1992). Dissolved methane profiles in the Great Salt Lake (salinity = ~ 333 g l^{-1}) are indicative of a source in the sediments (Baedecker, 1985; Schink et al., 1983), and bacterial formation of methane in the sediments from precursors like methionine has been briefly noted in the context of a review on a broader topic (Zeikus, 1983). The Dead Sea (salinity = ~ 340 g l^{-1}) has been the object of numerous investigations, although none that deal with methane (e.g., Cohen et al., 1983; Nissenbaum, 1975; Nissenbaum et al., 1972, 1990; Oren, 1983, 1988a, 1988b, 1990a, 1990b, 1993, 1994, 1995; Oren and Gurevich, 1994; Oren et al., 1984). Bacterial growth in the Dead Sea is severely constrained by the abundance of divalent cations (Cohen et al., 1983). Initial attempts to enrich for methanogens from the Dead Sea were unsuccessful (Oren, 1988b). Nonetheless, extremely halophilic methanogens have been isolated from lagoonal sediments of the Crimea, suggesting that such organisms may inhabit the sediments of other extremely hypersaline systems (Zhilina, 1986; Zhilina and Zavarzin, 1987).

In this investigation we compared the methanogenic capability of sediments from the Dead Sea, whose waters are at salt-saturation concentrations (> 300 g l^{-1}), with those from the less saline Solar Lake (salinity ≤ 180 g l^{-1}; Cohen et al., 1977). The organic-rich Solar Lake sediments were quite active and readily able to metabolize a diversity of ^{14}C-methylotrophic methanogenic precursors to methane and carbon dioxide. In contrast, Dead Sea sediments demonstrated a weak ability to form methane only from [^{14}C]-methanol, and not all samples were able to achieve this activity. However, the detection of even this small amount of methanogenic activity from Dead Sea materials constitutes the first report of methanogenesis in this extreme environment.

12.2 METHANOGENESIS IN DEAD SEA SEDIMENT SAMPLES

Dead Sea surface sediments (upper ~ 10 cm) underlying a benthic salt crust were collected by using a large bottom grab. Sediments were collected on two occasions. Samples were taken from the southwestern portion of the lake opposite Masada (site 4) on 29 February 1996 at water depths of 30 and 50 m. Only a small amount of sediment was recovered on this occasion. On 27 August 1996 samples were recovered from the northwestern corner of the lake offshore of a sewage drain pipe from East Jerusalem at water depths of 10 m (site 1; about 10 m from the shore), 22 m (site 2; about 20 m from the shore), 30 m (site 3; about 40 m from the shore), and for comparison, samples were taken again from the vicinity of site 4 opposite Masada (water depth = 18 m). Retrieval of sediment was more successful in the August trip. Sediments were stored at ~ 20°C in completely filled serum bottles or plastic jars and were shipped to the USGS laboratories in Menlo Park, CA.

Dead Sea sediment slurries were generated by placing 1.5 cm^3 of wet sediment into 13-ml serum bottles. After crimp-sealing the bottles with black butyl rubber stoppers, the gas phase was flushed with O_2-free N_2 for 5–10 min, 1.5 ml of deoxygenated (N_2 bubbled) Dead Sea water was injected, and the bottles were vortexed to generate a slurry. ^{14}C-labeled methanogenic substrates (0.5 ml) were injected into the slurries and the samples were incubated at room temperature, during which time headspace levels of $^{14}CH_4$ and $^{14}CO_2$ were quantified by gas chromatography/gas proportional counting (Culbertson et al., 1981). Because of the limited amount of sediment material collected, only two experiments were conducted with the samples collected in February. In experiment #1, sediment from 30 m, 50 m, and a Dead Sea water control (no sediment) were incubated with ^{14}C-methanol. In experiment #2, sediment slurries from 50 m were incubated with several potential ^{14}C-labeled methane precursors, including methionine, acetate, trimethylamine (TMA), methanol, and dimethylsulfide (DMS). In addition, an incubation was conducted with $^{14}CH_4$ to determine if the sediments had the capacity to oxidize methane. In the third experimental series, the sediments collected in August were all examined for their ability to metabolize [^{14}C]-methanol, and in addition, sediments from sites 1 and 4 were tested for their ability to metabolize ^{14}C-labeled DMS, acetate, and TMA. Incubations were also conducted with Na^{14}HCO$_3$ for the purpose of determining the partitioning of carbon dioxide between the liquid and vapor phases so that the amount of dissolved $^{14}CO_2$ formed from oxidation of the ^{14}C-precursors could be calculated. An additional experiment without radioisotope was conducted at all four sites to determine if the addition of glycerol to sediments could stimulate the production of methane. Methane in the headspaces of these slurries was determined by flame-ionization gas chromatography (Oremland et al., 1987). All sediments were incubated statically in the dark (~ 20°C).

12.2.1 February 1996 Samples

Production of $^{14}CH_4$ from ^{14}C-methanol was not observed, but its oxidation to $^{14}CO_2$ was evident in sediment from 30 and 50 m, although not in controls incubated with only Dead Sea water (Figure 12.1). The liquid phase levels of $^{14}CO_2$ were not

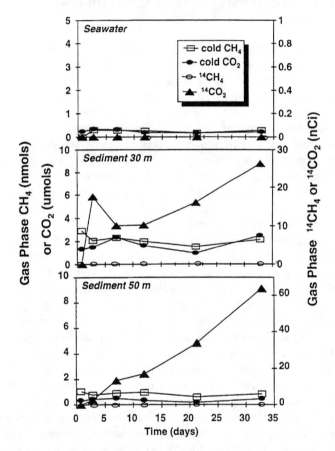

FIGURE 12.1 Metabolism of [^{14}C]-methanol (1.1 µCi; final concentration = 5.3 µM) by Dead Sea sediment and water collected in February, 1996. Results represent the average of two samples.

determined in this experiment, and final observed quantities in the vapor phase represented oxidation of ~ 3–6% of the amount of isotope added. Production of unlabeled methane or carbon dioxide was not observed. These results indicated that bacterial oxidation of methanol under anaerobic conditions can occur in the sediments, but not in the waters of the Dead Sea. Presumably this oxidation is linked to an electron acceptor such as nitrate, sulfate, or Fe^{3+}. No formation of $^{14}CH_4$ was evident during incubations of all of the other ^{14}C-precursors with 50 m sediment slurries (not shown), but both methanol and the 2-C of acetate were oxidized to

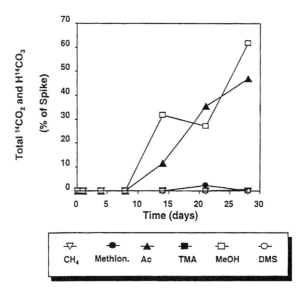

FIGURE 12.2 Time course of $^{14}CO_2$ production from various substrates with 50 m sediment collected from the Dead Sea in February, 1996. Activities and concentrations of the radioisotopes were as follows: methanol (0.3 µCi; 1.5 µM), DMS (2.0 µCi; 14 µM), TMA (1.9 µCi; 181 µM), [2-^{14}C]-acetate (1.9 µCi; 9.0 µM), [5-^{14}C]-methionine (1.9 µCi; 11 µM), and methane (1.8 µCi; 3.2 µM). Results represent the average of two samples.

$^{14}CO_2$, and oxidation represented metabolism of ~ 45% and ~ 63%, respectively, of the acetate and methanol added (Figure 12.2). No oxidation of ^{14}C-labeled DMS, methionine, or TMA was noted. Anaerobic $^{14}CH_4$ oxidation was not observed (Figure 12.2), although it occurs in moderately hypersaline soda lakes (Iversen et al., 1987; Oremland et al., 1993).

12.2.2 August 1996 Samples

The metabolism of [^{14}C]-methanol was examined at all sites (Figure 12.3). Some production of $^{14}CO_2$ was evident at all four sites, although levels formed were variable and in some cases were no longer detectable by the final sampling times (e.g., sites 2 and 3). There was obvious production of $^{14}CH_4$ only at sites 3 and 4, while none was noted at sites 1 and 2. Sites 3 and 4 metabolized 40%–50% of the added methanol. No methanogenic activity was detected at sites 1 and 4 with samples incubated with ^{14}C-labeled DMS, acetate, or TMA (not shown). However, after 36 days of incubation, a small amount of oxidation of the acetate (~ 1% of added) and TMA (~ 5% of added) were evident at site 4, but not at site 1. Neither site 1 nor 2 demonstrated production of unlabeled methane over 36-day incubations, but there was obvious methanogenic activity at sites 3 and 4 (not shown). Glycerol (10 mM) did not enhance the rate of methanogenesis with site 3 samples, but there was a twofold stimulation with the low activity samples of site 4. Glycerol has considerable importance as a substrate for heterotrophic bacteria in the water column of the Dead Sea (Oren, 1993). However, in the sediments, it is likely that bacterial metabolism

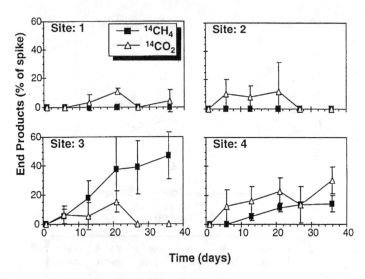

FIGURE 12.3 Metabolism of [^{14}C]-methanol (0.3 µCi; 1.4 µM) in sediments of the Dead Sea collected from all four sites in August, 1996. Results represent the mean of three slurries and bars indicate ± one standard deviation.

of glycerol does not readily yield methanogenic substrates that will stimulate methanogenesis (e.g., methanol).

Since we observed methanogenic activity at the offshore deep site of the sewage outfall transect (site 3), but at the two inshore sites (sites 1 and 2), it would appear that any organics supplied to the sediments from the outfall did not have an overt influence on methanogenic activity. This is reinforced by the observed activity at site 4. Therefore, it appears that some of the sediments of the Dead Sea have the ability to sustain methanogenesis from autochthonous sources.

12.3 METHANOGENESIS IN THE SOLAR LAKE

Sediment from the upper ~ 8 cm of a cyanobacterial mat in Solar Lake was recovered by hand and shipped in a completely filled and sealed plastic box. Solar Lake samples were treated similarly to those of the Dead Sea, but they were not slurried. Instead, solid plugs of sediment (2–3 cm^3) were placed in serum vials and the radioisotopes were injected after sealing. Dissolved methane in the water column of Solar Lake was sampled during September, 1996 when the water column was mixed, and during December, 1996 when the water column was stratified. Procedures for methane extraction have been published (Oremland et al., 1987). All experiments were initiated within a month of sample collection.

Production of ^{14}CH$_4$ was readily evident in Solar Lake samples labeled with [^{14}C]-methanol or [^{14}C]-TMA, but not for [^{14}C]-DMS, [2-^{14}C]-acetate, or H^{14}CO$_3^-$ (Figure 12.4). Production of radioactive gaseous products from DMS and TMA was evident after < 24 h incubation, while a lag of ~ 7 days occurred before [^{14}C]-methylmercury (MeHg) was degraded (Figure 12.4) or the oxidation of [^{14}C]-acetate was detected

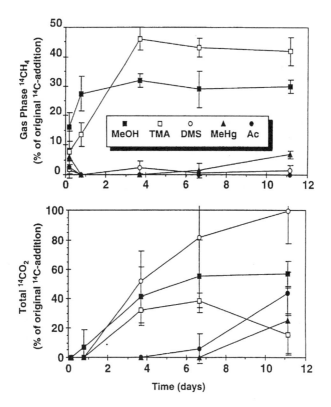

FIGURE 12.4 Metabolism of ^{14}C-labeled methane precursors to ^{14}CH$_4$ (top) or ^{14}CO$_2$ (bottom) by the Solar Lake benthic mat collected in May, 1996. Symbols represent the mean of three slurries and bars indicate ± one standard deviation. Activities and final concentrations of the radiotracers were as follows: methanol (0.2 µCi; 1.4 µM), TMA (0.2 µCi; 20.2 µM), DMS (0.2 µCi; 4.0 µM), [2-^{14}C]-acetate (0.2 µCi; 1.4 µM), and MeHg (0.2 µCi; 23.5 µM).

(Figure 12.4B). Methanogenic attack of MeHg has been observed in freshwater, estuarine, and soda-lake sediments (Oremland et al., 1991; 1995). Both [2-^{14}C]-acetate and [^{14}C]-DMS were metabolized only to ^{14}CO$_2$ (Figure 12.4B), suggesting an exclusive attack mediated by sulfate reducers. Solar Lake waters and sediment have extensive sulfate reduction (Fründ and Cohen, 1992; Jørgensen and Cohen, 1977; Jørgensen et al., 1979a, 1979b). A significant amount of ^{14}CO$_2$ was also associated with the metabolism of methanol and TMA. In a second experiment, a small amount (~ 2%) of the added [^{14}C]-DMS was associated with the formation of ^{14}CH$_4$, with the majority recovered as ^{14}CO$_2$ (Figure 12.5). The ^{14}C-DMS was added at a concentration four times higher than that used in the previous experiment. More of the added DMS pool can enter methanogenic, as opposed to sulfate-reduction, pathways, as the concentration of this substance rises in sediments, a situation similar to that for methanol (Kiene et al., 1986; King, 1984; Oremland et al., 1993). Because [^{14}C]-methylbromide undergoes strong nucleophilic attack by sulfide to form methane thiol and DMS (Oremland et al., 1994), it should have displayed a degradation pattern similar to DMS. While

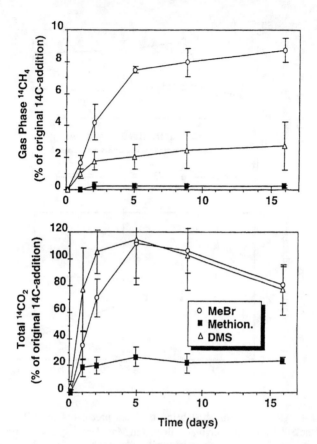

FIGURE 12.5 Metabolism of ^{14}C-labeled methane precursors to ^{14}CH$_4$ (top) or ^{14}CO$_2$ (bottom) by Solar Lake benthic mat collected in May, 1996. Symbols represent the mean of three slurries ± one standard deviation. Activities and concentrations of the radiotracers were as follows: MeBr (3.6 µCi; 18.7 µM), [5-^{14}C]-methionine (1.8 µCi; 10.3 µM), and DMS (2.1 µCi; 14.3 µM).

the patterns of ^{14}CO$_2$ production were similar for both precursors, there was about four times more ^{14}CH$_4$ formed from [^{14}C]-methylbromide than from [^{14}C]-DMS. Although this is probably related somehow to the kinetics of the nucleophilic attack on methylbromide by HS$^-$, we do not have an obvious explanation for this observation. Methionine can also act as a non-competitive substrate for methanogenesis in hypersaline sediments (Oremland et al., 1982) owing to its metabolism, which liberates methane thiol as a methane precursor. However, addition of [^{14}C]-methionine to Solar Lake sediments resulted only in the formation of a modest amount of ^{14}CO$_2$ (Figure 12.5).

Dissolved methane is present in the water column of Solar Lake (not shown). Concentrations during mixis were ~ 0.2 µM throughout the water column, while during stratification, they increased steadily with depth, reaching a concentration of 2.7 µM at 6 m. Previous studies have shown that methylotrophic methanogens are present in Solar Lake sediments (Giani et al., 1984), and that an outward flux of methane can be detected from cores recovered from these sediments (Conrad et al.,

1995). It is now clear that benthic methanogenesis is of sufficient magnitude to influence the chemistry of the water column during stratification.

12.4 CONCLUSION

The results of this survey indicate that some of the sediments of the Dead Sea harbor anaerobes capable of oxidizing methanol, TMA, and acetate, but not DMS or methionine. In addition, methanogenic activity from a methanol precursor was detected at two of the sites, which were also able to form unlabeled methane. Although the general pattern of methanogenesis from a methanol precursor is consistent with the "non-competitive" substrate theory (King, 1984; Oremland and Polcin, 1982), the restriction of methanogenesis to only methanol in the Dead Sea is perplexing, since bacteria in sediments from moderately hypersaline environments attack methylated amines and DMS in addition to methanol (Giani et al., 1984, 1989; King, 1988; Oremland et al., 1993). Furthermore, the metabolism of ^{14}C-methanol to methane in the Dead Sea sediments was slow, with only about half of the material mineralized after a month of incubation (Figure 12.3). In contrast, mineralization of [^{14}C]-methanol was evident within a few hours in Mono Lake sediments (Oremland and King, 1988), and DMS, acetate, and TMA were also rapidly mineralized (Oremland et al., 1993). Rate constants for degradation of ^{14}C-labeled TMA, methanol, and DMS were roughly 10 times higher in Solar Lake than in Dead Sea samples (Table 12.1). In addition, metabolism of the labeled substrates to ^{14}CH$_4$ and ^{14}CO$_2$ was essentially complete in Solar Lake samples (Figures 12.4 and 12.5). The use of only methylotrophic methanogenic substrates in Solar Lake samples is consistent with previous observations (Giani et al., 1984).

TABLE 12.1
Highest Measured Rate Constants for the Formation of Radiolabeled Gaseous Products from Methanol, TMA, and DMS in Dead Sea and Solar Lake Sediments

Sample	Substrate	k (^{14}CH$_4$)[a]	k (^{14}CO$_2$)[a]
Dead Sea[b]	methanol	0.02	0.07
Dead Sea[c]	TMA	0.00	0.03
Dead Sea[c]	DMS	0.00	0.00
Solar Lake	methanol	0.18	0.12
Solar Lake	TMA	0.11	0.10
Solar Lake	DMS	0.02	0.36

[a] k = fraction of substrate degraded day^{-1} calculated from the linear portions of ^{14}C-gas evolution time courses.
[b] Data from Site 3.
[c] Data from Site 4.

In conclusion, the substrate diversity and activity of Dead Sea methanogens is severely constrained when compared with moderately hypersaline environments. Furthermore, methanogenic activity in the Dead Sea appears "patchy," and was not evident in all the samples tested from around the lake. The reasons for this are not known, but further experiments to examine the possible adverse effects of high salinity and divalent cations on methanogenesis seem to be a logical departure point for future research.

ACKNOWLEDGMENTS

A. Oren acknowledges the Israeli Ministry of Infrastructure for financial support. We thank skipper Moti Gonen and his crew for assistance during the Dead Sea cruises. Y. Cohen acknowledges the German Ministry of Science and Technology (BMBF) for financial support in the framework of the Red Sea project.

REFERENCES

Baedecker, M.J. 1985. *Organic Material in Sediments of Great Salt Lake, Utah: Influence of Changing Depositional Environments.* Ph.D. thesis, George Washington University, Washington, DC.

Cohen, S., A. Oren, and M. Shilo. 1983. The divalent cation requirement of Dead Sea halobacteria. *Arch. Microbiol.* 136:184–190.

Cohen, Y., W.E. Krumbein, M. Goldberg, and M. Shilo. 1977. Solar Lake (Sinai). 1. Physical and chemical limnology. *Limnol. Oceanogr.* 22:597–609.

Conrad, R., P. Frenzel, and Y. Cohen. 1995. Methane emission from hypersaline microbial mats: lack of aerobic methane oxidation activity. *FEMS Microbiol. Ecol.* 16:297–306.

Culbertson, C.W., A.J.B. Zehnder, and R.S. Oremland. 1981. Anaerobic oxidation of acetylene by estuarine sediments and enrichment cultures. *Appl. Environ. Microbiol.* 41:396–403.

Dickins, H.D. and E.S. van Vleet. 1992. Archaebacterial activity in the Orca Basin determined by the isolation of characteristic isopranyl ether-like lipids. *Deep-Sea Res.* 39:521–536.

Fründ, C. and Y. Cohen. 1992. Diurnal cycles of sulfate reduction under oxic conditions in cyanobacterial mats. *Appl. Environ. Microbiol.* 58:70–77.

Giani, D., L. Giani, Y. Cohen, and W.E. Krumbein. 1984. Methanogenesis in the hypersaline Solar Lake (Sinai). *FEMS Microbiol. Lett.* 25:219–224.

Giani, D., D. Jannsen, V. Schostak, and W.E. Krumbein. 1989. Methanogenesis in a saltern in the Bretagne (France). *FEMS Microbiol. Lett.* 62:143–150.

Iversen, N., R. S. Oremland, and M. J. Klug. 1987. Big Soda Lake (Nevada). 2. Pelagic methanogenesis and anaerobic methane oxidation. *Limnol. Oceanogr.* 32:804–814.

Jørgensen, B.B. and Y. Cohen. 1977. Solar Lake (Sinai). 5. The sulfur cycle of the benthic cyanobacterial mats. *Limnol. Oceanogr.* 22:657–666.

Jørgensen, B.B., J.G. Kuenen, and Y. Cohen. 1979a. Microbial transformations of sulfur compounds in a stratified lake (Solar Lake, Sinai). *Limnol. Oceanogr.* 24:799–822.

Jørgensen, B.B., N.P. Revsbech, T.H. Blackburn, and Y. Cohen. 1979b. Diurnal cycle of oxygen and sulfide microgradients and microbial photosynthesis in a cyanobacterial mat sediment. *Appl. Environ. Microbiol.* 38:46–58.

Kiene, R.P., R.S. Oremland, A. Catena, L.G. Miller, and D.G. Capone. 1986. Metabolism of reduced methylated sulfur compounds by anaerobic sediments and by a pure culture of an estuarine methanogen. *Appl. Environ. Microbiol.* 52:1037–1045.
King, G.M. 1984. Utilization of hydrogen, acetate, and "non-competitive" substrates by methanogenic bacteria in marine sediments. *Geomicrobiol. J.* 3:275–306.
King, G.M. 1988. Methanogenesis from methylated amines in a hypersaline algal mat. *Appl. Environ. Microbiol.* 54:130–136.
Nissenbaum, A. 1975. The microbiology and biogeochemistry of the Dead Sea. *Microb. Ecol.* 2:139–161.
Nissenbaum, A., M.J. Baedecker, and I.R. Kaplan. 1972. Organic geochemistry of Dead Sea sediments. *Geochim. Cosmochim. Acta* 36:709–727.
Nissenbaum, A., M. Stiller, and A. Nishri. 1990. Nutrients in pore waters from Dead Sea sediments. *Hydrobiologia* 197:83–89.
Ollivier, B., P. Caumette, J.L. Garcia, and R.A. Mah. 1994. Anaerobic bacteria from hypersaline environments. *Microbiol. Rev.* 58:27–38.
Oremland, R.S. 1988. The biogeochemistry of methanogenic bacteria, in *Biology of Anaerobic Microorganisms,* A.J.B. Zehnder, Ed. John Wiley & Sons, New York. 641–706.
Oremland, R.S. and G.M. King. 1988. Methanogenesis in hypersaline environments, in *Microbial Mats: Physiological Ecology of Benthic Microbial Communities,* Y. Cohen and E. Rosenberg, Eds. American Society for Microbiology, Washington, DC. 180–190.
Oremland, R.S. and L.G. Miller. 1993. Biogeochemistry of natural gases in three alkaline, permanently stratified (meromictic) lakes, in *The Future of Energy Gases,* D. Howell, Ed. USGS Professional paper 1570, Washington, DC. 453–470.
Oremland, R.S. and S. Polcin. 1982. Methanogenesis and sulfate reduction: competitive and non-competitive substrates in estuarine sediments. *Appl. Environ. Microbiol.* 44:1270–1276.
Oremland, R.S., L. Marsh, and D.J. Des Marais. 1982. Methanogenesis in Big Soda Lake, Nevada: an alkaline, moderately hypersaline desert lake. *Appl. Environ. Microbiol.* 43:462–468.
Oremland, R.S., L.G. Miller, and M.J. Whiticar. 1987. Sources and flux of natural gases from Mono Lake, California. *Geochim. Cosmochim. Acta* 51:2915–2929.
Oremland, R.S., C.W. Culbertson, and M.R. Winfrey. 1991. Methylmercury decomposition in sediments and bacterial cultures: involvement of methanogens and sulfate-reducers in oxidative demethylation. *Appl. Environ. Microbiol.* 61:2745–2753.
Oremland, R.S., L.G. Miller, C.W. Culbertson, S.W. Robinson, R.L. Smith, D. Lovley, M.J. Whiticar, G.M. King, R.P. Kiene, N. Iversen, and M. Sargent. 1993. Aspects of the biogeochemistry of methane in Mono Lake and the Mono Basin of California, in *Biogeochemistry of Global Change: Radiatively Active Trace Gases,* R.S. Oremland, Ed. Chapman and Hall, New York. 704–741.
Oremland, R.S., L.G. Miller, and F.E. Strohmaier. 1994. Degradation of methyl bromide in anaerobic sediments. *Environ. Sci. Technol.* 28:514–520.
Oremland, R.S., L.G. Miller, P. Dowdle, T. Connell, and T. Barkay. 1995. Methylmercury oxidative degradation potentials in contaminated and pristine sediments of the Carson River, Nevada. *Appl. Environ. Microbiol.* 61:2745–2753.
Oren, A. 1983. *Halobacterium sodomense* sp. nov., a Dead Sea halobacterium with an extremely high magnesium requirement. *Int. J. Syst. Bacteriol.* 33:381–386.
Oren, A. 1988a. Anaerobic degradation of organic compounds at high salt concentrations. *Antonie van Leeuwenhoek* 54:267–277.

Oren, A. 1988b. The microbial ecology of the Dead Sea. *Adv. Microb. Ecol.* 10:193–229.
Oren, A. 1990a. Formation and breakdown of glycine betaine and trimethylamine in hypersaline environments. *Antonie van Leeuwenhoek* 58:291–298.
Oren, A. 1990b. The use of protein synthesis inhibitors in the estimation of the contribution of halophilic archaebacteria to bacterial activity in hypersaline environments. *FEMS Microbiol. Ecol.* 73:187–192.
Oren, A. 1993. Availability, uptake, and turnover of glycerol in hypersaline environments. *FEMS Microbiol. Ecol.* 12:15–23.
Oren, A. 1994. The ecology of extremely halophilic bacteria. *FEMS Microbiol. Rev.* 13:415–440.
Oren, A. 1995. Uptake and turnover of acetate in hypersaline environments. *FEMS Microbiol. Ecol.* 18:75–84.
Oren, A. and P. Gurevich. 1994. Production of D-lactate, acetate, and pyruvate from glycerol in communities of halophilic archaea in the Dead Sea and in saltern crystallizer ponds. *FEMS Microbiol. Ecol.* 14:147–156.
Oren, A., W.G. Weisburg, M. Kessel, and C.R. Woese. 1984. *Halobacteroides halobius* gen. nov., sp. nov., a moderately halophilic anaerobic bacterium from the bottom sediments of the Dead Sea. *Syst. Appl. Microbiol.* 5:58–70.
Schink, B., F.S. Lupton, and J.G. Zeikus. 1983. A radioassay for hydrogenase activity in viable cells and documentation of aerobic hydrogen consuming bacteria in extreme environments. *Appl. Environ. Microbiol.* 45:1491–1500.
Wiesenburg, D.A., J.M. Brooks, and B.B. Bernard. 1985. Biogenic hydrocarbon gases and sulfate reduction in the Orca Basin brine. *Geochim. Cosmochim. Acta* 49:2069–2080.
Zeikus, J.G. 1983. Metabolic communications between biodegradive populations in nature, in *Microbes in their Natural Environments,* H. Slater, E. Whittenbury, and J. Wimpenny, Eds. Soc. Gen. Microbiology Symposium 34. Cambridge University Press, London. 423–462.
Zhilina, T.N. 1986. Methanogenic bacteria from hypersaline environments. *Syst. Appl. Microbiol.* 7:216–222.
Zhilina, T.N. and G.A. Zavarzin. 1987. *Methanohalobium evestigatus,* nov. gen., nov. sp. An extremely halophilic methanogenic archaebacterium. *Dokl. Acad. Nauk USSR* 293:464–468.

Section IV

Ion Metabolism and Osmotic Regulation

13 The Molecular Mechanism of Regulation of the NhaA Na^+/H^+ Antiporter of *Escherichia coli*, a Paradigm for an Adaptation to Na^+ and H^+

Etana Padan

CONTENTS

13.1 Introduction 164
13.2 Antiporter Genes 164
13.3 Antiporter Proteins 165
13.4 pH Sensor of NhaA 166
13.5 Control of Expression of *nhaA* is Induced Specifically by
      Na^+ (Li^+) 169
13.6 The NhaR Regulator 169
13.7 The Molecular Mechanism of the Interaction Between
      Na^+, *nhaA*–DNA and NhaR 170
13.8 The Footprint of NhaR on *nhaA* is Na^+ Sensitive 170
13.9 The Involvement of H-Ns in the Regulation of *nhaA* 171
13.10 Conclusion 172
Acknowledgments 172
References 172

13.1 INTRODUCTION

H^+ and Na^+ are the most common ions, which very often become stressors to all cells, whether eukaryotic or prokaryotic. Hence, every cell is equipped with a mechanism adapting the cells to these ions. As with any adaptive mechanism, this includes a signaling device that senses the concentrations of the ions and alarms the cells when they are too high or too low, a transducer, and finally an effector machinery to protect the cells. Our aim is to understand, in molecular terms, all steps in this adaptive mechanism.

In the early '70s, P. Mitchell and colleagues (Mitchell, 1961; Mitchell and Moyle, 1967; West and Mitchell, 1974) discovered an activity in membranes that couples the fluxes of Na^+ and H^+. He coined the term Na^+/H^+ antiporters for the proteins responsible for this activity, and suggested that they are involved in the homeostasis of both Na^+ and H^+ in cells. Since then, Na^+/H^+ antiporter activity has indeed been found in the cytoplasmic membrane of most cells examined and in many organellar membranes (review in Padan and Schuldiner, 1994). One bacterium was found to lack Na^+/H^+ antiporter and accordingly was very restricted in its pH range of growth (Speelmans et al., 1993).

13.2 ANTIPORTER GENES

A clue to the molecular nature of the antiporters was obtained when we identified, mapped, and cloned two genes of *Escherichia coli*, *nhaA* (Goldberg et al., 1987; Karpel et al., 1988) and *nhaB* (Pinner et al., 1992), and deleted them from the chromosome, each separately and both together (Pinner et al., 1993). The single deletions showed that *nhaA* is the major indispensable gene for the adaptation to Na^+ and Li^+ and to withstand alkaline pH in the presence of Na^+, whereas *nhaB* is the housekeeping gene. The double deletion showed that without these genes there is no Na^+/H^+ antiporter activity left in the membrane, and these cells become highly sensitive to the ions (Pinner et al., 1993). This high Na^+ sensitivity has since made the double-deletion mutant most useful for cloning and expression of both homologous and heterologous antiporter genes (Padan and Schuldiner, 1996).

The DNA sequences of these two genes are predicted to encode proteins of very little homology but with overall similar secondary structure, comprising 12 TMS (trans membrane segments) connected with hydrophilic loops (Padan and Schuldiner, 1994, 1996; Rothman et al., 1996). This model characterizes many secondary transporters utilizing chemiosmotic energy as a driving force.

The model has recently been substantiated by determining the topology of NhaA across the membrane utilizing *pho*A fusions (Manoil and Beckwith, 1986), combined with epitope mapping and exposure to proteolysis from either side of the membrane (Rothman et al., 1996). Most interestingly, even at this low resolution of structure, we could identify one or even two very short TMS (VII and VIII) in NhaA (Figure 13.1 and section 13.4).

The Molecular Mechanism of Regulation of the NhaA Na+/H+ Antiporter

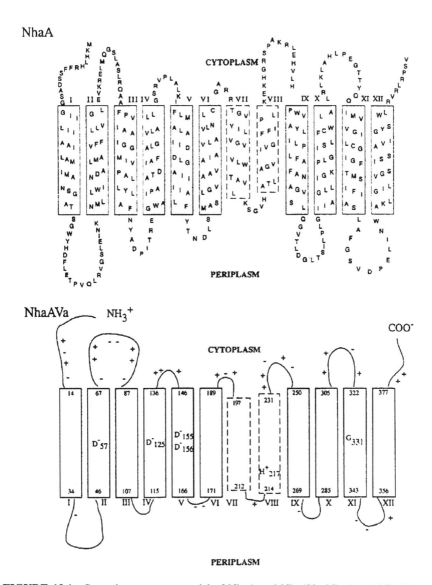

FIGURE 13.1 Secondary structure model of NhaA and NhaAVa. NhaA and NhaAVa are the Na+/H+ antiporters of *Escherichia coli* and *Vibrio alginolyticus* respectively. They share extensive homology (Nakamura et al., 1995), and their putative secondary structure is similar.

13.3 ANTIPORTER PROTEINS

Both the NhaA and NhaB proteins were identified, overexpressed, purified and reconstituted in a functional form in proteoliposomes. This was done in the past with classical biochemistry approaches (Taglicht et al., 1991; Pinner et al., 1994).

Today, the NhaA protein was engineered fused at its 5'-end to the strong *tac* promoter and at its 3'-end to polyhistidine, a His-tag (Olami et al., 1997). These allow for very strong expression and most efficient affinity purification of the protein on a Ni^{2+}-NTA column (Olami et al., 1997). With this procedure, 4 mg of pure protein is obtained per liter of cell culture.

The purified reconstituted antiporters were found to be as active as the native proteins in the membrane, in terms of kinetics, turnover numbers and bioenergetics, and thus opened the way to many direct biochemical studies (Padan and Schuldiner, 1996). These proved that a single polypeptide is sufficient to catalyze the exchange reaction, and that is the stoichiometry of the exchange reaction $2H^+/Na^+$ for NhaA (Taglicht et al., 1993) and $3\ H^+/2\ Na^+$ for NhaB (Pinner et al., 1994). Most importantly, NhaA but not NhaB was found to be highly sensitive to pH (Pinner et al., 1994; Taglicht et al., 1991). Interestingly, a similar dramatic pH sensitivity is characteristic of many antiporters as well as other transporters and proteins involved in pH regulation (Aronson, 1985; Bearson et al., 1997; Dell et al., 1994; Grinstein et al., 1992; Hall et al., 1995; Nakamura et al., 1995; Sekler et al., 1996; Slonczewski and Foster, 1996; Wakabayashi et al., 1992, 1994).

13.4 pH SENSOR OF NhaA

NhaA on its own responds to pH; its V_{max} is changed by three orders of magnitude over the pH range between 7 and 8.5 (Padan and Schuldiner, 1996; Taglicht et al., 1991). Hence, this molecule must be equipped with a "sensor" that directly senses pH, the environmental signal. In addition, it must have a transducer that transduces the signal into a change in V_{max}. Identifying the residues playing a role in the pH response of NhaA and their topology with respect to the face of the membrane has therefore become a major focus of our work.

A simple educated guess has been that such a role is played by a histidine residue, which in solution has a pK of 6, within the pH range of the NhaA response. Therefore, we replaced each of the 8 His of NhaA with Arg, and found that only H225 is important for the pH response (Gerchman et al., 1993). Its replacement with Arg (H225R) yielded an antiporter with a pH profile shifted by half a pH unit toward acidic pH. As a result, the mutated cells do not grow at alkaline pH in the presence of Na^+.

Histidine and arginine share certain properties; both bear a positive charge, are polar, and capable of hydrogen bonding. Which of these properties is important for the pH response? The fact that H225R is inactive at alkaline pH in the presence of Na^+ has provided a powerful tool to answer this question by looking for suppressor mutations. The mutant H225R-plasmid was randomly mutagenized by hydroxylamine and transformed into $\Delta nhaA\Delta nhaB$. Mutants that restore growth at alkaline pH in the presence of Na^+ were selected (Rimon et al., 1995). Among these mutants only revertants and first-site suppressants replacing H225 were found. The suppressants contained Cys and Ser (H225C and H225S respectively). Hence, polar groups rather than charge are important at position 225. Nevertheless, charge at position 225 affected the pH response; as opposed to the acidic shift caused by the positive charge of H225R, Asp (H225D) with a negative charge caused an alkaline shift and

Ala (H225A), which is neither hydrophilic nor charged, produced a carrier that does not respond to pH (Rimon et al., 1995).

Hence, although we do not know how H225 acts, it is intimately involved in the pH response of NhaA, and therefore, a study of its topology and reactivity has been undertaken. The fact that H225C is functional and responds to pH has provided the way (Olami et al., 1997) of using SH-reagents. Once a Cys is introduced to a strategic site in a protein, it can be used to probe the protein by site-directed modification (Akabas et al., 1992, 1994; Altenbach et al, 1990; Wu et al., 1995). For this approach, known as Cys scanning, a prerequisite is to construct a Cys-less antiporter to eliminate all native Cys residues that might complicate the interpretation of the results. Cys-less NhaA was constructed and found active and pH sensitive (Olami et al., 1997). These results showed that the native Cys of NhaA are dispensable. Next, a Cys-less NhaA with H225C was constructed, again found active and pH sensitive, and therefore ready for probing NhaA with SH reagents. In addition, His-tagged derivatives were constructed for directly probing the purified proteins reconstituted in proteoliposomes (Olami et al., 1997).

N-ethyl maleimide (NEM), the classical SH reagent, was used first. As expected from the dispensability of the native Cys, NEM had no effect on the wild-type or the Cys-less NhaA. However, it inhibited C-less H225C (Olami et al., 1997). To quantitate the degree of NEM-alkylation in the membrane, we used [^{14}C]-NEM and His-tagged derivatives of NhaA: wild-type, C-less and C-less H225. Membranes were exposed to [^{14}C]-NEM, and then the proteins, affinity purified on a Ni^{2+} column, were separated by SDS-PAGE and autoradiographed. As expected the C-less control was not labeled, but surprisingly, H225C was labeled at least tenfold stronger than all three native cysteines together. Accordingly, the C-less H225C reached saturation at 0.5 mM NEM at an alkyl/NhaA stoichiometry of 1, while the native cysteines were not saturated even at 2.5 mM (Olami et al., 1997).

This difference cannot be ascribed to the membrane's own impermeation to NEM, since right-side-out (RSO) and inside-out (ISO) membrane vesicles gave identical results. Given that deprotonation of the Cys is required for alkylation by NEM, we have suggested that the native cysteines reside in the protein in a hydrophobic environment that prevents deprotonation. On the other hand, H225C is exposed to an aqueous environment and therefore gets readily alkylated. Indeed, according to our model, the native Cys residues reside inside the TMS, whereas H225C is located at the edge of the shortest TMS facing the bulk (Figure 13.1). We therefore believe that quantitative probing of NEM alkylation of Cys replacements is a powerful tool to get information regarding both the aqueous vs. hydrophobic domains along a transport protein, and the residues of the transporter bordering TMS. Thus, the number of the TMS can be assessed (Olami et al., 1997). Very similar conclusions have been reached by Yamaguchi and colleagues using the TetA protein (Kimura et al., 1996).

Importantly, as shown previously (Akabas et al., 1994; Karlin, 1993), the probing of Cys replacements with impermeant SH reagents in membranes of known orientation, (ISO or RSO), allows us to determine the topology of a residue in a membrane protein, since impermeant SH-reagents attack only residues exposed to the medium. Table 13.1 shows that, whereas residues exposed to the periplasm will react with an

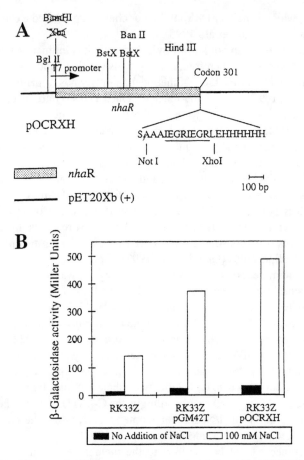

FIGURE 13.2 His-tagged NhaR, construction and activity. (A) pOCRXH is a plasmid-bearing *nha*R tagged at its C-terminus with sequences encoding two factor Xa cleavage sites (underlined) and six histidines. (B) To test the activity of the His-tagged NhaR, RK33Z, a strain bearing *nhaA'-'lacZ* translational fusion (Karpel et al., 1991) was transformed with pOCRXH or pGM42T encoding native NhaR. Cells were induced with 100 mM Na$^+$ at pH 7.5 and β-galactosidase activity determined (Karpel et al., 1991).

impermeant probe in RSO vesicles but not in ISO vesicles, residues exposed to the cytoplasm will exhibit the opposite behavior. RSO and ISO membrane vesicles of His-tagged Cys-less H225C were exposed to the impermeant reagents MTSET [methanethiosulfonate ethyl-trimethyl ammonium bromide (Akabas et al., 1992; Stauffer and Karlin, 1994)] and PCMBS (p-chloromercuribenzosulfonate), and then the H225C residues left free were titrated by alkylation with [^{14}C]-NEM as described above. As a control, the same protocol was conducted with permeant probes, MTSEA (methanethiosulfonate ethylammonium (Olami et al., 1997) and PCMB (p-chloromercurybenzoate). The results were straightforward, showing that H225C is periplasmic (Table 13.1).

TABLE 13.1
The Membrane Topology of H225, Which is Involved in the Response of NhaA to PH[a]

	NEM-reactive H225C (% of control) in the presence of:				
	NEM	PCMB	PCMBS	MTSEA	MTSET
RSO	100	4	5	0	5
ISO	100	8	54	0	47

[a] Right-side-out (RSO) and inside-out (ISO) membrane vesicles bearing His-tagged Cys-less H225C NhaA (Olami et al., 1997) were exposed to either permeant N-ethyl maleimide (NEM), p-chloro-mercurybenzoate (PCMB) and methanethiosulfonate ethylammonium (MTSEA) or impermeant p-chloromercuribenzo-sulfonate (PCMBS) and methanethiosulfonate ethyl-trimethyl (MTSET) reagents and then all free H226C residues left free were titrated by [^{14}C]NEM alkylation (Olami et al., 1997).

To determine the length of the TMS VIII, we performed the same protocol with E241, predicted to be located on the internal side of the membrane at the edge of TMS VIII opposing H225. Preliminary results indeed suggest that it faces the cytoplasm; it is exposed to impermeant probes only in ISO membrane vesicles. Taken together, these results support the model of NhaA (Rothman et al., 1996) and the existence of short TMS in this transporter. Although as yet we do not know the meaning of the short TMS, we know that this structure and the critical His225 are conserved among the NhaA family members; both were found in the recently cloned *nhaA* from *Vibrio alginolyticus* (Nakamura et al., 1995; Figure 13.1).

13.5 CONTROL OF EXPRESSION OF *nhaA* IS INDUCED SPECIFICALLY BY Na$^+$ (Li$^+$)

The next level of regulation of NhaA is expression at the gene level. Experiments with *nhaA'-'lacZ* protein fusion (Karpel et al., 1991) and northern analysis (Dover et al., 1996) showed that the environmental signals that turn on *nhaA* are Na$^+$ and Li$^+$, while alkaline pH potentiates the effect of the ions. This demonstrated for the first time that there is a unique regulatory network responding specifically to Na$^+$ and not to either osmolarity or ionic strength. Therefore, we focused on this novel system. Interestingly, a similar role has recently been assigned to Na$^+$ in the regulation of expression of the Na$^+$–ATPase of *Enterococcus hirae* (Murata et al., 1996).

13.6 THE NhaR REGULATOR

The regulatory gene of *NhaA* is *NhaR*, the downstream neighbor of *NhaA*. It encodes a positive regulator that works in trans; Δ*nhaR* confers Na$^+$ and Li$^+$ sensitivity in spite of the presence of *NhaA* (Rahav-Manor et al., 1992); when applied in multicopy on a plasmid, it dramatically increases expression of *NhaA'-'lacZ* protein fusion (Rahav-Manor et al., 1992). NhaR is homologous to the LysR family of positive regulators that are involved in the response of bacteria to various environmental stresses (Christman et al., 1989; Henikoff et al., 1988; Rahav-Manor et al., 1992). The immediate signal for the *NhaR*-dependent induction of *NhaA* is intracellular rather than extracellular Na^+. This was demonstrated by changing intracellular Na^+ without changing the extracellular Na^+. Multicopy *nhaA* decreased Na^+ and inhibited expression (Rahav-Manor et al., 1992). Deleting both *NhaA* and *NhaB* increased $[Na^+]_{in}$ and enhanced expression (Dover et al., 1996).

13.7 THE MOLECULAR MECHANISM OF THE INTERACTION BETWEEN Na^+, *nhaA*-DNA AND NhaR

A genetic approach undertaken previously (Carmel et al., 1996) showed that NhaR may harbor a Na^+ "sensor." E134G, a point mutation in *nhaR*, increased the sensitivity of the regulation to Na^+ and conferred resistance to Li^+ on the mutated cells. These results suggested that the interaction among Na^+, NhaR, and the target regulatory sequences of *nhaA* is direct. We have recently undertaken a biochemical approach to study this interaction with its purified components. NhaR was affinity purified to homogeneity by constructing a His-tagged NhaR that was found to be as active as the wild-type protein (Figure 13.2). Separation by gel filtration on a HPLC column showed one homogenous peak at 60 kDa (Carmel et al., 1997a). Both wild-type NhaR (Rahav-Manor et al., 1992) and its His-tagged derivative (Carmel et al., 1997a) were found to bind specifically to DNA sequences overlapping the promoter region of *nhaA*. The mobility of PCR-amplified fragments of *nhaA* DNA of various lengths was tested in the gel retardation assay in the presence or absence of His-tagged NhaR. Retarded mobility identified the regulatory sequences of *nhaA* recognized by His-tagged NhaR (Carmel et al., 1997a).

Interestingly, each fragment that bound the regulator showed at least one LysR consensus motive (Goethals et al., 1992; Schell, 1993). The smallest fragment overlapping all binding sequences was located between bp −120 and +14. It contained, consequently, three times the LysR consensus motive. Na^+ had no effect on the gel retardation pattern, suggesting that a change in the footprint rather than the affinity is affected by Na^+. This behavior is characteristic of the LysR family of regulators (Schell, 1993; Toledano et al., 1994).

13.8 THE FOOTPRINT OF NhaR ON *nhaA* IS Na^+ SENSITIVE

The DNAase I footprint of His-tagged NhaR on *nhaA* revealed a very long segment spanning about 90 bases, which aligned with the segment binding His-tagged NhaR and the three consensus motives of LysR family members (Goethals et al., 1992; Schell, 1993). Na^+ up to 100 mM had no effect on the DNAse I footprint. This negative result was found to be the outcome of the DNAase protection assay, which is not sensitive enough and limited to the minor groove of the DNA (Sasse-Dwight and Gralla, 1991). Thus, the DMS methylation protection footprint both *in vitro* and *in vivo* revealed the specific effect of Na^+ (Carmel et al., 1997b).

Dimethyl sulfate methylates G, and to a lesser extent A, mainly in the major groove of the DNA (Sasse-Dwight and Gralla, 1991). It was used to identify the footprint of His-tagged NhaR. The DNA bases that were not protected by the regulator were expected to be methylated by DMS, whereas the protected bases were not. After treatment with DMS, which can be conducted both *in vivo* and *in vitro*, the DNA was isolated and treated with piperidine, which splits every methylated base but not the non-methylated ones. Analysis of the products obtained *in vitro* by primer extension showed that bases -24, -29, -60, and -92 were protected by His-tagged NhaR. Whereas the NhaR-protection tested *in vitro* of bases -24, -29, and -92 was not affected by either 100 mM K^+ or 100 mM Na^+, protection of base -60 was differentially affected by the type of the ion. Na^+ but not K^+ (100 mM each) specifically exposed it to methylation and subsequent cleavage. The effective concentration of Na^+ affecting the footprint was found within the range expected from the intracellular concentration (10–20 mM). Remarkably, in line with the finding that the expression is markedly potentiated *in vivo* by pH (Karpel et al., 1991), the Na^+ interaction with base -60, but not with either bases -24 and -29, was dramatically affected by pH. At alkaline pH (7.5–8.5), but not at acidic pH (6.5), Na^+ was effective on NhaR/*nhaA* interaction at base -60.

A DMS methylation protection assay was also conducted *in vivo*. As expected from the *in vitro* results, bases -24 and -29 were protected but not affected by the type of the ion, while base -60 was protected in the presence of K^+ but not in the presence of Na^+. Most interestingly, base -92, which *in vitro* was protected by the regulator but unaffected by the presence of Na^+, was *in vivo* exposed by Na^+ but not by K^+. The Na^+ specific effects on the footprint of NhaR on *nhaA* required NhaR, since they were not observed in a $\Delta nhaR$ strain (Carmel et al., 1997b). We therefore suggest that Na^+ directly affects the interaction of NhaR with G^{-60} of *nhaA* but indirectly the interaction with G^{-92}. The latter probably requires either a certain topology of the DNA or another factor existing only *in vivo*. Taken together, this work suggests that NhaR is both the sensor and the transducer of the Na^+ signal that regulates expression of *nhaA* and undergoes a conformational change upon binding Na^+. This change is expressed directly in a decrease in the binding of NhaR to G^{-60} in a pH-dependent fashion. This is also manifested in the binding of NhaR to G^{-92} observed only *in vivo*. The G^{-92}/NhaR interaction suggests an involvement of another factor *in vivo*.

13.9 THE INVOLVEMENT OF H-NS IN THE REGULATION OF *nhaA*

In the hierarchy of regulation of gene expression the recently established global regulation occupies the highest level. It is conducted by genes unlinked to the target operons, which can be numerous. Although the global regulator *rpoS* was not found to be involved in *nhaA* regulation, we have recently established a connection between the Na$^+$-specific, NhaR-dependent, regulation of *nhaA* and H-NS, a DNA-binding protein and a global regulator (Dover et al., 1996). Thus, the expression of *nhaA'-'lacZ* was derepressed in strains bearing *hns* mutation, and transformation with a low-copy-number plasmid carrying *hns*$^+$ repressed expression and restored Na$^+$ induction. The derepression in *hns* strains was *nhaR* independent. Most interestingly, multicopy *nhaR*, which in an *hn*$^+$ background acted only as an Na$^+$-dependent positive regulator, acted as a repressor in an *hns* strain in the absence of Na$^+$ but was activated in the presence of the ion. Hence, an interplay between *nhaR* and *hns* in the regulation of *nhaA* was suggested. Although the mechanism of regulation mediated by H-NS is not known, it has been suggested to involve a change in the topology of the DNA (Tupper et al., 1994).

13.10 CONCLUSION

We have established that the NhaA Na$^+$/H$^+$ antiporter, the main system responsible for adaptation to Na$^+$ and alkaline pH (in the presence of Na$^+$) in *E. coli* and many other enteric bacteria, is under a very intricate control (Padan and Schuldiner, 1996). At the protein level it is regulated directly by pH, one of its regulatory signals. At the gene level, its transcription is dependent on NhaR, a positive regulator of the LysR family, and regulated by Na$^+$, the other environmental signal. Na$^+$ affects directly the NhaR/*nhaA* interaction by changing the footprint of NhaR on *nhaA* in a pH-dependent fashion. The expression of *nhaA* is also under global regulation of H-NS, which has been suggested to involve a change in the topology of the DNA in an unknown mechanism (Tupper et al., 1994). Since H$^+$ and Na$^+$ are the most common ions that challenge every cell, it is suggested that the pattern of regulation of *nhaA* found in *E. coli* is a paradigm for a mechanism and response of proteins and genes to these ions.

ACKNOWLEDGMENTS

The research in the author's laboratory is supported by grants from the Israel Science Foundation administered by the Israel Academy of Sciences and Humanities, and the German-Israeli Project Cooperation on Future-Oriented Topics (DIP).

REFERENCES

Akabas, M., D. Stauffer, M. Xu, and A. Karlin. 1992. Acetylcholine receptor channel structure probed in cysteine-substitution mutants. *Science* 258:307–310.

Akabas, M., C. Kaufmann, P. Archdeacon, and A. Karlin. 1994. Identification of acetylcholine receptor channel-lining residues in the entire M2 segment of the α-subunit: implications for the secondary structure and for the locations of the gate and a selectivity filter. *Neuron* 13:919–927.

Altenbach, C., T. Marti, H.G. Khorana, and W. Hubbell. 1990. Transmembrane protein structure: spin labelling of bacteriorhodopsin mutants. *Science* 248:1088-1092.

Aronson, P.S. 1985. Kinetic properties of the plasma membrane Na$^+$/H$^+$ exchanger. *Ann. Rev. Physiol.* 47:545–560.
Bearson, S., B. Bearson, and J.W. Foster. 1997. Acid stress responses in enterobacteria. *FEMS Microbiol. Lett.* 147:173–180.
Carmel, O., N. Dover, O. Rahav-Manor, P. Dibrov, D. Kirsch, S. Schuldiner, and E. Padan. 1997a. A single amino acid substitution (G134-Ala) in NhaR1 increases the inducibility by Na$^+$ of the product of *nhaA*, a Na$^+$/H$^+$ antiporter gene in *Escherichia coli*. *EMBO J.* 13:1981–1989.
Carmel, O., O. Rahav-Manor, N. Dover, B. Shaanan, and E. Padan. 1997b. The Na$^+$ specific interaction between the LysR-type regulator, NhaR and the *nhaA* gene, encoding the Na$^+$/H$^+$ antiporter of *Escherichia coli*. *EMBO J.* 16:5922–5929.
Christman, M., G. Storz, and B. Ames. 1989. OxyR, a positive regulator of hydrogen peroxide-inducible genes in *Escherichia coli* and *Salmonella typhimurium* is homologous to a family of bacterial regulatory proteins. *Proc. Natl. Acad. Sci. USA* 86:3484–3488.
Dell, C.L., M.N. Neely, and E.R. Olson. 1994. Altered pH and lysine signalling mutants of *cadC*, a gene encoding a membrane-bound transcriptional activator of the *Escherichia coli cad*BA operon. *Mol. Microbiol.* 14:7–16.
Dover, N., C. Higgins, O. Carmel, A. Rimon, E. Pinner, and E. Padan. 1996. Na$^+$-induced transcription of *nhaA*, which encodes an Na$^+$/H$^+$ antiporter in *Escherichia coli*, is positively regulated by *nha*R and affected by *hns*. *J. Bacteriol.* 178:6508-6517.
Gerchman, Y., Y. Olami, A. Rimon, D. Taglicht, S. Schuldiner, and E. Padan. 1993. Histidine 226 is part of the pH sensor of *NhaA*, a Na$^+$/H$^+$ antiporter in *Escherichia coli*. *Proc. Natl. Acad. Sci. USA* 90:1212–1216.
Goethals, K., M. van Motagu, and M. Holsters. 1992. Conserved motifs in a divergent *nod* box of *Azorhizobium caulinodans* ORS571 reveal a common structure in promoters regulated by LysR-type proteins. *Proc. Natl. Acad. Sci. USA* 89:1646–1650.
Goldberg, B.G., T. Arbel, J. Chen, R. Karpel, G.A. Mackie, S. Schuldiner, and E. Padan. 1987. Characterization of Na$^+$/H$^+$ antiporter gene of *E. coli*. *Proc. Natl. Acad. Sci. USA* 84:2615–2619.
Grinstein, S., M. Woodside, C. Sardet, J. Pouyssegur, and D. Rotin. 1992. Activation of the Na$^+$/H$^+$ antiporter during cell volume regulation. Evidence for a phosphorylation-independent mechanism. *J. Biol. Chem.* 267:23823–23828.
Hall, H.K., K.L. Karem, and J.W. Foster. 1995. Molecular responses of microbes to environmental pH stress. *Adv. Microb. Physiol.* 37:229–272.
Henikoff, S., G. Haughn, J. Calvo, and J. Wallace. 1988. A large family of bacterial activator proteins. *Proc. Natl. Acad. Sci. USA* 85:6602–6606.
Karlin, A. 1993. Structure of nicotinic acetylcholine receptors. *Curr. Opin. Neurobiol.* 3:299–309.
Karpel, R., Y. Olami, D. Taglicht, S. Schuldiner, and E. Padan. 1988. Sequencing of the gene *ant*, which effects the Na$^+$/H$^+$ antiporter activity in *Escherichia coli*. *J. Biol. Chem.* 263:10408–10414.
Karpel, R., T. Alon, G. Glaser, S. Schuldiner, and E. Padan. 1991. Expression of a sodium proton antiporter (NhaA) in *Escherichia coli* is inducted by Na$^+$ and Li$^+$ ions. *J. Biol. Chem.* 266:21753–21759.
Kimura, T., M. Suzuki, T. Sawai, and A. Yamaguchi, A. 1996. Determination of a transmembrane segment using cysteine-scanning mutants of transposon Tn10-encoded metal-tetracycline/H$^+$ antiporter. *Biochemistry* 35:15896–15899.
Manoil, C. and J. Beckwith. 1986. Tn *pho*A: a probe for protein export signals. *Proc. Natl. Acad. Sci. USA* 82:8129–8133.

Mitchell, P. 1961. Coupling of phosphorylation to electron and hydrogen transfer by a chemiosmotic type of mechanism. *Nature* 191:144–146.
Mitchell, P. and J. Moyle. 1967. Respiration-driven proton translocation in rat liver mitochondria. *Biochem. J.* 105:1147–1162.
Murata, T., I. Yamato, K. Igarashi, and Y. Kaninuma. 1996. Intracellular Na^+ regulates transcription of the *ntp* operon encoding a vacuolar-type Na^+ translocating ATPase in *Enterococcus hirae*. *J. Biol. Chem.* 271:23661–23666.
Nakamura, T., Y. Komano., and T. Unemoto. 1995. Three aspartic residues in membrane-spanning regions of Na^+/H^+ antiporter from *Vibrio alginolyticus* play a role in the activity of the carrier. *Biochim. Biophys. Acta* 1230:170–176.
Olami, Y., A. Rimon, Y. Gerchman, A. Rothman, and E. Padan. 1997. Histidine 225, a residue of the NhaA-Na^+/H^+ antiporter of *Escherichia coli* is exposed and faces the cell exterior. *J. Biol. Chem.* 272:1761–1768.
Padan, E. and S. Schuldiner. 1994. Molecular physiology of Na^+/H^+ antiporters, molecular devices that couple the Na^+ and H^+ circulation in cells. *Biochim. Biophys. Acta* 1185:129–151.
Padan, E. and S. Schuldiner. 1996. Bacterial Na^+/H^+ antiporters–molecular biology, biochemistry and physiology, in *Transport Processes in Eukaryotic and Prokaryotic Organisms*, W.N. Konings, H.R. Kaback, and J.S. Lolkema, Eds. Elsevier, Amsterdam. 501–531.
Pinner, E., E. Padan, and S. Schuldiner. 1992. Cloning, sequencing and expression of *nhaB* gene encoding a Na^+/H^+ antiporter in *Escherichia coli*. *J. Biol. Chem.* 267:11064–11068.
Pinner, E., Y. Kotler, E. Padan, and S. Schuldiner. 1993. Physiological role of NhaB, a specific Na^+/H^+ antiporter in *Escherichia coli*. *J. Biol. Chem.* 268:1729–1734.
Pinner, E., E. Padan, and S. Schuldiner. 1994. Kinetic properties of NhaB, a Na^+/H^+ antiporter from *E. coli*. *J. Biol. Chem.* 269:26274–26479.
Rahav-Manor, O., O. Carmel, R. Karpel, D. Taglicht, G. Glaser, S. Schuldiner, and E. Padan. 1992. NhaR, a protein homologous to a family of bacterial regulatory proteins (LysR), regulates *nha*A, the sodium proton antiporter gene in *Escherichia coli*. *J. Biol. Chem.* 267:10433–10438.
Rimon, A., Y. Gerchman, Y. Olami, S. Schuldiner, and E. Padan. 1995. Replacements of histidine 226 of NhaA-Na^+/H^+ antiporter of *Escherichia coli*–Cysteine (H226C) or serine (H226S) retain both normal activity and pH sensitivity, aspartate (H226D) shifts the pH profile toward basic pH, and alanine (H226A) inactivates the carrier at all pH values. *J. Biol. Chem.* 270:26813–26817.
Rothman, A., E. Padan, and S. Schuldiner. 1996. Topological analysis of NhaA, a Na^+/H^+ antiporter from *Escherichia coli*. *J. Biol. Chem.* 271:32288–32292.
Sasse-Dwight, S. and J. Gralla. 1991. Footprinting protein-DNA complexes *in vivo*. *Meth. Enzymol.* 208:146–168.
Schell, M. 1993. Molecular biology of the LysR family of transcriptional regulators. *Ann. Rev. Microbiol.* 47:597–626.
Sekler, I., S. Kobayashi, and R.R. Kopito. 1996. A cluster of cytoplasmic histidine residues specifies pH dependence of the AE2 plasma membrane anion exchanger. *Cell* 86:929–935.
Slonczewski, J.L. and J.W. Foster. 1996. pH-regulated genes and survival at extreme, in *Escherichia coli* and *Salmonella, Cellular and Molecular Biology*, 2nd ed, F.C. Neidhardt, R. Curtiss III, J.L. Ingraham, E.C.C. Lin, K.B. Low, Jr., B. Magasanik, W.S. Reznikoff, M. Riley, M. Schaechter, and H.E. Umbarger, Eds. ASM Press, Washington, DC. 1539–1549.

Speelmans, G., B. Poolman, T. Abee, and W.N. Konings. 1993. Energy transduction in the thermophilic anaerobic bacterium *Clostridium fervidus* is exclusively coupled to Na$^+$ ions. *Proc. Natl. Acad. Sci. USA* 90:7975-7979.

Stauffer, D. and A. Karlin. 1994. The electrostatic potential of the acetylcholine binding sites in the nicotinic receptor probed by reactions of binding-site cysteines with charged methanethiosulfonates. *Biochemistry* 33:6840–6849.

Taglicht, D., E. Padan, and S. Schuldiner. 1991. Overproduction and purification of a functional Na$^+$/H$^+$ antiporter coded by *nhaA (ant)* from *Escherichia coli*. *J. Biol. Chem.* 266:11289–11294.

Taglicht, D., E. Padan, and S. Schuldiner. 1993. Proton-sodium stoichiometry of NhaA, an electrogenic antiporter from *Escherichia coli*. *J. Biol. Chem.* 268:5382–5387.

Toledano, M., I. Kullik, F. Trinh, P. Baird, T. Schneider, and G. Storz. 1994. Redox dependent shift of OxyR-DNA contacts along an extended DNA-binding site: a mechanism for differential promoter selection. *Cell* 78:397–909.

Tupper, A., T. Owen-Hughes, D. Ussery, D. Santos, J. Ferguson, J. Sidebotham, J. Hinton, and C. Higgins. 1994. The chromatin-associated protein H-NS alters DNA topology *in vitro*. *EMBO J.* 13:258–268.

Wakabayashi, S., C. Sardet, P. Fafournoux, and J. Pouyssegur. 1992. The Na$^+$/H$^+$ antiporter cytoplasmic domain mediates growth factor signals and controls "H$^+$–sensing." *Proc. Natl. Acad. Sci. USA* 89:2424–2428.

Wakabayashi, S., B. Bertrand, M. Shigekawa, P. Fafournoux, and J. Pouyssegur. 1994. Growth factor activation and H$^+$-sensing of the Na$^+$/H$^+$ exchanger isoform 1 (Nhe1) – evidence for an additional mechanism not requiring direct phosphorylation. *J. Biol. Chem.* 269:5583–5588.

West, I.C. and P. Mitchell. 1974. Proton/sodium antiport in *E. coli*. *Biochem. J.* 144:87-90.

Wu, J., S. Fillingos, and H.R. Kaback. 1995. Dynamics of lactose permease of *Escherichia coli* determined by site-directed chemical labeling and fluorescence spectroscopy. *Biochemistry* 34:8257–8263.

14 The Biochemistry and Genetics of the Synthesis of Osmoprotective Compounds in Cyanobacteria

Martin Hagemann, Arne Schoor,
Stefan Mikkat, Uta Effmert, Ellen Zuther,
Kay Marin, Sabine Fulda, Josef Vinnemeier,
Anja Kunert, Carsten Milkowski,
Christian Probst, and Norbert Erdmann

CONTENTS

14.1 Introduction 177
14.2 Biosynthesis of Sucrose and Trehalose 179
14.3 Biosynthesis of Glucosylglycerol 179
14.4 Cloning of Genes Involved in Glucosylglycerol Synthesis 181
14.5 Biosynthesis of Glycine Betaine 183
14.6 Future Prospects 184
Acknowledgment 184
References 184

14.1 INTRODUCTION

Cyanobacteria form a group of evolutionary old bacteria, capable, like chloroplasts of higher plants, of performing oxygen-evolving photosynthesis. This chapter will deal with their acclimation to different salt concentrations. In this respect, they behave like most other prokaryotic and eukaryotic cells. During adaptation to salt stress, a balanced water potential is achieved by the accumulation of osmoprotective compounds–compatible solutes (Brown, 1976), while excess inorganic ions are extruded from the cells.

In 1980, the heteroside glucosylglycerol [2-O-(α-D-glucopyranosyl)-glycerol] (GG) was found as the first osmoprotectant in the marine strain *Synechococcus* NM100 (Borowitzka et al., 1980). Later, the disaccharides sucrose and trehalose and the quaternary ammonium compounds glycine betaine and glutamate betaine were discovered as other principal osmoprotective compounds in cyanobacteria (Figure 14.1).

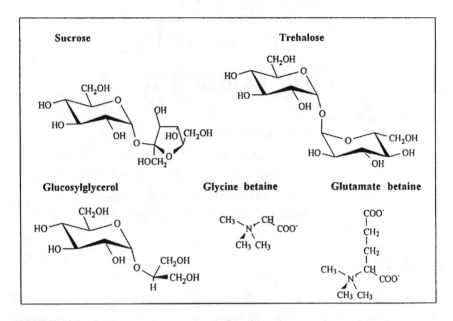

FIGURE 14.1 Osmoprotective compounds found in salt-stressed cyanobacteria.

The examination of a large number of cyanobacterial strains allowed them to be assigned to three salt resistance groups that differ distinctly in the principal osmoprotective substance they accumulate. The least halotolerant strains accumulate sucrose or trehalose and can tolerate up to 0.7 M NaCl. Moderately halotolerant cyanobacteria accumulate GG. Their tolerance limit is 1.8 M NaCl. The highest tolerance is exhibited by halophilic strains that accumulate glycine- or glutamate betaine. These strains tolerate salt concentrations up to 2.7 M. No link has been discovered between the kind of osmoprotective substance accumulated and the taxonomic affiliation (Reed et al., 1986).

The correlation between the salt tolerance limit and the type of osmoprotectant used indicates that betaines are more compatible than disaccharides, while heterosides are in between (Reed et al., 1986). Warr et al. (1988) found a decreased compatibility according to the proposed halotolerance groups in the order glycine betaine, polyol derivatives and disaccharides with the glutamine synthetase from different cyanobacteria. However, comparing different enzymes and several inhibitory treatments *in vitro*, a dependence on the protective effect from the test conditions became obvious (Galinski, 1993). In another approach, GG seemed to be slightly more effective than trehalose to compensate the salt sensitivity of a *Synechocystis* mutant defective in osmolyte synthesis, while sucrose showed only very little effect because it was metabolized by this strain (Mikkat et al., 1997). In summary, some experimental data confirm the view that the degree of compatibility of an osmoprotective compound might establish a certain salt tolerance, but additionally it seems very likely that factors like ion-export capacity or membrane composition also contribute to the salt resistance limit of a special strain.

14.2 BIOSYNTHESIS OF SUCROSE AND TREHALOSE

Most of the cyanobacteria accumulating sucrose or trehalose were originally isolated from freshwater biotopes, supporting the view that these substances lead only to a relatively low salt tolerance. However, sucrose and trehalose were also found as the principal osmoprotectant in some cyanobacteria of marine origin. The biochemical pathway as well as the genes and proteins involved in their salt-induced synthesis are only poorly investigated in cyanobacteria.

In plants, sucrose biosynthesis could be realized by two pathways, mainly by cooperation of sucrose-phosphate synthase (SPS, UDP-glucose: D-fructose-6-phosphate glucosyltransferase) and sucrose-6-phosphate phosphohydrolase (SPP) and probably in addition by sucrose synthase (UDP-glucose: fructose-2-glucosyltransferase) (Huber and Huber, 1996). SPS activity has been recently shown in the cyanobacterial strain *Anabaena* sp. PCC 7119 (Porchia and Salerno, 1996). The biosynthesis of sucrose in the salt-stressed cyanobacterium *Synechococcus* sp. PCC 7942 also involves the cooperation of SPS and SPP. This biosynthesis could be activated by adding NaCl to extracts from control cells (Hagemann, unpublished results).

Stress-induced trehalose synthesis has been extensively characterized in bacteria (Lippert et al., 1993; Strøm and Kaasen, 1993) and yeast. Trehalose was built up from UDP-glucose and glucose-6-phosphate via the intermediate trehalose-6-phosphate by the cooperation of trehalose-6-phosphate synthase and phosphatase activities. Recently, an alternative biosynthetic pathway for trehalose has been found in thermophilic archaea, where starch was successively degraded to trehalose (Maruta et al., 1996). In cyanobacteria, the biosynthetic pathway of trehalose has not been investigated so far, but analogous to *Escherichia coli* and the synthesis of sucrose, it should probably be made from UDP-glucose and glucose-6-phosphate.

14.3 BIOSYNTHESIS OF GLUCOSYLGLYCEROL

The osmoprotective compound GG is characteristic for cyanobacteria of moderate halotolerance and has been extensively studied using the model strain *Synechocystis* sp. PCC 6803. GG is synthesized from ADP-glucose and glycerol-3-phosphate via the intermediate glucosylglycerol-phosphate (GGP) by the cooperation of GGP-synthase (GGPS) and GGP-phosphatase (GGPP) (Hagemann and Erdmann, 1994). The GGPS was found to be strictly dependent on ADP-glucose and divalent cations

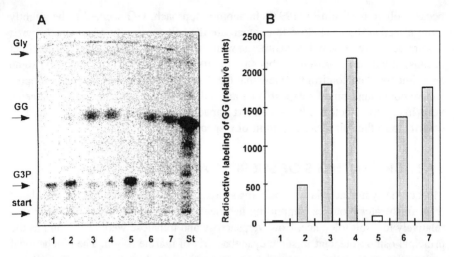

FIGURE 14.2 Regulation of the activity of glucosylglycerol-synthesizing enzymes from *Synechocystis* sp. PCC 6803 by NaCl. (A) TLC separation of radiolabeled reaction products of the enzyme assays. Gly = glycerol; GG = glucosylglycerol; G3P = glycerol-3-phosphate. (B) Results of quantitative estimation of radioactivity incorporated into GG using a phosphoimager. Crude protein extracts were obtained from control cells (tests 1-3) and from cells shocked for 2 h by 684 mM NaCl (lanes 4-7). The cells were homogenized with NaCl-free buffer (lanes 1, 2, 5, 6, 7) or 200 mM NaCl-containing buffer (lanes 3, 4). Enzyme activity was measured without NaCl in the assays (lanes 1, 5), in the presence of 100 mM NaCl (lane 6) or 200 mM NaCl (lanes 2, 3, 4, 7). For details of the assays see Hagemann and Erdmann (1994).

(Mg^{2+}). The dependence on NaCl of GGPS and GGPP represents one of the most interesting features of this enzyme system (Figure 14.2). About 200 mM NaCl in the enzyme assay were found to promote maximal activities in crude protein extracts (Hagemann et al., 1996b). GGPS and GGPP activities were inhibited in protein extracts from salt-adapted cells after homogenization and assay under NaCl-free conditions, and could be restored after adding NaCl to the assays. Furthermore, in protein extracts obtained from low-salt-grown cells containing no internal GG, the activities of GGPS and GGPP could be activated merely by adding enhanced NaCl concentrations to the buffers used for homogenization and/or assay (Hagemann and Erdmann, 1994; Hagemann et al., 1996a). Other salts were also found to be effective in the activation of GGPS. It was observed that the Na^+ rather than the Cl^- seems to be the effective stimulant (Schoor et al., unpublished results).

The experiments outlined above allowed three important conclusions:

- The GG-synthesizing enzyme system is present but is inactive in cells growing in basal medium, which is immediately activated by salt *in vivo* and *in vitro*. This salt adaptation strategy seems to be very useful in the case of *Synechocystis*, because a dramatic decrease of the overall protein synthesis was observed during the first hours after the salt shock (Hagemann et al., 1990). The only activation of the expression of the genes encoding for GGPS and GGPP would lead to a much slower adaptation

rate compared with the biochemical activation of already existing enzymes. However, besides the dominant regulation of the enzyme activities of GGPS and GGPP on the biochemical level, a slight increase in their gene expression could also be found (Hagemann et al., 1996b, 1997b).
- The salt-dependent activation and inactivation of GGPS and GGPP is completely reversible. This eliminates the possibility that the enzymes were activated by partial proteolysis, as was found for the salt-induced activation of enzymes involved in the biosynthesis of the structural related compound isofloridoside (galactosylglycerol) in the Chrysophyceaean alga *Poterioochromonas malhamensis* (Kauss, 1987). Changes of the phosphorylation state or in the conformation of the proteins are mechanisms that might be responsible for the reversible activation/deactivation of GGPS and GGPP. Salt-induced changes of the protein phosphorylation have been found in *Synechocystis* (Hagemann et al., 1993). This corresponds with the results that showed the regulation of SPS activity by the protein phosphorylation state in plants according to the environmental conditions (Huber and Huber, 1996).
- The signal transduction events leading to active GG biosynthesis do not necessarily require intact membrane structures. In general, the initial event in the translation of the signal salt stress into the cellular response is still a matter for discussion. Several models include changes of features of the cytoplasmic membrane as the primary key signal (Csonka and Hanson, 1991). For the activation of GG synthesis, it can be assumed that high salt concentrations interact with cytoplasmic proteins (GGPS and GGPP or regulatory proteins), and this seems to be sufficient to translate the environmental signal. Such an activation of enzymes related to salt adaptation was also found for trehalose-6-phosphate synthase (Giæver et al., 1988; Lippert et al., 1993), photosystem I (Jeanjean et al., 1993) and cytochrome oxidase (Gabbay-Azaria et al., 1989).

From the results represented above we suggested that the concentration of NaCl might play an important direct regulatory role in triggering the salt adaptation (Hagemann et al., 1996b). The role of K^+-glutamate in the regulation of the activities of proteins and genes involved in synthesis and uptake of osmoprotective compounds in enteric bacteria has been discussed (Booth and Higgins, 1990). Furthermore, the remarkable high salt tolerance of GGPS and GGPP might ensure that almost all fixed carbon is used for GG synthesis in salt-shocked cells, since enzymes that might compete for carbon are inhibited by salt at the same time.

14.4 CLONING OF GENES INVOLVED IN GLUCOSYLGLYCEROL SYNTHESIS

For cloning of genes involved in GG synthesis we have generated several salt-sensitive mutants of *Synechocystis* (Hagemann and Zuther, 1992). Out of 18 mutants, nine were defective in GG synthesis. These mutants accumulated enhanced amounts of sucrose, while in the wild type only traces were detectable. The amounts of sucrose in mutants were not sufficient to compensate the loss of GG and to ensure adaptation to high salinity (Hagemann and Zuther, 1992). External supplied sucrose was not able to complement such mutants, while GG and trehalose did (Mikkat et al., 1997).

In one salt-sensitive mutant of *Synechocystis* the GGPP activity was impaired, leading to the accumulation of GGP. The integration of an antibiotic resistance gene used for mutagenesis led to a deletion of about 1.1 kb affecting two reading frames (Hagemann et al., 1996a). One ORF was identical with the *stpA* gene (salt tolerance protein A) isolated from another salt-sensitive mutant of *Synechocystis*. It was assumed that the StpA protein might play a regulatory role in salt-treated cells (Onana et al., 1994). After characterization of single mutants, overexpression of StpA in *E. coli* and biochemical characterization of the isolated protein, it could be shown that *stpA* encodes for the GGPP (Hagemann et al., 1997b). Surprisingly, the StpA protein obtained after overexpression in *E. coli* was also active under low salt conditions, while in extracts from *Synechocystis*, enhanced salt concentrations were necessary. This indicates that the GGPP itself probably does not depend on higher salt concentrations, but rather regulatory factors seem to be influenced by the salinity.

The availability of several salt-sensitive mutants of *Synechocystis* impaired in GG-synthesis and the completely known genome sequence of this strain (Kaneko et al., 1996) will soon help us to also identify and clone the gene encoding GGPS. The region affected in another GG-defect mutant was already cloned and sequenced. A deletion of about 10 kb was identified. The deleted ORFs will be checked by the generation of single mutants for the GGPS encoding gene (Hagemann et al., unpublished results).

Recently, it was shown that besides the capability to synthesize GG *de novo*, *Synechocystis* cells are able to take up GG actively from the surrounding medium (Mikkat et al., 1996). The activity of this transporter was restricted to GG, sucrose and trehalose (Figure 14.3) and became enhanced fourfold in salt-adapted cells (Mikkat et al., 1996). The increase of transport activity seemed to be based on increased transcription, since in northern blot experiments an increase of about the same magnitude was detected for the mRNA encoding a protein of this uptake system (Hagemann et al., 1997a). During the genetic analysis of salt-sensitive mutants of *Synechocystis* an ORF was sequenced showing similarities to ATP-binding subunits of ABC transporters (Hagemann et al., 1996a). Subsequent analysis revealed that this ORF (named *ggtA*) encodes for one subunit of the GG transport system, since a directed mutant in *ggtA* was unable to transport GG but remained salt tolerant. Furthermore, the *ggtA* mutant lost significant amounts of GG into the medium, indicating that the main function of this transport system is directed to the reuptake of GG and other osmoprotectants leaked through the cytoplasmic membrane (Hagemann et al., 1997a). The GG transporter belongs to the group of binding protein-dependent transport systems (ABC transporters). In contrast to the ABC transporter responsible for glycine betaine uptake in *E. coli* encoded by *proU* (Csonka and Hanson, 1991), the genes encoding for the substrate-binding protein and the permease were not organized together with the ATP-binding protein in one operon in *Synechocystis* (Hagemann et al., 1997a).

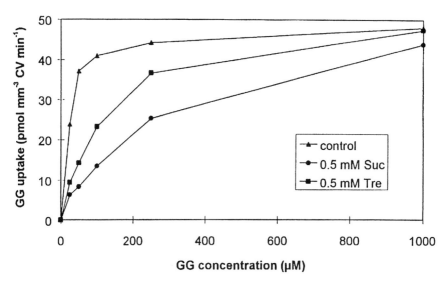

FIGURE 14.3 Uptake of glucosylglycerol (GG) by cells of *Synechocystis* sp. PCC 6803 grown in basal medium (2 mM NaCl). The GG transport rate was measured using different GG concentrations (control, ▲). The affinity of the GG transporter for sucrose and trehalose is shown by the reduced uptake rate in the presence of 0.5 mM sucrose (●) or trehalose (■). For details of the assays see Mikkat et al. (1996).

14.5 BIOSYNTHESIS OF GLYCINE BETAINE

Glycine betaine is the osmoprotectant that is characteristic for cyanobacteria from hypersaline environments (Reed et al., 1986). The osmoprotectant glutamate betaine was described only in two *Calothrix* strains (Mackay et al., 1984). In cyanobacteria able to synthesize glycine betaine, an active uptake system has been found for this compound (Moore et al., 1987). The affinity of the glycine betaine transport system is about 10 times higher than the GG-transporter of *Synechocystis* (Mikkat et al., 1996), but it should also serve the recovery of glycine betaine that has diffused out of the cell.

The enzymes involved in glycine betaine synthesis have not been characterized in cyanobacteria. It was reported that the oxidation of choline by a betaine-aldehyde dehydrogenase represents a necessary step in glycine betaine synthesis in *Spirulina subsalsa* (Gabbay-Azaria and Tel-Or, 1993). In salt-stressed higher plants, glycine betaine is mainly synthesized via choline in chloroplasts (Hanson et al., 1985). *E. coli* and other bacteria depend in their glycine betaine synthesis on the uptake of choline, which is then metabolized to glycine betaine by choline dehydrogenase (BetA) and betaine-aldehyde dehydrogenase (BetB) (Csonka and Hanson, 1991). Another pathway was found in halophilic bacteria starting from glycine, which is stepwise methylated (Galinski, 1995).

The choline oxidase gene from *Arthrobacter globiformis* (Deshnium et al., 1995) and the whole *bet*-operon (*betA* and *betB*) including a gene encoding a choline uptake protein from *E. coli* (Nomura et al., 1995) were transferred to the low-salt-tolerant cyanobacterial strain *Synechococcus* sp. PCC 7942. In both cases, glycine betaine accumulation was observed, leading to a slightly improved salt tolerance. This relatively low efficiency of glycine betaine in *Synechococcus* could be explained by its limited amount. Alternatively, in *Synechococcus*, processes other than the amount and kind of osmoprotectant could be limiting for its salt resistance. Nevertheless, these experiments represent a big step forward in the analyses of cyanobacterial salt adaptation by a transgene approach.

14.6 FUTURE PROSPECTS

Much knowledge about salt adaptation of cyanobacteria has been accumulated over the last decades and is awaiting application in plant sciences. For successful improvement of plant stress tolerance it is necessary to get a comprehensive look inside the complex events of salt adaptation. In our opinion, three points should be given special attention:

1. Further characterization of processes that may contribute to salt adaptation in order to select the most promising elements. Until now, research has been mainly focused on investigations of osmoprotectants. Processes like Na^+-exclusion, membrane composition and ion channels have not received the attention they deserve.
2. Isolation of genes encoding proteins involved in the main processes. Much progress could be made in selecting the genes involved in GG biosynthesis. However, a promising field might be the elucidation of the molecular background of glycine betaine and trehalose synthesis.
3. Investigation of the regulatory mechanisms in salt-treated cyanobacteria. At the moment, regulatory processes (signal recognition and transduction) are more or less unknown. However, information on these processes will be indispensable to understanding cyanobacterial salt adaptation and to use such mechanisms after transgene expression in plants.

ACKNOWLEDGMENT

Since 1991, the research on cyanobacterial salt adaptation in Rostock has been generously supported by grants from the Deutsche Forschungsgemeinschaft.

REFERENCES

Booth, I.R. and C.F. Higgins. 1990. Enteric bacteria and osmotic stress: intracellular potassium glutamate as a secondary signal of osmotic stress? *FEMS Microbiol. Rev.* 75:239–246.

Borowitzka, L.J., S. Demmerle, M.A. Mackay, and R.S. Norton. 1980. Carbon-13 nuclear magnetic resonance study of osmoregulation in a blue green alga. *Science* 210:650–651.

Brown, A.D. 1972. Microbial water stress. *Bacteriol. Rev.* 40:803–808.

Csonka, L.N. and A.D. Hanson. 1991. Prokaryotic osmoregulation: genetics and physiology. *Ann. Rev. Microbiol.* 45:569–606.

Deshnium, P., A.D. Los, H. Hayashi, L. Mustardy, and N. Murata. 1995. Transformation of *Synechococcus* with a gene for choline oxidase enhances tolerance to salt stress. *Plant Mol. Biol.* 29:897–907.

Gabbay-Azaria, R. and E. Tel-Or. 1993. Mechanisms of salt tolerance in cyanobacteria, in P.P. Gresshoff, Ed., *Plant Responses to the Environment.* CRC, Boca Raton. 123–132.

Gabbay-Azaria, R., M. Schonfeld, and E. Tel-Or. 1989. Cytochrome oxidase and H^+-ATPase activities in plasma-membranes of the marine cyanobacterium *Spirulina subsalsa* and their possible role in salt tolerance, in *Plant Membrane Transport,* J. Dainty, M.I. De Michelis, E. Marre, and F. Rasi-Caldogno, Eds. Elsevier Science, Amsterdam. 687–688.

Galinski, E.A. 1993. Compatible solutes of halophilic eubacteria: molecular principles, water-solute interaction, stress protection. *Experientia* 49:487–496.

Galinski, E.A. 1995. Osmoadaptation in bacteria. *Adv. Microb. Physiol.* 37:273–328.

Giæver, H.M., O.B. Styrvold, I. Kaasen, and A.R. Strøm. 1988. Biochemical and genetic characterization of osmoregulatory trehalose synthesis in *Escherichia coli. J. Bacteriol.* 170:2841–2849.

Hagemann, M. and N. Erdmann. 1994. Activation and pathway of glucosylglycerol synthesis in the cyanobacterium *Synechocystis* sp. PCC 6803. *Microbiology* 140:1427–1431.

Hagemann, M. and E. Zuther. 1992. Selection and characterization of mutants of the cyanobacterium *Synechocystis* sp. PCC 6803 unable to tolerate high salt concentrations. *Arch. Microbiol.* 158:429–434.

Hagemann, M., D. Golldack, J. Biggins, and N. Erdmann. 1993. Salt-dependent protein phosphorylation in the cyanobacterium *Synechocystis* PCC 6803. *FEMS Microbiol. Lett.* 113:205–210.

Hagemann, M., S. Richter, and E. Zuther. 1996a. Characterization of a glucosylglycerol-phosphate-accumulating, salt-sensitive mutant of the cyanobacterium *Synechocystis* sp. strain PCC 6803. *Arch. Microbiol.* 166: 83–91.

Hagemann, M., A. Schoor, and N. Erdmann. 1996b. NaCl acts as a direct modulator in the salt adaptive response: salt-dependent activation of glucosylglycerol synthesis *in vivo* and *in vitro. J. Plant Physiol.* 149:746–752.

Hagemann, M., S. Richter, and S. Mikkat. 1997a. The *ggtA* gene encodes a subunit of the transport system for the osmoprotective compound glucosylglycerol in *Synechocystis* sp. strain PCC 6803. *J. Bacteriol.* 179:714–720.

Hagemann, M., A. Schoor, R. Jeanjean, E. Zuther, and F. Joset. 1997b. The *stpA* gene from *Synechocystis* sp. strain PCC 6803 encodes the glucosylglycerol-phosphate phosphatase involved in cyanobacterial osmotic response to salt shock. *J. Bacteriol.* 179:1727–1733.

Hagemann, M., L. Wölfel, and B. Krüger. 1990. Alterations of protein synthesis in the cyanobacterium *Synechocystis* sp. PCC 6803 after a salt shock. *J. Gen. Microbiol.* 136:1393–1399.

Hanson, A.D., A.M. May, R. Grumet, J. Bode, G.C. Jamieson, and D. Rhodes. 1985. Betaine synthesis in chenopods: localization in chloroplasts. *Proc. Natl. Acad. Sci. USA* 82:3678–3682.

Huber, S.C. and J.L. Huber. 1996. Role and regulation of sucrose-phosphate synthase in higher plants. *Ann. Rev. Plant Physiol. Plant Mol. Biol.* 47:431–444.

Jeanjean, R., H.C.P. Matthijs, B. Onana, M. Havaux, and F. Joset. 1993. Exposure of the cyanobacterium *Synechocystis* PCC 6803 to salt stress induces concerted changes in respiration and photosynthesis. *Plant Cell Physiol.* 34:1073–1079.

Kaneko, T., S. Sato, H. Kotani, A. Tanaka, E. Asamizu, Y. Nakamura, N. Miyajima, M. Hirosawa, M. Sugiura, S. Sasamoto, T. Kimura, T. Hosouchi, A. Matsuno, A. Muraki, N. Nakazaki, K. Nruo, S. Okumura, S. Shimpo, C. Takeuchi, T. Wada, A. Watanabe, M. Yamada, M. Yasuda, and S. Tabata 1996. Sequence analysis of the genome of the unicellular cyanobacterium *Synechocystis* sp. strain PCC 6803. II. Sequence determination of the entire genome and assignment of potential protein-coding regions. *DNA Res.* 3:109–136.
Kauss, H. 1987. Some experiments of calcium-dependent regulation in plant metabolism. *Ann. Rev. Plant Physiol.* 38:47–72.
Lippert, K., E.A. Galinski, and H.G. Trüper. 1993. Biosynthesis and function of trehalose in *Ectothiorhodospira halochloris*. Antonie van Leeuwenhoek 63:85–91.
Mackay, M.A., R.S. Norton, and L.J. Borowitzka. 1984. Organic osmoregulatory solutes in cyanobacteria. *J. Gen. Microbiol.* 130:2177–2191.
Maruta, K., H. Mitsuzumi, T. Nakada, M. Kubota, H. Chaen, S. Fukuda, T. Sugimoto, and M. Kurimoto. 1996. Cloning and sequencing of a cluster of genes encoding novel enzymes of trehalose biosynthesis from thermophilic archaebacterium *Sulfolobus acidocaldarius*. *Biochim. Biophys. Acta* 1291:177–181.
Mikkat, S., M. Hagemann, and A. Schoor. 1996. Active transport of glucosylglycerol is involved in salt adaptation of the cyanobacterium *Synechocystis* sp. strain PCC 6803. *Microbiology* 142:1725–1732.
Mikkat, S., U. Effmert, and M. Hagemann. 1997. Uptake and use of the osmoprotective compounds trehalose, glucosylglycerol, and sucrose by the cyanobacterium *Synechocystis* sp. PCC 6803. *Arch. Microbiol.* 167:112–118.
Moore, D.J., R.H. Reed, and W.D.P. Stewart. 1987. A glycinebetaine transport system in *Aphanothece halophytica* and other glycinebetaine-synthesizing cyanobacteria. *Arch. Microbiol.* 147:399–405.
Nomura, M., M. Ishitani, T. Tabake, A. K. Rai, and T. Tabake. 1995. *Synechococcus* sp. PCC 7942 transformed with *Escherichia coli bet* genes produces glycinebetaine from choline and acquires resistance to salt stress. *Plant Physiol.* 107:703–708.
Onana, B., R. Jeanjean, and F. Joset. 1994. A gene, *stpA*, involved in the establishment of salt tolerance in the cyanobacterium *Synechocystis* PCC 6803. *Russian Plant Physiol.* 41:1176–1183.
Porchia, A.C. and G.L. Salerno. 1996. Sucrose biosynthesis in a prokaryotic organism: presence of two sucrose-phosphate synthases in *Anabaena* with remarkable difference compared with the plant enzymes. *Proc. Natl. Acad. Sci. USA* 93:13600–13604.
Reed, R.H., L.J. Borowitzka, M.A. Mackay, J.A. Chudek, R. Foster, S.R.C. Warr, D.J. Moore, and W.D.P. Stewart. 1986. Organic solute accumulation in osmotically stressed cyanobacteria. *FEMS Microbiol. Rev.* 39:51–56.
Strøm, A.R. and I. Kaasen. 1993. Trehalose metabolism in *Escherichia coli*: stress protection and stress regulation of gene expression. *Mol. Microbiol.* 8:205–210.
Warr, S.R.C., R.H. Reed, and W.D.P. Stewart. 1988. The compatibility of osmotica in cyanobacteria. *Plant Cell Environ.* 11:137–142.

15 Compatible Solutes: Ectoine Production and Gene Expression

Erwin A. Galinski and Petra Louis

CONTENTS

15.1 Introduction 187
15.2 Occurrence of Ectoines in Halophilic and Halotolerant
 Microorganisms 188
15.3 Two Model Production Strategies 189
 15.3.1 Ectoine from *Halomonas elongata* 189
 15.3.2 Hydroxyectoine from *Marinococcus* M52 193
15.4 Biosynthetic Pathways 194
15.5 Construction and Characterization of Ect⁻ Mutants of
 Halomonas elongata 194
15.6 Genomic Organization of Ectoine Genes from
 Marinococcus halophilus 195
15.7 Osmoregulated Expression in *Escherichia coli* 196
15.8 Conclusion 197
Acknowledgments 198
References 198

15.1 INTRODUCTION

Microorganisms exposed to high salt concentrations are faced with a dual challenge—high ionic strength and low water activity. Brines are therefore effectively dry environments and, as water is freely permeable across the membrane, this situation will lead to subsequent dehydration of "normal" organisms that are unable to adapt. This is precisely why salt is often used for the preservation of food (e.g. olives, fish, sheep cheese etc.).

Contrary to the situation in extremely halophilic halobacteria (Archaea), which tolerate salt in the cytoplasm, halophilic phototrophic and halotolerant aerobic eubacteria (Bacteria) have developed a different strategy of adaptation (Eisenberg et al., 1992; Galinski, 1995; Oren, 1994; Rengpipat et al., 1988). They exclude salt from the cytoplasm and, instead, synthesize or accumulate non-ionic organic osmolytes.

While cell wall, outer membrane and outer face of the cytoplasmic membrane still experience high salinity (with the need for salt adaptation), the interior of the cells remains relatively salt-free. This is, for example, demonstrated by the fact that typical cytoplasmic enzymes (in contrast to the so-called halophilic enzymes) behave like "normal" enzymes and are strongly inhibited by salt (Oren and Gurevich, 1993; Rengpipat et al., 1988).

The solutes responsible for osmotic equilibrium have been named compatible solutes (Brown, 1976, 1990) because of their compatibility with cellular metabolism—even at very high concentrations (1–2 M). It would, however, be incorrect to assume that these solutes have only an osmoregulatory function. While their presence surely enables the cell to maintain osmotic equilibrium, the water activity nevertheless remains as low inside as outside. As the hydration shell of enzymes is sensitive toward changes in water activity, it was proposed that compatible solutes, in addition, have a stabilizing effect on biomolecules in a low water environment (Galinski, 1993, 1995). Hence, potential biotechnological applications for the preservation of enzymes and whole cells have become a major focus of attention (Galinski and Tindall, 1992; Lippert and Galinski, 1992; Louis et al., 1994; Ventosa and Nieto, 1995).

It is, accordingly, not surprising that efforts are undertaken to establish a biotechnological production of compatible solutes, to elucidate their molecular function and to understand the expression and regulation of genes responsible for the biosynthesis of these solutes. However, investigations at the genetic level are still at an early stage with halophilic-compatible solute-producing eubacteria. Due to the natural competence of cyanobacteria, this phototrophic group of organisms was the first to claim attention, mainly with respect to synthesis and regulation of the glucosylglycerol-synthesizing system of moderately halophilic species (Hagemann et al., Chapter 14). Halophilic and halotolerant chemoheterotrophic producers of compatible solutes, on the other hand, have long been neglected because of a lack of basic genetic methods. Fundamental work of A. Ventosa and others on the moderately halophilic *Chromohalobacter-Deleya-Halomonas* group (Mellado et al., 1995b) finally started a systematic genetic exploration. Characterization of plasmids and construction of shuttle vectors (Fernandez-Castillo et al., 1992; Louis and Galinski 1997a; Mellado et al., 1995a, 1995c; Vargas et al., 1995) was paralleled by the development of a transposon mutagenesis system for *Halomonas elongata* by Kunte and Galinski (1995), which was based on the finding that antibiotics susceptibility, gene transfer and transposition can be achieved at the lower end of the salinity range of extremely halotolerant eubacteria. Until recently, the genes for *de novo* biosynthesis of compatible solutes had not been elucidated at the molecular level.

15.2 OCCURRENCE OF ECTOINES IN HALOPHILIC AND HALOTOLERANT MICROORGANISMS

The best known compatible solute is certainly glycine betaine. It is produced by a large number of halophilic phototrophic bacteria, of both the oxygenic and anoxygenic type (Imhoff, 1986; Mackay et al., 1984; Reed et al., 1984; Severin et al.,

1992) as well as by halophilic methanogenic Archaea (Lai et al., 1991; Robertson et al., 1990). In addition, it could be demonstrated that the ability to accumulate and use glycine betaine as a compatible solute is widespread among all kinds of halophilic and halotolerant bacteria (Imhoff and Rodriguez-Valera, 1984; Severin et al., 1992; Wohlfarth et al., 1990). In many cases, choline can also function as a precursor that is accumulated and subsequently oxidized to yield glycine betaine (Canovas et al., 1996; Landfald and Strøm, 1986; Smith et al., 1988). This ability (conversion of choline into betaine) is often called betaine synthesis. To avoid confusion with a true *de novo* biosynthesis, we suggest that the term choline oxidation should be used in this context. Among the aerobic chemoheterotrophic bacteria, however, other solutes seem to play a prominent part. The predominant compatible solutes of this group of eubacteria are proline, a number of N-acetylated diamino acids and the ectoines (Galinski, 1995).

As presented in Table 15.1, the ectoines (Figure 15.1), first discovered in the phototrophic bacterial genus *Ectothiorhodospira* (Galinski et al., 1985), compose the most characteristic and widespread class of compounds. Still, it should be noted that, so far, only a few bacterial lineages have been investigated systematically, mainly the Proteobacteria, the Firmicutes (Gram-positive bacteria) and Cyanobacteria. Within the latter group, as yet, no ectoines have been detected, and other branches harbor too few halophilic representatives to allow any conclusions. Ectoines, which in the light of the above are very common solutes of chemoheterotrophic Proteobacteria and Gram-positive bacteria, can be seen as cyclic forms of the greater family of N-acetylated diamino acids.

15.3 TWO MODEL PRODUCTION STRATEGIES

As the limited availability of ectoines has for a long time restricted further investigations, we have placed emphasis on the development of efficient production techniques. Two organisms were chosen, one for the production of L-ectoine (*Halomonas elongata* DSM 2081^T) and another for S,S-β-hydroxyectoine (*Marinococcus* M52). As will be shown below, two unique strategies were established that may have great potential for the production of not only ectoines but also other natural compounds from halophilic and halotolerant bacteria.

15.3.1 Ectoine from *Halomonas elongata*

As the synthesis of ectoines, like that of other compatible solutes, is strictly osmoregulated, their cytoplasmic concentration increases gradually with the salinity of the medium and reaches a maximum of approximately 2 M (equivalent to 20%–30% of the cells' dry weight [cdw]) in the case of *Halomonas elongata*. Further reasons to choose this organism as a potential production strain were its high growth rate in synthetic medium, its broad salt tolerance and a special response toward dilution stress.

While typical moderately halophilic microorganisms like *Marinococcus* species (see below) display a rather broad salinity optimum, *H. elongata* DSM 2081^T has its maximum growth rate in a narrow range between 3%–5% salt but is still able to

TABLE 15.1
Summary of Bacterial Species Known to Synthesize Ectoines (L-Ectoine and/or S,S-β-Hydroxyectoine)

Anoxygenic Phototrophic Bacteria (Purple Bacteria)

Ectothiorhodospira halochloris (DSM 1059[T])	Galinski et al., 1985
Ectothiorhodospira halophila (DSM 244[T])	Galinski et al., 1985
Ectothiorhodspira abdelmalekii (DSM 2110[T])	Galinski et al., 1985
Rhodospirillum salinarum (BN 40)	Galinski, 1986
Rhodovulum sulfidophilum (DSM 1374[T])	Galinski, 1986
(basonym Rhodobacter sulfidophilus)	

Aerobic Chemoheterotrophic Proteobacteria

Arhodomonas aquaeolei (DSM 8974)	unpublished
Chromohalobacter marismortui (ATCC 17056)	Severin et al., 1992
Halomonas elongata (ATCC 33173[T])	Wohlfarth et al., 1990
Halomonas eurihalina (ATCC 99336)	del Moral et al., 1994
(basonym Volcaniella eurihalina)	
Halomonas halmophila (CCM 2833[T])	Wohlfarth et al., 1990
(basonym Flavobacterium halmephilum)	
Halomonas halodenitrificans (DSM 735[T])	Severin et al., 1992
(basonym Paracoccus halodenitrificans)	
Halomonas halophila (CCM 3662[T])	Wohlfarth et al., 1990
(basonym Deleya halophila)	
Halomonas salina (ATCC 49509)	del Moral et al., 1994
(basonym Deleya salina)	
Halomonas variabilis (DSM 3051[T])	Severin et al., 1992
(basonym Halovibrio variabilis)	
Marinomonas sp. (Prado et al 1991)	del Moral et al., 1994
Pseudomonas halophila (DSM 3050[T])	Severin et al., 1992
Pseudomonas halosaccharolytica (CCM 2851)	Severin et al., 1992
Salinivibrio costicola (CCM 2811)	Severin et al., 1992
(basonym Vibrio costicola)	
Vibrio alginolyticus (DSM 2171)	Severin et al., 1992
Vibrio fischeri (DSM 507[T], DSM 7151)	Schmitz and Galinski., 1996
Vibrio harveyi (DSM 6904, DSM 2165)	Schmitz and Galinski., 1996

Aerobic Chemoheterotrophic Gram-positive Bacteria

High GC

Brachybacterium tyrofermentans (DSM 10673)	unpublished
Brachybacterium alimentarium (DSM 10672)	unpublished
Brevibacterium casei (DSM 20657[T])	Frings et al., 1993
Brevibacterium epidermidis (DSM 20660[T])	Frings et al., 1993
Brevibacterium linens (DSM 29425[T])	Frings et al., 1993
Brevibacterium linens (CNRZ 211)	Bernard et al., 1993
Brevibacterium iodinum (DSM 20626[T])	Frings et al., 1993
Micrococcus varians var. halophilus (CCM 3316)	Severin et al., 1992

TABLE 15.1 (CONTINUED)
Summary of Bacterial Species Known to Synthesize Ectoines (L-Ectoine and/or S,S-β-Hydroxyectoine)

Nesterenkonia halobia (DSM 20541^T)	Severin et al., 1992
(basonym *Micrococcus halobius*)	
Nocardiopsis alba subsp. *alba* (DSM 43119)	unpublished
Nocardiopsis alba subsp. *prasina* (DSM 43845^T)	unpublished
Nocardiopsis alborubida (DSM 40465^T)	unpublished
Nocardiopsis dassonvillei (DSM 43111^T)	unpublished
Nocardiopsis listeri (DSM 40297^T)	unpublished
Nocardiopsis lucentensis (DSM 44048)	Yassin et al., 1993
Streptomyces griseolus (DSM 40067^T)	Severin et al., 1992
Streptomyces parvulus (ATCC 12434)	Inbar and Lapidot, 1988
Coccus 19 (Valderrama et al 1991)	del Moral et al., 1994
Coccus 28 (Valderrama et al 1991)	del Moral et al., 1994
Low GC	
Bacillus haloalkaliphilus (DSM 5271^T)	Müller, 1991
(former *Bacillus* WN13^T)	
Bacillus halophilus (DSM 4771)	Müller, 1991
Bacillus pantothenticus (DSM 26^T)	Müller, 1991
Bacillus pasteurii (DSM 33)	Müller, 1991
Bacillus sp. (DSM 578)	Müller, 1991
Bacillus 30 (Bejar et al 1992)	del Moral et al., 1994
Bacillus 49 (Bejar et al 1992)	del Moral et al., 1994
Halobacillus halophilus (DSM 2266^T)	Severin, 1993
(basonym *Sporosarcina halophila*)	
Marinococcus halophilus (DSM 20408^T)	Severin et al., 1992
Marinococcus albus (DSM 20748^T)	Severin et al., 1992
Marinococcus M52	Severin et al., 1992

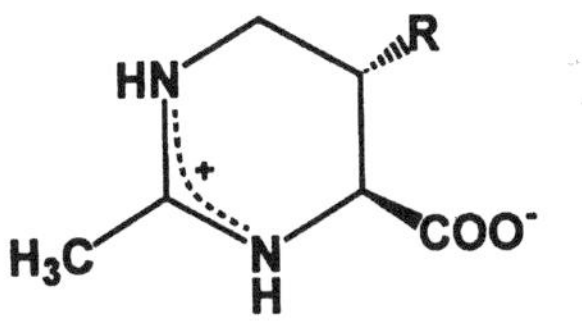

L-Ectoine (R=H)
S,S-β-Hydroxyectoine (R=OH)

FIGURE 15.1 Molecular formula of L-ectoine (1,4,5,6-tetrahydro-2-methyl-4-pyrimidine carboxylic acid) and S,S-β-hydroxyectoine.

grow well at 20% salt and higher. This organism should, therefore, be classified as a marine organism with an exceptionally broad salt tolerance. Aiming at the large-scale production of ectoine, and encouraged by the observation that some halophiles (and even *Escherichia coli*) release compatible solutes in response to a dilution stress

(Fischel and Oren, 1993; Lamark et al., 1992; Schleyer et al., 1993; Tschichholz and Trüper, 1990), we searched for a producer strain with the following properties: rapid extrusion of ectoine during dilution, high tolerance toward osmotic shock in both directions, and effective resynthesis of the lost solutes under high salt conditions. Rapid dilution, as for example by heavy rainfall or flooding, and slow increase in salinity due to evaporation are natural events typical for environments of fluctuating salinity. The halotolerant Gram-negative bacterium *H. elongata*, which was first isolated from a solar saltern (Vreeland et al., 1980), proved to be the ideal candidate (Sauer, 1995; Wohlfarth et al., 1990). Consequently, this organism was used to develop a novel process for the production of ectoine, called "bacterial milking" (Sauer and Galinski, 1998).

This process, which is briefly described in Figure 15.2, can in theory be repeated indefinitely (nine cycles were performed in the authors' laboratory with no sign of deterioration). It is based on a near 100% biomass recycling and can effectively be seen as a semi-continuous extraction of ectoine, which provides a raw-product solution (50 mM ectoine) containing mainly salt as a major impurity. During each cycle, approximately 65% of the cells' compatible solutes were gained, providing a maximum yield of 155 mg ectoine per cycle per g cdw. The solute was obtained with highest purity following one chromatographic step and subsequent crystallization. "Bacterial milking" for ectoine production has been upscaled to pilot plant level and is currently used for the commercial production

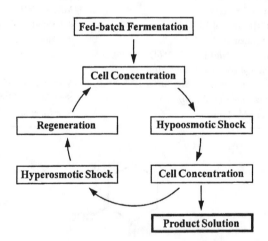

FIGURE 15.2 Flow diagram of the "bacterial milking" process using *H. elongata*. A synthetic glucose mineral salt medium at a salinity of 15% NaCl is used for batch and fed-batch cultivation with exponential feeding strategy. The final high cell-density culture of 40 g cdw l^{-1} is reduced to approximately one-fifth of its volume using a cross-flow filtration unit (concentration step). By refilling the bioreactor with distilled water, the salinity is rapidly reduced to 3% NaCl, which triggers the release of ectoine into the medium. A second cross-flow filtration step serves the purpose of harvesting the ectoine containing product solution and thus remove the dilution medium. Addition of hypersaline fermentation medium (to a final concentration of 15% NaCl) subsequently induces resynthesis of the lost solutes and returns the process to the beginning.

of ectoines at BITOP, Gesellschaft für biotechnische Optimierung mbH, Witten, Germany. The rapid-release phenomenon as the underlying principle for "bacterial milking" is at present under investigation. The recent discovery of mechanosensitive channels in *E. coli* (MscL, MscS), which open when cell turgor exceeds a critical level, might possibly help to explain this unusual property (Martinac et al., 1987; Sukharev et al., 1994).

15.3.2 Hydroxyectoine from *Marinococcus* M52

An economically feasible chemical synthesis for S,S-β-hydroxyectoine, with its two chiral configurations, appears to be out of reach. Therefore, we developed a biotechnological process for the production of the compound, using *Marinococcus* M52 (Frings et al., 1995). This strain was isolated by Wohlfarth (1993) from brackish waters near Agadir, Morocco, and was classified in the genus *Marinococcus* because of DNA-DNA hybridization experiments (Jahnke, 1994). The organism, which is able to synthesize both ectoine and hydroxyectoine, fully converts ectoine into hydroxyectoine when it enters the stationary phase. Hence, there is no need for a separation of the compounds. In addition, the final content of hydroxyectoine is remarkably high (15% of cdw when grown at 10% salt).

An exponential feeding strategy similar to the one applied with *Halomonas elongata* failed with *Marinococcus* M52 (μ_{max} = 0.27 h^{-1}) because this species, like other *Marinococcus* species, typically displayed an ever-decreasing growth rate with increasing biomass. The assumption that unknown growth factors or other metabolic byproducts become growth limiting during the fermentation process was experimentally confirmed by exchanging the growth medium at a later stage. With this combination of a conventional fed-batch strategy and subsequent medium exchange by cross-flow filtration, it took approximately five days to reach a density of 50 g l^{-1} cdw (Frings et al., 1995). As a consequence of these observations, *Marinococcus* M52 was also grown in a dialysis bioreactor with a cuprophane membrane (Märkl, 1989; Märkl et al., 1990), where cells are constantly dialyzed against fresh growth medium. Here the maximum growth rate was maintained over the entire batch phase, followed by a reduced rate during the fed-batch procedure. At a cell density of 100 g l^{-1} cdw, the authors still observed a growth rate of 0.07 h^{-1}, equivalent to a doubling time of less than 10 h (Krahe et al., 1996). A final density of 132 g l^{-1} (almost four times higher than with a conventional fed-batch strategy) was reached in only two thirds of the fermentation time (60 h instead of 90 h). As the commercial application of the dialysis cultivation technique—for the time being—still depends on the development of novel scale-up procedures, there is clearly a need to track down the growth-limiting factor and further enhance growth-medium development.

The prime advantage of *Marinococcus* M52 as a production strain is that, at stationary phase, hydroxyectoine is the sole product, and that downstream processing, in particular removal of salt, is extremely simple (Frings et al., 1995). Contrary to the described "bacterial milking" above, Gram-positive *Marinococcus* cells display a different behavior when exposed to hypoosmotic shock. Intracellularly accumulated hydroxyectoine is not excreted into the medium. Due to water influx, the cells increase in volume (swelling) but are protected from disintegration by a seemingly strong cell wall. The physiological response to retain solutes, rather than excrete them during a dilution procedure, allows salts to be removed by a simple washing procedure and to gain a salt-free raw material for the extraction of solutes. Thus, a pure product can be gained following only three steps of downstream processing: Soxhlet extraction, removal of pigments by a hydrophobic absorber resin and crystallization (Frings et al., 1995). As a further consequence of the "bacterial washing" technique, it is now possible to directly apply crude (salt-free) cell extracts for *in vitro* stabilization studies.

15.4 BIOSYNTHETIC PATHWAYS

The biosynthetic pathway for ectoine has been elucidated at enzymological level in the two halophilic eubacteria *Ectothiorhodospira halochloris* and *Halomonas elongata* (Galinski and Trüper, 1994; Peters et al., 1990; Tao et al., 1992). It comprises three steps, the first being the conversion of aspartate semialdehyde, an intermediate in the amino acid metabolism, to L-2,4-diaminobutyric acid. This is followed by acetylation to Nγ-acetyldiaminobutyric acid. Finally, ectoine is formed in a cyclic condensation reaction (Figure 15.3). The biosynthetic pathway for hydroxyectoine is still unsolved; the substrate for the hydroxylation reaction, in particular, still needs to be identified. ^{13}C-NMR labeling experiments performed by Sauer (1995) provided strong evidence that ectoine or Nγ-acetyldiaminobutyric acid are likely precursors.

15.5 CONSTRUCTION AND CHARACTERIZATION OF ECT⁻ MUTANTS OF *HALOMONAS ELONGATA*

The first genetic studies regarding ectoine synthesis were performed by Min-Yu et al. (1993) who cloned and sequenced the gene for L-ectoine synthase in a *Halomonas* sp. following determination of the N-terminal sequence of the purified enzyme.

Using a transposon mutagenesis system established by Kunte and Galinski (1995) for the Gram-negative moderate halophile *H. elongata,* a salt-sensitive mutant (*H. elongata* SAA4) that was able to grow only to 4% NaCl in a minimal medium (unpublished results) was obtained. This mutant had lost the ability to synthesize ectoine and hydroxyectoine, but accumulated high amounts of glutamate and glutamine at lower salinities to maintain osmotic equilibrium. Furthermore, L-2,4-diaminobutyric acid, an intermediate in the biosynthetic pathway for ectoine, could be detected. We therefore concluded that, for the second enzyme of the ectoine biosynthetic pathway, the transposon has inserted into the gene L-2,4-diaminobutyric acid Nγ-acetyltransferase, which uses L-2,4-diaminobutyric acid as a substrate. The supplementation of *H. elongata* SAA4 with Nγ-acetyl-L-2,4-diaminobutyric acid, the respective product of the destroyed gene, restored its ability to synthesize ectoine.

The addition of ectoine not only restored the organism's salt tolerance, but also its ability to synthesize the hydroxylated derivative hydroxyectoine. This observation provides additional evidence that hydroxyectoine is synthesized directly from ectoine or the immediate precursor rather than by a separate pathway.

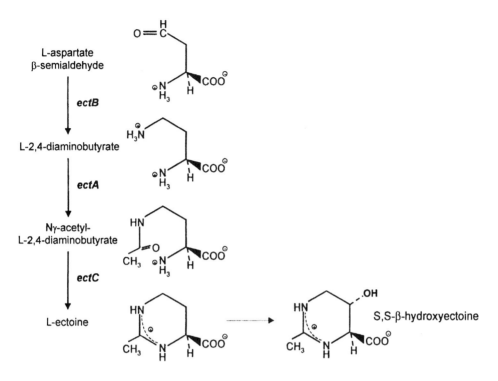

FIGURE 15.3 Biosynthetic pathway for ectoine based on enzymological studies. *ectB*: L-2,4-diaminobutyric acid transaminase, *ectA*: L-2,4-diaminobutyric acid Nγ-acetyltransferase, *ectC*: L-ectoine synthase. The proposed formation of hydroxyectoine (dotted line) from ectoine is at present still hypothetical.

15.6 GENOMIC ORGANIZATION OF ECTOINE GENES FROM *MARINOCOCCUS HALOPHILUS*

The genes for the biosynthesis of ectoine were cloned from the Gram-positive moderate halophile *Marinococcus halophilus* by construction of a genomic DNA library in *E. coli* and screening for clones with enhanced salt tolerance (Louis and Galinski, 1997b). The clone *E. coli* (pOSM11), which was able to grow up to 5% NaCl in a minimal medium (maximal salt tolerance of parent strain: 3% NaCl), had obtained the ability to synthesize ectoine. Sequencing of the recombinant DNA fragment of *E. coli* (pOSM11) revealed four major open reading frames oriented in the same direction and predicted to encode proteins of 172, 427, 129, and 110 amino acids, with deduced molecular masses of 19385, 47192, 14796, and 13142 Da, respectively, followed by the 5'-end of another open reading frame (*orfB*) (Figure 15.4). As the *lacZ* promoter of the vector was oriented in the opposite direction, transcription of these reading frames from the *lacZ* promoter could be excluded.

Sequence comparisons with known proteins as well as physiological examinations of suitable subclones revealed the physiological function of these open reading frames (Louis and Galinski, 1997b). The gene *ectA* encodes the diaminobutyric acid

pOSM11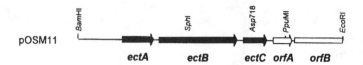

FIGURE 15.4 Map of the sequenced DNA fragment of *M. halophilus* (pOSM11). Position of the open reading frames in bps: *ectA*: 722-1240, *ectB*: 1329-2612, *ectC*: 2716-3105, *orfA*: 3216-3548, *orfB*: 3584-4351 (truncated). (From Louis and Galinski 1997b, with permission.)

acetyl transferase, *ectB* the diaminobutryic acid transaminase and *ectC* the ectoine synthase required for ectoine biosynthesis, whereas *orfA* and *orfB* apparently are not involved in this pathway.

15.7 OSMOREGULATED EXPRESSION IN *ESCHERICHIA COLI*

Remarkably, the intracellular ectoine concentration in *E. coli* (pOSM11) harboring the ectoine genes from *M. halophilus* rose with elevated salt concentrations of the medium (Figure 15.5). As no ectoine could be detected extracellularly, ectoine seemed to be synthesized in an osmoregulated manner.

The question arises as to how these genes originating from a halophilic Gram-positive bacterium are expressed in an osmoregulated way in *E. coli*. The search

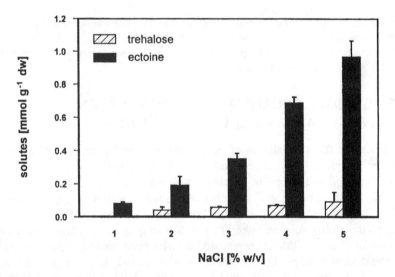

FIGURE 15.5 Correlation between the intracellular solute concentration of *E. coli* (pOSM11) and medium salinity during growth in minimal medium MM63 as measured by isocratic HPLC. Means and standard deviations of three independent experiments. Glutamate levels did not exceed 0.12 mmol g^{-1} dw regardless of the salt concentration of the medium. The trehalose pool of *E. coli* XL1-Blue in the presence of 3% NaCl was 0.35 mmol g^{-1} cdw. (From Louis and Galinski 1997b, with permission.)

for promoter consensus sequences revealed possible promoter sequences for σ^{70}-dependent promoters upstream of the second gene *ectB*. However, no expression of this gene could be observed in physiological examinations with a subclone lacking the putative promoter region upstream of the first gene *ectA*. Thus, the biological significance of these putative promoters remains obscure. A search for promoter consensus sequences of the osmoresponsive promoters for the compatible solute transport systems *proU*, *proP* and *opuA* (Kempf and Bremer, 1995; Mellies et al., 1994, 1995) revealed no matches. As several osmoregulated genes in *E. coli*, including the genes for the biosynthesis of trehalose, are known to be under the control of σ^S-dependent promoters (Gordia and Gutierrez, 1996; Manna and Gowrishankar, 1994; Mellies et al., 1995; Strøm and Kaasen, 1993), a search for these consensus sequences (according to Strøm and Kaasen, 1993) was also performed, but no matches were found. Instead, a sequence similar to the consensus for σ^B of *Bacillus subtilis* known to transcribe the so-called "general stress proteins," which are induced by various stimuli such as salt stress, ethanol treatment and starvation (Hecker et al., 1996), was found about 300 bps upstream of the first gene *ectA*. Deleting this region resulted in a subclone that was still able to synthesize ectoine, but no longer in an osmoregulated fashion. Part of the promoter region (about 200 bps upstream of *ectA*) has a very high AT content (above 70% compared with 53.6% of total DNA of *M. halophilus* [Hao et al., 1984]) possibly forming curved DNA, which is known to be preferentially bound by the histone-like protein H-NS. An involvement of this protein in the osmoregulated expression of other genes of *E. coli* is known (Gowrishankar and Manna, 1996; Owen-Hughes et al., 1992), so it is possibly also involved in the osmoregulated expression of the ectoine genes from *M. halophilus* in *E. coli*. We expect that rather global regulatory mechanisms such as histone-like DNA-binding proteins; the DNA topology, which also responds to environmental factors; or the intracellular potassium concentration (Booth and Higgins, 1990; Meury and Kohiyama, 1992) will help to explain the osmoregulated expression of cloned genes from only distantly related organisms.

15.8 CONCLUSION

While current biotechnological production techniques provide us with the necessary quantities of ectoine and hydroxyectoine to pursue comprehensive application studies, future developments may rely on production techniques using recombinant overproducing strains, preferably *E. coli*. In addition, it is intriguing to speculate, but still needs to be demonstrated, that a moderate degree of salt tolerance can be achieved in all kinds of organisms by simply transferring the ectoine genes as described above. Concerning general aspects of osmoregulation, we hope to gain further insights by cloning and sequencing of the ectoine genes of mutant *H. elongata* SAA4. In addition, comparison of the regulating sequences of the ectoine synthesis genes from both halophiles (*H. elongata* and *M. halophilus*) will hopefully lead to a deeper understanding of their osmoregulated expression and possibly also reveal fundamental similarities with general stress response mechanisms currently explored in *E. coli*.

ACKNOWLEDGMENTS

Most of the studies reported from the biotechnology group at the Institute for Microbiology & Biotechnology in Bonn were carried out by Ph.D. and diploma students; their industrious efforts and intellectual impact is gratefully acknowledged. In addition, special thanks are due to Marlene Stein for her invaluable help with the creation of artwork. Finally, we would like to express our gratitude for financial support by the European Commission (Bio2-CT93-0274, Bio4-CT96-0488) and the Deutsche Forschungsgemeinschaft (Ga393/1+3).

REFERENCES

Bejar, M.V., E. Quesada, M.C. Gutiérrez, A. del Moral, M.J. Valderrama, A. Ventosa, F. Ruiz-Berraquero, and A. Ramos-Cormenzana. 1992. Numerical taxonomy of moderately halophilic Gram-positive endospore-forming rods. *Syst. Appl. Microbiol.* 14:223–228.

Bernard, T., M. Jebbar, Y. Rassouli, S. Himdi-Kabbab, J. Hamelin, and C. Blanco. 1993. Ectoine accumulation and osmotic regulation in *Brevibacterium linens*. *J. Gen. Microbiol.* 139:129–136.

Booth, I.R. and C.F. Higgins. 1990. Enteric bacteria and osmotic stress: intracellular potassium glutamate as a secondary signal of osmotic stress? *FEMS Microbiol. Rev.* 75:239–246.

Brown, A.D. 1976. Microbial water stress. *Bacteriol. Rev.* 40:803–846.

Brown, A.D. 1990. *Microbial Water Stress Physiology: Principles and Perspectives.* Wiley, Chichester.

Canovas, D., C. Vargas, L.N. Csonka, A. Ventosa, and J.J. Nieto. 1996. Osmoprotectants in *Halomonas elongata*: high-affinity betaine transport system and choline-betaine pathway. *J. Bacteriol.* 178:7221–7226.

del Moral, A., J. Severin, A. Ramos-Cormenzana, H.G. Trüper, and E.A. Galinski. 1994. Compatible solutes in new moderately halophilic isolates. *FEMS Microbiol. Lett.* 122:165–172.

Eisenberg, H., M. Mevarech, and G. Zaccai. 1992. Biochemical, structural, and molecular genetic aspects of halophilism. *Adv. Prot. Chem.* 43:1–62.

Fernandez-Castillo, R., C. Vargas, J.J. Nieto, A. Ventosa, and F. Ruiz-Berraquero. 1992. Characterization of a plasmid from moderately halophilic eubacteria. *J. Gen. Microbiol.* 138:1133–1137.

Fischel, U. and A. Oren. 1993. Fate of compatible solutes during dilution stress in *Ectothiorhodospira marismortui*. *FEMS Microbiol. Lett.* 113:113–118.

Frings, E., H.J. Kunte, and E.A. Galinski. 1993. Compatible solutes in representatives of the genera *Brevibacterium* and *Corynebacterium*: occurrence of tetrahydropyrimidines and glutamine. *FEMS Microbiol. Lett.* 109:25–32.

Frings, E., T. Sauer, and E.A. Galinski. 1995. Production of hydroxyectoine: high cell-density cultivation and osmotic downshock of *Marinococcus* strain M52. *J. Biotechnol.* 43:53–61.

Galinski, E.A. 1986. *Salzadaptation durch kompatible Solute bei halophilen Phototrophen Bakterien.* Ph.D. thesis, University of Bonn.

Galinski, E.A. 1993. Compatible solutes of halophilic eubacteria: molecular principles, water-solute interaction, stress protection. *Experientia* 49:487–496.

Galinski, E.A. 1995. Osmoadaptation in bacteria. *Adv. Microb. Physiol.* 37:273–328.

Galinski, E.A. and B.J. Tindall. 1992. Biotechnological prospects for halophiles and halotolerant microorganisms, in *Molecular Biology & Biotechnology of Extremophiles,* R.H. Herbert and R. Sharp, Eds. Blackie & Son, Glasgow. 76–114.

Galinski, E.A. and H.G. Trüper. 1994. Microbial behaviour in salt-stressed ecosystems. *FEMS Microbiol. Rev.* 15:95–108.

Galinski, E.A., H.P. Pfeiffer, and H.G. Trüper. 1985. 1,4,5,6-Tetrahydro-2-methyl-4-pyrimidinecarboxylic acid, a novel cyclic amino acid from halophilic phototrophic bacteria of the genus *Ectothiorhodospira. Eur. J. Biochem.* 149:135–139.

Gordia, S. and C. Gutierrez. 1996. Growth-phase-dependent expression of the osmotically inducible gene *osmC* of *Escherichia coli* K-12. *Mol. Microbiol.* 19:729–736.

Gowrishankar, J. and D. Manna. 1996. How is osmotic regulation of transcription of the *Escherichia coli proU* operon achieved? *Genetica* 97:363–378.

Hao, M.V., M. Kocur, and K. Komagata. 1984. *Marinococcus* gen. nov., a new genus for motile cocci with meso-diaminopimelic acid in the cell wall; and *Marinococcus albus* sp. nov. and *Marinococcus halophilus* (Novitsky and Kushner) comb. nov. *J. Gen. Appl. Microbiol.* 30:449–459.

Hecker, M.,W. Schumann, and U. Völker. 1996. Heat-shock and general stress response in *Bacillus subtilis. Mol. Microbiol.* 19:417–428.

Imhoff, J. F. 1986. Osmoregulation and compatible solutes in eubacteria. *FEMS Microbiol. Rev.* 39:57–66.

Imhoff, J.F. and F. Rodriguez-Valera. 1984. Betaine is the main compatible solute of halophilic eubacteria. *J. Bacteriol.* 160:478–479.

Inbar, L. and A. Lapidot. 1988. The structure and biosynthesis of new tetrahydropyrimidine derivatives in actinomycin D producer *Streptomyces parvulus. J. Biol. Chem.* 263:16014–16022.

Jahnke, K.D. 1994. A modified method of quantitative colorimetric DNA-DNA hybridization on membrane filters for bacterial identification. *J. Microbiol. Meth.* 20:223–230.

Kempf, B. and E. Bremer. 1995. OpuA, an osmotically regulated binding protein-dependent transport system for the osmoprotectant glycine betaine in *Bacillus subtilis. J. Biol. Chem.* 270:16701–16713.

Krahe, M., G. Antranikian, and H. Märkl. 1996. Fermentation of extremophilic microorganisms. *FEMS Microbiol. Rev.* 18:271–285.

Kunte, H.J. and E.A. Galinski. 1995. Transposon mutagenesis in halophilic eubacteria: conjugal transfer and insertion of transposon Tn5 and Tn1732 in *Halomonas elongata. FEMS Microbiol. Lett.* 128:293–299.

Lai, M.C., K.R. Sowers, D.E. Robertson, M.F. Roberts, and R.P. Gunsalus. 1991. Distribution of compatible solutes in the halophilic methanogenic archaebacteria. *J. Bacteriol.* 173:5352–5358.

Lamark, T., O.B. Styrvold, and A.R. Strøm. 1992. Efflux of choline and glycine betaine from osmoregulating cells of *Escherichia coli. FEMS Microbiol. Lett.* 96:149–154.

Landfald, B. and A.R. Strøm. 1986. Choline-glycine betaine pathway confers a high level of osmotic tolerance in *Escherichia coli. J. Bacteriol.* 165:849–855.

Lippert, K. and E.A. Galinski. 1992. Enzyme stabilization by ectoine-type compatible solutes: protection against heating, freezing and drying. *Appl. Microbiol. Biotechnol.* 37:61–65.

Louis, P. and E.A. Galinski. 1997a. Identification of plasmids in the genus *Marinococcus* and complete nucleotide sequence of plasmid pPL1 from *Marinococcus halophilus. Plasmid* 38:107–114.

Louis, P. and E.A. Galinski. 1997b. Characterization of genes for the biosynthesis of the compatible solute ectoine from *Marinococcus halophilus* and osmoregulated expression in *Escherichia coli*. *Microbiology* 143:1141–1149.
Louis, P., H.G. Trüper, and E.A. Galinski. 1994. Survival of *Escherichia coli* during drying and storage in the presence of compatible solutes. *Appl. Microbiol. Biotechnol.* 41:684–688.
Mackay, M.A., R.S. Norton, and L.J. Borowitzka. 1984. Organic osmoregulatory solutes in cyanobacteria. *J. Gen. Microbiol.* 130:2177–2191.
Manna, D. and J. Gowrishankar. 1994. Evidence for involvement of proteins HU and RpoS in transcription of the osmoresponsive *proU* operon in *Escherichia coli*. *J. Bacteriol.* 176:5378–5384.
Märkl, H. 1989. Folien und Membranen als neue Elemente im Fermenterbau. *Forum Mikrobiol.* 12:234–237.
Märkl, H., M. Lechner, and F. Götz. 1990. A new dialysis fermentor for the production of high concentrations of extracellular enzymes. *J. Ferment. Bioeng.* 69:244–249.
Martinac, B., M. Buechner, A.H. Delcour, J. Adler, and C. Kung. 1987. Pressure-sensitive ion channels in *Escherichia coli*. *Proc. Natl. Acad. Sci. USA* 84:2297–2301.
Mellado, E., J.A. Asturias, J.J. Nieto, K.N. Timmis, and A. Ventosa. 1995a. Characterization of the basic replicon of pCM1, a narrow-host-range plasmid from the moderate halophile *Chromohalobacter marismortui*. *J. Bacteriol.* 177:3443–3450.
Mellado, E., E.R.B. Moore, J.J. Nieto, and A. Ventosa. 1995b. Phylogenetic interferences and taxonomic consequences of 16S ribosomal DNA sequence comparison of *Chromohalobacter marismortui*, *Volcaniella euryhalina*, and *Deleya salina* and reclassification of *V. euryhalina* as *Halomonas euryhalina* comb. nov. *Int. J. Syst. Bacteriol.* 45:712–716.
Mellado, E., J.J. Nieto, and A. Ventosa. 1995c. Construction of novel shuttle vectors for use between moderately halophilic bacteria and *Escherichia coli*. *Plasmid* 34:157–164.
Mellies, J., R. Brems, and M. Villarejo. 1994. The *Escherichia coli proU* promoter element and its contribution to osmotically signaled transcription activation. *J. Bacteriol.* 176:3638–3645.
Mellies, J., A. Wise, and M. Villarejo. 1995. Two different *Escherichia coli proP* promoters respond to osmotic and growth phase signals. *J. Bacteriol.* 177:144–151.
Meury, J. and M. Kohiyama. 1992. Potassium ions and changes in bacterial DNA supercoiling under osmotic stress. *FEMS Microbiol. Lett.* 99:159–164.
Min-Yu, L., H. Ono, and M. Takano. 1993. Gene cloning of ectoine synthase from *Halomonas* sp. *Ann. Rep. Int. Centre Coop. Res. Biotechnol. Japan* 16:193–200.
Müller, E. 1991. *Kompatible Solute und Prolingewinnung bei halophilen und halotoleranten Bacilli*. Ph.D. thesis, University of Bonn.
Oren, A. 1994. The ecology of the extremely halophilic archaea. *FEMS Microbiol. Rev.* 13:415–440.
Oren, A. and P. Gurevich. 1993. The fatty acid synthetase complex of *Haloanaerobium praevalens* is not inhibited by salt. *FEMS Microbiol. Lett.* 108:287–290.
Owen-Hughes, T.A., G.D. Pavitt, D.S. Santos, J.M. Sidebotham, C.S.J. Hulton, J.C.D. Hinton, and C.F. Higgins. 1992. The chromatin-associated protein H-NS interacts with curved DNA to influence DNA topology and gene expression. *Cell* 71:255–265.
Peters, P., E.A. Galinski, and H.G. Trüper. 1990. The biosynthesis of ectoine. *FEMS Microbiol. Lett.* 71:157–162.

Prado, B., A. del Moral, E. Quesada, R. Rios, M. Monteoliva-Sanchez, V. Campos, and A. Ramos-Cormenzana. 1991. Numerical taxonomy of moderately halophilic Gram-negative rods isolated from the Salar del Atacama, Chile. *Syst. Appl. Microbiol.* 14:275–281.

Reed, R.H., J.A. Chudek, R. Foster, and W.D.P. Stewart. 1984. Osmotic adjustment in cyanobacteria from hypersaline environments. *Arch. Microbiol.* 138:333–337.

Rengpipat, S., S.E. Lowe, and J.G. Zeikus. 1988. Effect of extreme salt concentrations on the physiology and biochemistry of *Halobacteroides acetoethylicus*. *J. Bacteriol.* 170:3065–3071.

Robertson, D., D. Noll, M.F. Roberts, J. Menaia, and R.D. Boone. 1990. Detection of the osmoregulator betaine in methanogens. *Appl. Environ. Microbiol.* 56:563–565.

Sauer, T. 1995. *Untersuchungen zur Nutzung von Halomonas elongata für die Gewinnung kompatibler Solute*. Ph.D. thesis, University of Bonn.

Sauer, T. and E.A. Galinski. Bacterial milking: a novel bioprocess for the production of compatible solutes. *Biotechnol. Bioeng.* (in press).

Schleyer, M., R. Schmid, and E.P. Bakker. 1993. Transient, specific and extremely rapid release of osmolytes from growing cells of *Escherichia coli* K-12 exposed to hypoosmotic shock. *Arch. Microbiol.* 160:424–431.

Schmitz, R.P.H. and E.A. Galinski. 1996. Compatible solutes in luminescent bacteria of the genera *Vibrio*, *Photobacterium* and *Xenorhabdus* (*Photorhabdus*): occurrence of ectoine, betaine and glutamate. *FEMS Microbiol. Lett.* 142:195–201.

Severin, J. 1993. *Kompatible Solute und Wachstumskinetik bei halophilen aeroben, heterotrophen Eubakterien*. Ph.D. thesis, University of Bonn.

Severin, J., A. Wohlfarth, and E.A. Galinski. 1992. The predominant role of recently discovered tetrahydropyrimidines for the osmoadaptation of halophilic eubacteria. *J. Gen. Microbiol.* 138:1629–1638.

Smith, L.T., J.A. Pocard, T. Bernard, and D. LeRudulier. 1988. Osmotic control of glycine betaine biosynthesis and degradation in *Rhizobium meliloti*. *J. Bacteriol.* 170:3142–3149.

Strøm, A.R. and I. Kaasen. 1993. Trehalose metabolism in *Escherichia coli*: stress protection and stress regulation of gene expression. *Mol. Microbiol.* 8:205–210.

Sukharev, S.I., P. Blount, B. Martinac, F.R. Blattner, and C. Kung. 1994. A large-conductance mechanosensitive channel in *E. coli* encoded by *mscL* alone. *Nature* 368:265–268.

Tao, T., N. Yasuda, H. Ono, A. Shinmyo, and M. Takano. 1992. Purification and characterization of 2,4-diaminobutyric acid transaminase from *Halomonas* sp. *Ann. Rep. Int. Centre Coop. Res. Biotechnol. Japan* 15:187–199.

Tschichholz. I. and H.G. Trüper. 1990. Fate of compatible solutes during dilution stress in *Ectothiorhodospira halochloris*. *FEMS Microbiol. Ecol.* 73:181–186.

Valderrama, M.J., B. Prado, A. del Moral, R. Rios, A. Ramos-Cormenzana, and V. Campos. 1991. Numerical taxonomy of moderately halophilic Gram-positive cocci isolated from the Salar de Atacama (Chile). *Microbiologia SEM* 7:35–41.

Vargas, C., R. Fernandez-Castillo, D. Canovas, A. Ventosa, and J.J. Nieto. 1995. Isolation of cryptic plasmids from moderately halophilic eubacteria of the genus *Halomonas*. Characterization of a small plasmid from *H. elongata* and its use for shuttle vector construction. *Mol. Gen. Genet.* 246:411–418.

Ventosa, A. and J.J. Nieto. 1995. Biotechnological applications and potentialities of halophilic microorganisms. *World J. Microbiol. Biotechnol.* 11:85–94.

Vreeland, R.H., C.D. Litchfield, E.L. Martin, and E. Elliot. 1980. *Halomonas elongata*, a new genus and species of extremely salt-tolerant bacteria. *Int. J. Syst. Bacteriol.* 30:485–495.

Wohlfarth, A. 1993. *Neue Naturstoffe aus halophilen heterotrophen Eubakterien.* Ph.D. thesis, University of Bonn.

Wohlfarth, A., J. Severin, and E.A. Galinski. 1990. The spectrum of compatible solutes in heterotrophic halophilic eubacteria of the family *Halomonadaceae. J. Gen. Microbiol.* 136:705–712.

Yassin, A. F., E.A. Galinski, A. Wohlfarth, K.D. Jahnke, K.P. Schaal, and H.G. Trüper. 1993. A new actinomycete species, *Nocardiopsis lucentensis* sp. nov. *Int. J. Syst. Bacteriol.* 43:266–271.

16 New Insights into the Extreme Salt Tolerance of the Unicellular Green Alga *Dunaliella*

Irena Gokhman, Morly Fisher, Uri Pick, and Ada Zamir

CONTENTS

16.1 Introduction 203
16.2 Salt-Induced Plasma Membrane Proteins 204
    16.2.1 General Scheme 204
    16.2.2 TTf (P150), an Internally Triplicated Transferrin-Like Protein 204
    16.2.3 Relationship of TTf to Iron Uptake 206
    16.2.4 Dca, a 60 kda Carbonic Anhydrase 207
    16.2.5 Regulation of Dca Accumulation 207
    16.2.6 TTf and Dca Common Features 209
16.3 Fatty Acid Elongase 210
16.4 Conclusion 211
Acknowledgments 212
References 213

16.1 INTRODUCTION

The genus *Dunaliella* includes unicellular green algae whose natural habitats range from marine ecosystems to hypersaline water bodies. Several *Dunaliella* species are able to proliferate over nearly the entire range of salt concentrations while maintaining a relatively low internal salinity. The algal cells achieve osmotic balance by intracellular accumulation of iso-osmotic levels of glycerol, a solute compatible with protein/enzyme functions (Avron, 1986). Glycerol-mediated osmotic adjustment is necessarily an essential feature of salt tolerance, yet other mechanisms must also be operating to permit the outstanding salt-adaptability of *Dunaliella*. Particular attention has been drawn to mechanisms responsible for ionic homeostasis, and potential components of such mechanisms have been described.

To identify additional mechanisms underlying salt tolerance, we chose an approach that was unbiased except for the assumption that proteins involved in such mechanisms preferentially accumulate at high salinities.

The identification of two salt-induced proteins revealed biochemical mechanisms that operate to alleviate indirect constraints imposed by high salinity. Analyses on the transcript level suggested the involvement of membrane lipid components.

A particularly significant outcome of this study is the bringing to light of a potentially new class of halotolerant proteins that remain functional over nearly the entire range of salinities.

16.2 SALT-INDUCED PLASMA MEMBRANE PROTEINS

16.2.1 GENERAL SCHEME

Total proteins from cells grown continuously in media with salinities ranging from 0.5 to 3.5 M NaCl, or plasma membrane fractions derived from such cells, were resolved by SDS-PAGE and examined for proteins whose accumulation increased with rising salinity. The patterns were compared and candidate proteins were purified to homogeneity, subjected to partial amino acid sequencing and used to elicit specific antibodies. These antibodies were subsequently used to identify, quantitate and localize the cognate proteins (Fisher et al., 1994; Sadka et al., 1991) and as tools in cDNA cloning (Fisher et al., 1996, 1997).

16.2.2 TTF (P150), AN INTERNALLY TRIPLICATED TRANSFERRIN-LIKE PROTEIN

The level of a ~150 kDa protein (p150) rose from barely detectable in cells grown in 0.5 M NaCl to a very pronounced level in cells grown in 3.5 M salt (Sadka et al., 1991). Moreover, in cells transferred from low to high salt, an increase in p150 roughly coincided with resumption of cell division, which occurred 15–20 h following the hyperosmotic shock. The protein was characterized biochemically and by immunoelectron microscopy as a major plasma membrane component.

The cDNA for p150 was cloned from a cDNA expression library with the aid of anti-p150 antibodies (Fisher et al., 1997). The ~4.4 kb cDNA for p150 (as corroborated by the match between predicted and determined amino acid sequences) codes for a 1274 amino acid protein. The predicted amino acid sequence of p150 includes three internally homologous segments of ~350 to 400 amino acids each ($\alpha 1$, $\alpha 2$ and β repeats ordered N to C; $\alpha 1$ is more than 60% identical to $\alpha 2$), separated by relatively short connecting sequences. A unique ~100 amino acids C-terminal sequence contains a 14-amino acid stretch, of which nine are acidic. Database searches (Figure 16.1) revealed that p150 closely resembled transferrins, a family of mostly ~80 kDa proteins so far identified only in various body fluids of animals (Aisen and Listowsky, 1980; Baker et al., 1987). Among the members of this family are serum transferrin, lactoferrin and ovotransferrin. Unlike the internally triplicated p150, all transferrins described so far are internally duplicated, i.e., contain two homologous segments. In the comparison shown in Figure 16.1, repeat $\alpha 1$ was

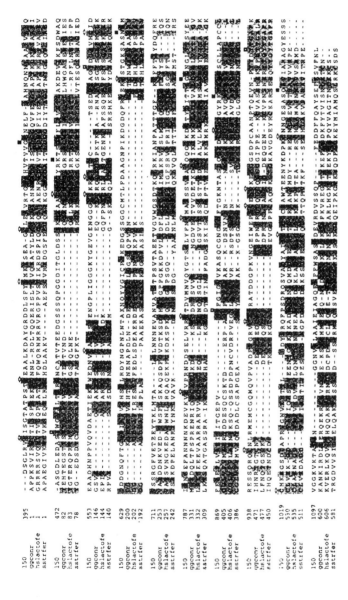

FIGURE 16.1 Alignment of p150 with different transferrins. To enable the alignment of the triplicated p150 with the duplicated animal transferrins, the α1 repeat of p150 was omitted. ggconr, chicken ovotransferrin; hslactofe, human lactoferrin; sstrfer, *Salmo salar* (Atlantic salmon) transferrin. Residue designations: diamonds, conserved Cys; filled boxes, canonical iron or anion liganding residues not conserved in p150; empty boxes, canonical iron and anion liganding residues conserved in p150.

omitted to permit alignment. The functionally best-characterized members of the transferrin family are the serum transferrins, major iron binding proteins that are essential for iron delivery to reticulocytes and other types of cells (Baker and Lindley, 1992). The serum ferri-transferrin is internalized by target cells via receptor-mediated endocytosis, and releases its bound iron after the internal acidification of the vesicles.

While most transferrins are soluble proteins, several membrane-associated transferrins have been described in recent years. The 97 kDa melanotransferrin, p97, first identified on the surface of human melanoma cells, is a typical example (Kennard et al., 1995).

The internally duplicated sequence of animal transferrins is reflected in the three-dimensional structure composed of two similar lobes as determined for human lactoferrin and rabbit serum transferrin (Baker and Lindley, 1992). In these proteins, each lobe bound a single Fe^{2+} cation synergistically with a single HCO_3^- (or CO_3^{2-}) anion. Based on the crystal structure (Baker and Lindley, 1992) the amino acid residues serving as iron ligands are Asp-60, Tyr-92, Tyr-192 and His-253 in the N-lobe, and Asp-395, Tyr-435, Tyr-528 and His-597 in the C-lobe (residue numbers as in human lactoferrin) (Figure 16.1). Conserved and non-conserved replacements of corresponding residues in other transferrins have been observed (Baldwin, 1993). In the three repeats of p150, the Asp is replaced by Gly, the two Tyr residues are conserved, and the His is replaced by Asn in $\alpha 1$ and $\alpha 2$ and by Gln in β. Residues involved in HCO_3^- (CO_3^{2-}) binding in the N- and C-lobes of animal transferrins are Arg-121 and Arg-465, respectively. In all three repeats of p150 these Arg residues are replaced by Lys which, interestingly, is invariably preceded by an Arg residue. Of the Cys residues in animal transferrins, mainly involved in disulfide bond formation, most are conserved in p150 (Baker and Lindley, 1992).

Altogether, the algal protein retains many of the structural features characteristic of animal transferrins. Based on its structural characteristics as well as additional evidence described below p150 was designated as TTf (triplicated transferrin). Northern blot hybridization analyses indicated that the salt-induced accumulation of p150 was transcriptionally regulated (Fisher et al., 1997).

16.2.3 Relationship of TTf to Iron Uptake

The similarity of TTf to animal iron carrier proteins suggested this algal protein is involved in an iron uptake mechanism. Accordingly, we concluded that the induction of p150 in high salinity might compensate for a decline in the effective iron availability under such conditions. Such limitation could result, for example, from the effect of salt on Fe^{3+} solubility or the interference by salt in the iron uptake machinery. Support for a role of TTf in iron uptake was provided by the finding that TTf accumulation is induced even in relatively low salinities if the growth medium is depleted of iron. Furthermore, in cells grown with $^{59}Fe^{3+}$, p150 becomes radioactively labeled, demonstrating the ability of this protein to bind iron *in vivo* (Fisher et al., 1997).

The mode of action of TTf might resemble that of the membrane-anchored melanotransferrin that has been shown recently to act in iron uptake via a mechanism other than receptor-mediated endocytosis of Fe-transferrin (Jefferies et al., 1996; Kennard et al., 1995). The details of this mechanism are still largely unknown, except for the distinction between an initial iron binding step and a subsequent energy-dependent iron internalization.

Similar to melanotransferrin, TTf in intact cells is pronase-sensitive, implying that the algal transferrin-like protein is exposed to the extracellular milieu. In further studies, an assay for iron uptake into whole *D. salina* cells was developed and used to demonstrate that (1) specific Fe^{3+} uptake was saturable and dependent on HCO_3^-/CO_3^{2-}, (2) The uptake consists of an energy-independent binding step followed by an energy-dependent internalization; (3) there is a general correspondence between Fe^{3+} uptake and the level of TTf.

The availability of iron is a major limiting factor in phytoplankton photosynthesis and hence biomass production. Yet, little is known regarding the detailed mechanisms of iron acquisition by algae. The utilization of animal-type transferrins for iron acquisition may not be unique for *Dunaliella* as suggested by the presence of immunologically cross-reacting proteins in other marine macro- and micro-algae (data not shown).

16.2.4 Dca, A 60 kDa Carbonic Anhydrase

Another protein accumulating proportionately to rising salinity is of 60 kDa (Fisher et al., 1994). Immunoelectron microscopy localized the protein to the cell surface. Following the general protocol outlined above, a full-length cDNA for p60 was cloned (Fisher et al., 1996). The cDNA encoded a 589 amino acid protein (Figure 16.2). Comparison with the N-terminal sequence determined for the purified protein indicated that the nascent polypeptide was processed by removal of the N-terminal 54 amino acid residues. In view of the plasma membrane localization of p60, the N-terminal processing is presumably related to the membrane targeting or membrane insertion of the protein.

The protein sequence is composed of an internal duplication. Each of the repeats shows clear-cut homology to animal carbonic anhydrases, as well as two nearly identical periplasmic carbonic anhydrases from the freshwater alga *Chlamydomonas reinhardtii* (Figure 16.2). Among the conserved amino acids are the Zn-liganding His residues, as well as the residues forming the hydrogen-bond network to Zn-bound solvent molecules. Whole *D. salina* cells exhibited carbonic anhydrase activity that was proportional to the level of p60. Further, carbonic anhydrase activity was demonstrated with purified p60. Based on this evidence, p60 was named Dca (Duplicated carbonic anhydrase) (Fisher et al., 1996).

16.2.5 Regulation of Dca Accumulation

The accumulation of Dca in high salinities can be understood as a response compensating for the limitation of CO_2 availability under these conditions (Sass and Ben-Yaakov, 1977). Since CO_2 is thought to be the major carbon source that enters the algal cells, its availability is a critical factor in photosynthesis and — since *Dunaliella* is an obligatory phototroph — also limits its proliferation. The possibility

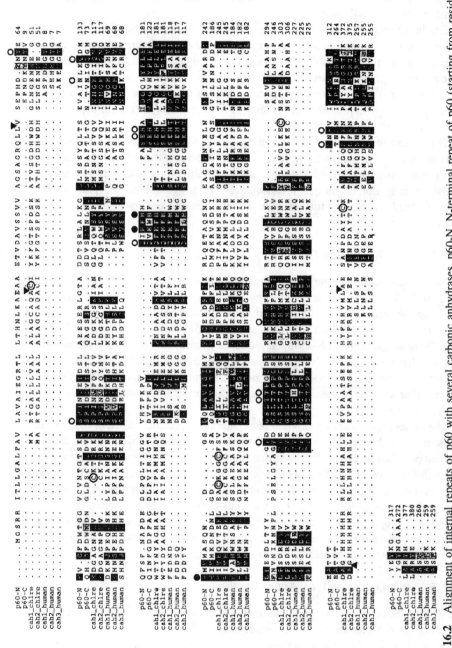

FIGURE 16.2 Alignment of internal repeats of p60 with several carbonic anhydrases. p60-N, N-terminal repeat of p60 (starting from residue 318); cah chlre, *Chlamydomonas reinhardtii* carbonic anhydrases I, II, III. Arrowheads, sites of proteolytic cleavage in p60-N, cah1 and cah2 of *C. reinhardtii*. Filled circles, Zn-liganding His residues; open circles, residues forming H-bond network to Zn-bound solvent molecules; circled residues, cysteines involved in disulfide bond formation.

that the induction of Dca accumulation was a response to CO_2 limitation, rather than to high salinity per se, was examined under conditions limiting CO_2 availability in relatively low salinity. Removal of bicarbonate from the medium, as well as raising the pH from 7.0 to 9.0, resulted in Dca accumulation (Fisher et al., 1996). In both high salt and under CO_2 limitation, Dca accumulated in parallel to its mRNA and hence regulation operates on the transcriptional level (Fisher et al., 1996).

16.2.6 TTF AND DCA COMMON FEATURES

Dca and TTf are structurally and functionally distinct proteins, yet they share several important characteristics. Hydropathy plots indicate that both proteins are largely hydrophilic throughout their entire length, and contain no sequence obviously matching a potential membrane-spanning domain. Furthermore, the proteins can be detached from the membrane and solubilized under conditions too mild for the solubilization of membrane-integral proteins. It is likely that both proteins are anchored via a post-translationally added modification.

A general similarity in the gross structure is the presence of internal repeats: in comparison with animal carbonic anhydrases and transferrins, each of the *Dunaliella* counterparts contains one extra repeat. The functional significance of this structural feature is not clear.

TTf and Dca function on the surface of the algal cells and are extracellularly exposed. Although the level of both proteins increases with external salinity, they are present on the cells even in relatively low salt, i.e., 0.5 M NaCl. It is therefore necessary that these proteins retain activity in a very wide range of salinities. The broad salt tolerance of Dca was demonstrated in assays comparing the activity of Dca and the surface carbonic anhydrase from *Chlamydomonas reinhardtii* (Fisher et al., 1996). In these assays, Dca activity did not change more than twofold in NaCl concentrations ranging from 0.2 to 2.2 M. In other experiments, evidence was obtained that TTf retained activity at 3.5 M NaCl (unpublished results).

It is intriguing to compare Dca and TTf with enzymes from extreme halophilic Archaea. These organisms achieve osmotic balance by intracellular accumulation of potassium salts, hence their enzymes invariably function in very high salinities. Such enzymes require high salt concentrations for structural integrity and activity (Eisenberg, 1992) and exhibit unique structural features (Dym et al., 1994; Frolow et al., 1996). Because *Dunaliella* retains a low intracellular ion concentration, internal *Dunaliella* proteins do not generally encounter high salinities. However, *Dunaliella* proteins that are extracellularly exposed need to remain active throughout the entire salinity range where the algae proliferate. Thus, one may consider Dca and TTf as representatives of a new class of halotolerant proteins.

The structural basis for halotolerance is still obscure. Yet, an outstanding distinguishing feature common to both proteins is the excess of acidic over basic amino acids. In a survey including 16 different transferrins (Figure 16.3) and more than 40 carbonic anhydrases of both eukaryotic and prokaryotic origins (Figure 16.4), the algal counterparts showed the highest ratio of acidic to basic residues without a drastic variation in the total contents of charged residues. The preponderance of acidic residues is a hallmark of proteins from extreme halophiles. The role of the

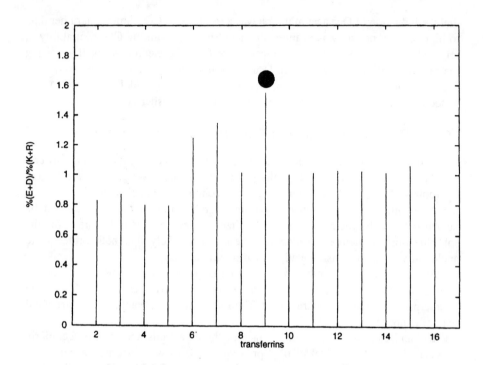

FIGURE 16.3 Ratio of acidic (E+D) over basic (R+K) amino acid residues in transferrin sequences included in GenBank. The line corresponding to the TTf is indicated by a filled circle.

negatively charged surface in creating a protective hydration shell around halophilic malate dehydrogenase and ferredoxin is discussed in Chapter 18. The outstanding high ratio of acidic to basic residues in Dca and TTf suggests these proteins may have preserved some features of proteins from extreme halophiles. Yet, the algal proteins differ in their tolerance for, rather than dependence on, salt.

16.3 FATTY ACID ELONGASE

Functions induced under high salinities were searched by looking for cDNAs corresponding to mRNAs accumulating preferentially at high salinities. A cDNA clone isolated in such a search specified a protein showing close similarity to recently described higher plant β-ketoacyl synthases acting in the synthesis of very long-chain fatty acids (VLCFAs). VLCFAs are widespread in plants and are found in cuticular waxes as well as seed oils. In higher plants, the first step in the 2-carbon addition cycle in VLCFA formation is the condensation of malonyl-CoA with a long-chain acyl-CoA to yield CO_2 and a β-ketoacyl-CoA. β-ketoacyl synthases involved in VLCFA in jojoba (Lassner et al., 1996) and *Arabidopsis* (James et al., 1995) have been recently characterized and cloned. The protein encoded by the cDNA cloned from salt-induced *D. salina* shows similarity to these enzymes along most of its length. A stretch of ~50 amino acids residues in the algal enzyme shows similarity

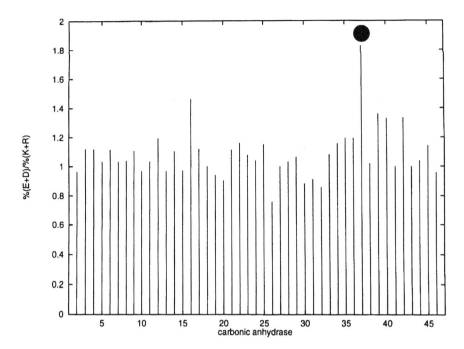

FIGURE 16.4 Ratio of acidic (E+D) over basic (R+K) amino acid residues in different carbonic anhydrase sequences included in GenBank. The line corresponding to Dca is indicated by a filled circle.

to segments in chalcone synthase enzymes from several plants as well as trihydroxystilbene synthase. Other enzymes condensing malonyl-CoA with various other acyl-CoAs were also shown to include a similar 50 amino acid motif (James et al., 1995).

The regulation of the β-ketoacyl synthase gene (*kcs*) homolog from *D. salina* was monitored by northern blot hybridization of mRNA extracted from cells at different times after transfer from 0.5 to 3.5 M NaCl. The results (Figure 16.5) show that (1) the hyperosmotic shock halts cell division for several hours, and (2) the mRNA level starts to increase and reaches a peak shortly before cell division is resumed. The transcript level then gradually declines. Possibly, despite this decline, the salt-adapted cells still harbor an elevated level of enzyme that could result in an increase in VLCFA contents in the plasma membrane or other membraneous compartments in the cell. By affecting permeability, these changes might facilitate salt adaptation. Another possibility is that the transient induction reflects cell cycle control of VLCFA biosynthesis, which becomes evident due to possible synchronization of cell division in the cells recovering from the hyperosmotic shock.

16.4 CONCLUSION

The studies reported in this communication draw attention to a rather overlooked aspect of salt stress, i.e., the reduced nutrient availability under high salinities. The most prominent plasma membrane proteins in cells grown in high salt appear to be

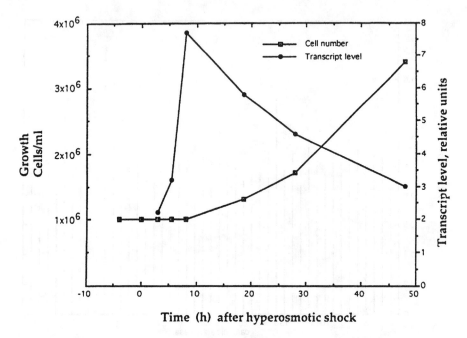

FIGURE 16.5 Transcript analysis of the gene encoding an analog of β-ketoacyl synthase acting in fatty acid elongation in higher plants. Total RNA isolated from *D. salina* at different times after transfer of the cells from 0.5 M to 3.5 M NaCl was analyzed by northern blot hybridization using as probe the entire cDNA clone described in the text. Transcript levels were determined by densitometry and expressed relative to a standard RNA. Cell counts were taken for each sample analyzed.

required for optimal uptake of CO_2 and iron. The unique functional and structural characteristics of these proteins lead us to propose the existence of halotolerant proteins that retain activity throughout practically the entire range of salinities.

The activation of a gene encoding an enzyme likely to function in the biosynthesis of very long-chain fatty acids raises interesting possibilities regarding salt-adaptive alterations in membrane lipid components.

ACKNOWLEDGMENTS

The authors are indebted to all the lab members who contributed to the studies reported, and to Dr. Michal Harel for compiling the data shown in Figures 16.3 and 16.4. These studies were partly funded by a grant to A. Zamir from the Minerva Foundation, Munich, Germany and the Minerva Willstatter Center for Research in Photosynthesis, Rehovot. The support of the Soref Foundation to A. Zamir is gratefully acknowledged.

REFERENCES

Aisen, P. and I. Listowsky. 1980. Iron transport and storage proteins. *Ann. Rev. Biochem.* 49:357–393.
Avron, M. 1986. The osmotic components of halotolerant algae. *Trends Biochem.* 11:5–6.
Baker, E.N. and P.F. Lindley. 1992. New perspectives on the structure and function of transferrins. *J. Inorg. Chem.* 47:147–160.
Baker, E.N., S.V. Rumball, and B.F. Anderson. 1987. Transferrins: insights into structure and function from studies on lactoferrin. *Trends Biochem. Sci.* 12:350-353.
Baldwin, G.S. 1993. Comparison of transferrin sequences from different species. *Comp. Biochem. Physiol.* 106B:203–218.
Dym, O., M. Mevarech, and J.L. Sussman. 1994. Structural features that stabilize halophilic malate dehydrogenase from an Archaebacterium. *Science* 267:1344–1346.
Eisenberg, H. 1992. Biochemical, structural and molecular genetic aspects of halophilism. *Adv. Prot. Chem.* 43:1–62.
Fisher, M., U. Pick, and A. Zamir. 1994. A salt-induced 60-kilodalton plasma membrane protein plays a potential role in the extreme halotolerance of the alga *Dunaliella*. *Plant Physiol.* 106:1359–1365.
Fisher, M., I. Gokhman, U. Pick, and A. Zamir. 1996. A salt-resistant plasma-membrane carbonic anhydrase is induced by salt in *Dunaliella salina*. *J. Biol. Chem.* 271:17718–17723.
Fisher, M., I. Gokhman, U. Pick, and A. Zamir. 1997. A structurally novel transferrin-like protein accumulates in the plasma membrane of the unicellular green alga *Dunaliella salina* grown in high salinities. *J. Biol. Chem.* 272:1565–1570.
Frolow, F., M. Harel, J.L. Sussman, M. Mevarech, and M. Shoham. 1996. Insights into protein adaptation to a saturated salt environment from the crystal structure of halophilic 2Fe-2S ferredoxin. *Nature Struct. Biol.* 3:452–458.
James, D.W., E. Lim, J. Keller, I. Plooy, E. Ralston, and H.K. Dooner. 1995. Directed tagging of the *Arabidopsis* fatty acid elongation (FAE1) gene with the maize transposon activator. *The Plant Cell* 7:309–319.
Jefferies, W.A., R. Gabathuler, S. Rothenberger, M. Food, and M.L. Kennard. 1996. Pumping iron in the '90s. *Trends Cell Biol.* 6:223–228.
Kennard, M.L., D.R. Richardson, R. Gabathuler, P. Ponka, and W.A. Jefferies. 1995. A novel iron uptake mechanism mediated by GPI-anchored human p97. *EMBO J.* 14:4178–4186.
Lassner, M.W., L.K. Mardizabal, and M.G. Metz. 1996. A jojoba β-ketoacyl-CoA synthase cDNA complements the canola fatty acid elongation mutation in transgenic plants. *The Plant Cell* 8:281–292.
Sadka, A., S. Himmelhoch, and A. Zamir. 1991. A 150 kDa cell surface protein is induced by salt in the halotolerant green alga *Dunaliella salina*. *Plant Physiol.* 95:822–831.
Sass, E. and S. Ben-Yaakov. 1977. The carbonate system in hypersaline solutions: Dead Sea brines. *Mar. Chem.* 5:183–199.

Section V

Biochemistry and Molecular Biology

17 What is Halophilic and What is Archaeal?

Donn J. Kushner

CONTENTS

17.1 Introduction 217
17.2 A Historical Note 217
17.3 Halophilic and Other Archaea 219
17.4 Journey to the Interior of Halophiles 222
Acknowledgments 223
References 223

17.1 INTRODUCTION

This topic, like others on halophiles, covers a very wide range: geology, hypersaline oceanography, microbial ecology, physiology, molecular biology and evolution. A broad look at what might still be considered an out of the way subject shows us how microorganisms that live in really strange and usually restrictive environments can introduce us to so many different aspects of our world and life in it. The late George Wald once stated that an intensive study of one special aspect of biology, in his case the biochemistry of vision, opened out to embrace the whole of the universe. This can be said of the halophiles as well, the more so as some of them also have a visual pigment.

17.2 A HISTORICAL NOTE

Halophilic bacteria, both bacterial (eubacterial) and archaeal (archaebacterial)* have been around for a long time, as have stories about them. Recent work (Norton et al., 1993; Vreeland and Powers, Chapter 5) has shown the presence of halobacteria in salt deposits hundreds of millions of years old and raised the fascinating possibility that these bacteria have been there all the time. I once speculated (Kushner, 1991) that Lot's wife, in a spot not too far from this conference, was the spiritual or physical ancestor of red halophiles. The Cities of the Plain have given their names to at least

---

* There is still much to be said for the original terms "archaebacteria" and "eubacteria," because "bacterium" was originally a structural description, derived from the Greek word for "rod." Readers should be reminded that both Archaea and Bacteria are prokaryotic cells.

two species of halophilic Archaea. The wise Chinese, possibly before the time of Sodom and Gomorrah, described the appearance of a red color in their salterns as an indicator of the proper time to draw off evaporating seawater into a smaller pond (Baas Becking, 1931). It is usually considered that the red color was caused by halophilic Archaea. The possibility that a red bloom of *Dunaliella* was responsible has not been seriously suggested, but perhaps should not be dismissed out of hand, since such a red, or reddish-orange bloom can sometimes appear before the deeper red or reddish purple of halobacteria. Of course, the red halophiles, whose detailed taxonomy is now understood at a much deeper level than the old designations of *Halobacterium* and *Halococcus*, have a very special place in the study of halophiles and in the affections of those who study them. I have long been interested in extreme environments, and have reviewed different aspects of the subject at intervals (Kushner, 1964, 1966, 1968, 1971, 1978, 1980, 1985, 1986, 1991, 1993a, 1993b). My own work with halobacteria began in 1961 in the laboratory of Dr. Norman Gibbons at the National Research Council of Canada in Ottawa.

Canadian interest in these bacteria started as an aspect of food microbiology, and had a somewhat Canadian "flavor," being related to the red rotting of salted fish. A classic paper was that of Harrison and Kennedy (1922) at Macdonald College on Montreal Island, an institute that some years later provided a scientific home for Dr. R.A. MacLeod, who made so many fundamental contributions to studies of the physiology of marine bacteria, especially to their requirement for Na^+ for growth and transport (MacLeod, 1980).

Work on marine bacteria has dealt mainly with the need for Na^+ rather than the ability of marine bacteria to tolerate high salt concentrations. The latter problem seems rather outside the experience of most marine bacteria, but in fact, a large proportion can tolerate high salt concentrations, in some cases up to 30% NaCl (Forsyth et al., 1971).

Such observations show that the rather clear-cut scheme of classification, based on slight, moderate and extreme halophiles (Kushner, 1978, 1993b), which I and others have used, must be taken flexibly. Marine bacteria may indeed be "slight halophiles," but they can also be rather salt-tolerant ones.

One sees a tendency in past work to not define the lower salt range of growth very exactly. This lower range is subject to genetic and environmental variation, whose determinants are not well understood.

One example of this is a change in the salt requirement of *Actinopolyspora halophila* (Gochnauer et al., 1975). The strain first isolated could not grow at less than 12% NaCl; however, an erythromycin-resistant mutant could grow in 6% NaCl (Johnson et al., 1986). This curious finding remains unexplained.

Another change in the lower level of NaCl requirement was found some years ago in *Marinococcus halophilus*. This bacterium (at the time known as *Planococcus halophilus*) behaved like a moderate halophile at temperatures of 25°C and higher, requiring at least 0.5 M NaCl for growth; NaCl could not be replaced by KCl or sucrose. However, at 20°C, cells grew (in a complex medium that must have contained some Na^+) without further NaCl addition; at this temperature, the bacteria would have been classified as somewhat salt-tolerant, but not halophilic (Novitsky and Kushner, 1975, 1976). What could have caused this change in salt requirement?

Do the bacteria have a temperature-sensitive transport system that specifically requires NaCl for protection?

Temperature-sensitive mutants of *Escherichia coli* can be protected by osmolytes. The so-called "osmophilic yeasts" have long been known to tolerate, but not require, high salt concentrations. However, some species require NaCl at 37°C but not at 30°C, so they are "tolerant" at one temperature and "philic" at another (reviewed by Kushner, 1978).

A recent study (Cánovas et al., 1996) of osmoprotectants in a halophilic strain of the normally salt-tolerant *Halomonas elongata* showed that glycine betaine, in addition to stimulating growth at higher NaCl concentrations, did the same at lower concentrations, effectively permitting the cells to be less salt dependent. Presumably, glycine betaine stimulates growth at high salt concentrations by acting as a compatible solute, but it is less clear why it permits cells to do without salt.

In general, more-complex media extend both the upper and the lower salt range for growth. This has been shown for *Salinivibrio costicola* (then known as *Vibrio costicola*) (Forsyth and Kushner, 1970), *Marinococcus* (formerly *Planococcus*) *halophilus* (Novitsky and Kushner, 1975) and more recently for *Haloferax volcanii*, in all cases, the more complex media enable cells to grow in lower salt concentrations, though usually the precise minima have not been established. *Hf. volcanii* can grow down to 0.7 M NaCl in a complex medium (compared with 1.5 M in a minimal medium). Its minimal NaCl requirement for growth can be lowered to 0.35 M by the presence of 50% D_2O, which increases hydrophobic interactions (Kauri and Kushner, 1990; V. Randhawa and D. J. Kushner, unpublished results). At one time, it was said that the advantage of working with halophiles is that nothing else grows in media of high salt concentrations. This rather sloppy attitude has generated some interesting contaminants. *Actinopolyspora halophila* was first found as a contaminant of saline growth media (Gochnauer et al., 1975). *Marinococcus* (*Planococcus*) *halophilus* was a contaminant that overgrew a culture from India that at one time had been a halophilic Archaeon isolated from salted hides (Novitsky and Kushner, 1975, 1976). The much studied salt-tolerant contaminant from the National Research Council of Canada, NRCC 21227, has recently been named, together with Ba_1, isolated from the Dead Sea, as *Halomonas canadensis* and *Halomonas israelensis*, respectively (Huval et al., 1995).

17.3 HALOPHILIC AND OTHER ARCHAEA

The most-studied halophiles are the archaeal ones, and vice versa. There are a number of reasons for this, including the very interesting properties of the halophilic Archaea, and the fact that they are often rather easier to study than thermophilic and methanogenic Archaea. High salt concentrations can cause problems. These have been largely overcome in protein purification. They still present difficulties if one wishes to use electrophoretic mobilities to study DNA/protein interactions in high salt environments (Forterre and Elie, 1993), and this limited our further exploration of an apparent example of repression of histidase formation (Kauri et al., 1993).

Still, halophilic Archaea are favored creatures for study, and we must ask which of their properties relate to their halophilic nature, which to their archaeal nature

TABLE 17.1
Properties and Structures of Halophiles and Archaea

Halophilic Archaea	Other Archaea	Halophilic Bacteria	Relevant for Halophilic Life?
Non-peptidoglycan walls	Yes	No	No
Ether-linked lipids	Yes	No	No
Bacteriorhodopsin	No	No	Maybe
Halorhodopsin	No	No	Maybe
Swimming behavior	?[a]	?	No
Gas vesicles	?	Yes	Maybe
Diphthamide in Ed-Tu	Yes	No	No
Salt-dependent enzymes	(No)[b]	Some	Yes
Salt-dependent phages	Yes	(No)	Yes
Special ribosomal proteins	(No)	No	Yes
Antibiotic profiles	Yes	No	No
Multi-subunit RNA polymerases	Yes	No	No
Multi-subunit enzymes	Yes	Yes	No

[a] ? = Limited knowledge.
[b] (No) = Some exceptions to the general rule.

and which to neither. Some of the properties of halophilic Bacteria and Archaea, and other Archaea are summarized in Table 17.1. The following can be taken as speculations and questions that I think should be asked, rather than as definitive answers. Which of these properties are closely related physiologically to halophilic life, especially life in high salt concentrations? Which properties, though very fascinating in themselves, may be unrelated to such life, or may be due to the archaeal nature of this group of halophiles, or may be just a specific property of a group of bacteria which, like other bacteria, can have curious attributes without apparent reason?

Many of the most interesting, and most studied, properties of halophilic Archaea seem more related to their archaeal than to their halophilic nature. The peculiarities of the Archaea are very well known, and are summarized in the book edited by Kates et al. (1993), especially in the chapters by Amils et al., Danson, Forterre and Elie, Kandler and König, Kates, Lanyi, Oesterhelt and Marwan, Skulachev, and Zillig et al. (as well as in chapters in this volume). Though there is a tendency to identify many of these properties with the halophilic bacteria that possess them, it is uncertain that most of them are essential or even especially suited for halophilic life.

Though very early workers considered that the evolutionary precursors of the extreme halophiles were more normal bacteria that had acquired salt tolerance, all the molecular evidence points to a close relationship among other Archaea (some of which might be quite halophilic themselves [Kushner, 1993a]). The wide variety of halophilic Bacteria and their relation to non-halophilic Bacteria suggests that evolutionary changes in salt responses are fairly easily gained over much of the microbial world. As has been seen, physiological and perhaps genetic changes affecting salt requirement occur fairly readily.

The lack of a peptidoglycan cell wall does not seem to be a special adaptation to life in high salt concentrations; all Archaea lack such walls. Though the rod-shaped and pleomorphic members of the *Halobacteriaceae* are very soft-bodied and often have glycoprotein outer layers, members of the genus *Halococcus* have very strong walls indeed.

Ether lipids, characteristic of all the Archaea, may well offer advantages to those genera growing in hot acid conditions, but probably do not do so to cells growing in high salt concentrations. Extremely salt-tolerant Bacteria, which can grow in just as high salt concentrations, have mainly ester-linked lipids.

What about bacteriorhodopsin and halorhodopsin, some of the most intensively studied and fascinating of the systems of these creatures? These proton and Cl⁻ pumps (which have other properties as well) are found only in halophilic Archaea, though apparently not in all genera and species. They don't need high salt for activity, since both can be prepared by dialysis of cells against distilled water, or detergent treatment of cells under low salt conditions. They function when incorporated into artificial vesicles, usually in low salt concentrations that would not permit the growth or survival of their parent bacteria (Duschl et al., 1988; Lanyi, 1993; Skulachev, 1993).

It is still easier to pose than to answer questions about the relevance of these pigments for halophilic life. As is well known, bacteriorhodopsin is produced under conditions of limited oxygen, such as would occur in highly saline water. It is produced under natural conditions (reviewed in Kushner, 1985), though there may still be some question as to the overall competitive advantage possessed by those halobacteria that produced bacteriorhodopsin over those that do not.

One of the interesting characteristics of halophilic Archaea is that they have internal chloride concentrations as high as those outside. This is rather rare for other aerobic microorganisms, in which chloride ions can be quite toxic, though the few studies of anaerobic halophiles suggest that some at least have high internal Cl⁻ ions, which are not toxic to enzyme systems (Kamekura and Kushner, 1984; Kushner and Kamekura, 1988). If the halorhodopsins are really involved in maintaining high internal Cl⁻ concentrations in halophilic Archaea, a correlation should be found between the distribution of halorhodopsin systems and such high internal concentrations. We must also ask how halophilic Archaea can grow in the dark, or does such growth lead to lower internal Cl⁻ concentrations?

What about the other special aspects of halophilic Archaea, their phototaxis, their curious swimming behavior, their gas vesicles, the sensitivity of elongation factor to diphtheria toxin? Gas vesicles could well offer an advantage to halobacteria growing in oxygen-limited conditions, but such vesicles are found in many cyanobacteria and other Bacteria. The vesicles of halobacteria, moreover, are stable in the absence of salt, which certainly makes them easier to study, but possibly less "halophilic." The rather dignified way of changing direction, involving back and forth movement rather than the alternate runs and twiddles of enterobacteria, has no obvious connection with growth in high salt concentrations. That Ef-Tu of halobacteria is sensitive to subunit A of diphtheria toxin is a curiosity that they share with other Archaea but is of no obvious relevance for life in salt — or any other aspects of archaeal life.

The multi-subunit enzymes, such as citrate synthase, are found in a number of Bacteria; again, no special pattern seems to be involved in the halophilic Archaea. Multi-subunit RNA polymerases are characteristic of all Archaea, and do not seem to have a special halophilic aspect (except presumably for salt dependence, characteristic of most enzymes of the halobacteria). Archaea are insensitive to the β-lactamases for obvious reasons; the lack of sensitivity to antibiotics that affect protein synthesis in Bacteria seems to be a characteristic of the Archaea shared by the halobacteria. It is true that high salt concentrations can lower sensitivity of Bacteria to certain antibiotics (Coronado et al., 1995; Goel et al., 1996), but there is nothing to indicate that the insensitivity of halobacteria is due to their salt environment, rather than to their archaeal nature. The ribosomal proteins are, of course, special in the halobacteria, as are so many other proteins. The need of high salt concentrations, especially KCl, is characteristic of *in vitro* protein synthesis in the halobacteria, but not in other Archaea. It is curious, however, that at least one of the ribosomal proteins of methanogens can be incorporated into the ribosomes of halobacteria in reconstitution experiments, which suggests similarities between them, shown by other work on archaeal ribosomes (Amils et al., 1993; Köpke et al., 1990).

The bacteriophages of halobacteria are indeed distinctive, being almost instantly inactivated in the absence of high salt concentrations. Those of halophilic Bacteria are relatively stable for long periods in the absence of salt, though their multiplication is very dependent on the salt environment of their bacterial hosts (Calvo et al., 1988; Goel et al., 1996; Kauri et al., 1991).

17.4 JOURNEY TO THE INTERIOR OF HALOPHILES

Our own work with phages aimed to search for an RNA phage as a source of mRNA for studies of *in vitro* protein synthesis. So far, most such studies in halophilic bacteria have used the artificial messenger poly-uridylic acid (Amils et al. 1993; Kushner and Kamekura, 1988). We were able to use the RNA of the R-17 phage of *Escherichia coli* in the *in vitro* system of *Salinivibrio* (formerly *Vibrio*) *costicola*, studying the sensitivity of this system to Cl^- ions (Choquet and Kushner, 1990). However, thus far, the only phages of halophilic Bacteria (as of halobacteria) have been DNA phages. Another interest in phages of halophilic Bacteria has been their potential use as indicators of internal ionic conditions inside the bacteria. It was thought that *in vitro* phage assembly systems might give accurate clues to the true internal ionic conditions inside such bacteria — a very big question (Kushner, 1988) that is far from being solved. Very briefly, though it seems accepted that internal solutes are at least osmotically equivalent to external ones for microorganisms growing in high salt media, the sum of the known internal solutes usually doesn't add up to the external ones. More than that, in order to grow, cells must have positive turgor pressure, presumably osmotic, which would involve higher inside than outside concentrations. Evidence for such turgor pressure appears from the results of penicillin on *S. costicola*, when even cells growing in high NaCl swell greatly after exposure to penicillin (Peerbaye and Kushner, 1993).

Would ionic conditions needed for phage assembly give a true indicator of internal ionic conditions? Not necessarily. It is sobering to realize that the conditions for best reassembly of ribosomes of halobacteria, involving high $(NH_4)_2SO_4$ concentrations, are certainly not those that actually exist within the cells (Amils et al, 1993; Sanchez et al., 1996). The real internal environments of halophilic Bacteria and Archaea remain fascinating mysteries whose solutions should tempt us still further.

ACKNOWLEDGMENTS

The work reported here from my own laboratory was supported over the years by the National Research Council and the National Scientific and Engineering Research Council of Canada. I am indebted to many past students, post doctoral fellows and colleagues, a number of whom were present at this meeting.

REFERENCES

Amils, R., P. Cammarano, and P. Londei. 1993. Translation in Archaea, in *The Biochemistry of Archaea (Archaebacteria)*, M. Kates, D.J. Kushner, and A.T. Matheson, Eds. New Comprehensive Biochemistry, Vol. 26. Elsevier, Amsterdam. 393–438.

Baas-Becking, L.G.M. 1931. Historical notes on salt and salt manufacture. *Sci. Month.* 32:434–446.

Calvo, M., A.G. de la Paz, V. Bejar, E. Quesada, and A. Ramos-Cormenzana. 1988. Isolation and characterization of phage F9-11 from a lysogenic *Deleya halophila* strain. *Curr. Microbiol.* 17:49–53.

Cánovas, D., C. Vargas, L. Csonka, A. Ventosa, and J.J. Nieto. 1996. Osmoprotectants in *Halomonas elongata*: high-affinity betaine transport system and choline-betaine pathway. *J. Bacteriol.* 178:7221–7226.

Choquet, C. and D.J. Kushner. 1990. Use of natural messenger RNAs in cell-free protein synthesizing systems of the moderate halophile *Vibrio costicola*. *J. Bacteriol.* 172:3462–3468.

Coronado, M.-J., C. Vargas, H.J. Kunte, E.A. Galinski, A. Ventosa, and J.J. Nieto. 1995. Influence of salt concentration on the susceptibility of moderately halophilic bacteria to antimicrobials and its potential use for genetic transfer studies. *Curr. Microbiol.* 31:365–371.

Danson, M.J. 1993. Central metabolism of the Archaea, in *The Biochemistry of Archaea (Archaebacteria)*, M. Kates, D.J. Kushner, and A.T. Matheson, Eds. New Comprehensive Biochemistry, Vol. 26. Elsevier, Amsterdam. 1–24.

Duschl, A., M.A. McCloskey, and J.K. Lanyi. 1988. Functional reconstitution of halorhodopsin: properties of halorhodopsin-containing proteoliposomes. *J. Biol. Chem.* 263:17016–17022.

Forsyth, M.P. and D.J. Kushner. 1970. Nutrition and distribution of salt response in populations of moderately halophilic bacteria. *Can. J. Microbiol.* 16:253–261.

Forsyth, M.P., D.B. Shindler, M.B. Gochnauer, and D.J. Kushner. 1971. Salt tolerance of intertidal marine bacteria. *Can. J. Microbiol.* 17:825–828.

Forterre, P. and C. Elie. 1993. Chromosome structure, DNA topoisomerases, and DNA polymerases in archaebacteria (Archaea), in *The Biochemistry of Archaea (Archaebacteria)*, M. Kates, D.J. Kushner, and A.T. Matheson, Eds. New Comprehensive Biochemistry, Vol. 26. Elsevier, Amsterdam. 325-366.

Gochnauer, M.B., G.G. Leppard, P. Komaratat, M. Kates, T. Novitsky, and D.J. Kushner. 1975. Isolation and characterization of *Actinopolyspora halophila*, gen. et sp. nov., an extremely halophilic actinomycete. *Can. J. Microbiol.* 21:1500-1511.

Goel, U., T. Kauri, H. Ackermann, and D.J. Kushner. 1996. A moderately halophilic vibrio from a Spanish saltern, and its lytic bacteriophage. *Can. J. Microbiol.* 42:1015-1023.

Harrison, F.C. and M.E. Kennedy. 1922. The red discolouration of cured codfish. *Trans. R. Soc. Canada* Section V:101-110.

Huval, J.H., R. Latta, R. Wallace, D.J. Kushner, and R.H. Vreeland. 1995. Description of two new species of *Halomonas*: *Halomonas israelensis* sp. nov. and *Halomonas canadensis* sp. nov. *Can. J. Microbiol.* 41:1124-1131.

Johnson, K.G., P.H. Lanthier, and M.B. Gochnauer. 1986. Studies of two strains of *Actinopolyspora halophila*, an extremely halophilic actinomycete. *Arch. Microbiol.* 143:370-378.

Kamekura, M. and D.J. Kushner. 1984. Effect of chloride and glutamate ions on *in vitro* protein synthesis by the moderate halophile, *Vibrio costicola*. *J. Bacteriol.* 160:385-390.

Kandler, O. and H. König. 1993. Cell envelopes of Archaea: Structure and chemistry, in *The Biochemistry of Archaea (Archaebacteria)*, M. Kates, D.J. Kushner, and A.T. Matheson, Eds. New Comprehensive Biochemistry, Vol. 26. Elsevier, Amsterdam. 223-260.

Kates, M. 1993. Membrane lipids of Archaea, in *The Biochemistry of Archaea (Archaebacteria)*, M. Kates, D.J. Kushner, and A.T. Matheson, Eds. New Comprehensive Biochemistry, Vol. 26. Elsevier, Amsterdam. 223-260.

Kates, M., D.J. Kushner, and A.T. Matheson, Eds. 1993. *The Biochemistry of Archaea (Archaebacteria)*. New Comprehensive Biochemistry, Vol. 26. Elsevier, Amsterdam.

Kauri, T. and D.J. Kushner. 1990. Nutrition of the halophilic archaebacterium *Haloferax volcanii*. *Syst. Appl. Microbiol.* 13:14-18.

Kauri, T., H.-W. Ackermann, U. Goel, and D.J. Kushner. 1991. A bacteriophage of a moderately halophilic bacterium. *Arch. Microbiol.* 156:435-438.

Kauri, T., I. Ahonkhai, M. Desjardins, and D.J. Kushner. 1993. Histidase activity of the halophilic archaebacterium, *Haloferax volcanii*. *Microbios* 76:245-249.

Köpke, A.K.E., C. Paulke, and H.S. Gewitz. 1990. Overexpression of the methanococcal ribosomal protein L12 in *Escherichia coli* and its incorporation into halobacterial 50S subunits yielding active ribosomes. *J. Biol. Chem.* 265:6436-6440.

Kushner, D.J. 1964. Microbial resistance to harsh and destructive environmental conditions. *Exp. Chemother.* 2:113-168.

Kushner, D.J. 1966. Mass culture of red halophilic bacteria. *Biotechnol. Bioengin.* 8:237-245.

Kushner, D.J. 1968. Halophilic Bacteria. *Adv. Appl. Microbiol.* 10:73-99.

Kushner, D.J. 1971. Influence of solutes and ions on microorganisms, in *The Inhibition and Destruction of the Microbial Cell*, W.B. Hugo, Ed. Academic, London. 259-283.

Kushner, D.J., Ed. 1978. Life in high salt and solute concentrations: halophilic bacteria, in *Microbial Life in Extreme Environments*. Academic, London. 318-368.

Kushner, D.J. 1980. Extreme environments, in D.C. Ellwood, J.N. Hedger, M.J. Latham, J.M. Lynch, and J.H. Slater, Eds., *Contemporary Microbial Ecology*. Academic, London. 29-54.

Kushner, D.J. 1985. The *Halobacteriaceae*, in *The Bacteria, Vol. 8. Archaebacteria*. Academic, C.R. Woese and R.S. Wolfe, Eds. New York. 171-214.

Kushner, D.J. 1986. Molecular adaptation of enzymes, metabolic systems and transport systems in halophilic bacteria. *FEMS Microbiol. Rev.* 39:121–127.

Kushner, D.J. 1988. What is the "true" internal environment of halophilic bacteria? *Can. J. Microbiol.* 34:482–486.

Kushner, D.J. 1991. Halophiles of all kinds: what are they up to now and where did they come from? in *General and Applied Aspects of Halophilic Microorganisms,* F. Rodriguez-Valera, Ed. Plenum, New York. 63–71.

Kushner, D.J. 1993a. Microbial life in extreme environments, in *Aquatic Microbiology: an Ecological Approach,* T.E. Ford, Ed. Blackwell Scientific, Cambridge, MA. 383–407.

Kushner, D.J. 1993b. Growth and nutrition of halophilic bacteria, in *The Biology of Halophilic Bacteria,* R.H. Vreeland and L. I. Hochstein, Eds. CRC, Boca Raton. 87–103.

Kushner, D.J. and M. Kamekura. 1988. Physiology of halophilic eubacteria, in *Halophilic Bacteria,* F. Rodríguez-Valera, Ed., Vol. 1. CRC, Boca Raton. 109–138.

Lanyi, J.K. 1993. Ion transport rhodopsins (bacteriorhodopsin and halorhodopsin): structure and function, in *The Biochemistry of Archaea (Archaebacteria),* M. Kates, D.J. Kushner, and A.T. Matheson, Eds. New Comprehensive Biochemistry, Vol. 26. Elsevier, Amsterdam. 189–208.

MacLeod, R.A. 1980. Observations on the role of inorganic ions in the physiology of marine bacteria, in *Saline Environments. Proceedings of the Japanese Conference on Halophilic Microbiology,* H. Morishita and M. Masui, Eds. Nakanishi Printing Co., Kyoto. 5–29.

Norton, C.F., T.J. McGenity, and W.D. Grant. 1993. Archaeal halophiles (halobacteria) from two British salt mines. *J. Gen. Microbiol.* 139:1077–1061.

Novitsky, T.J. and D.J. Kushner. 1975, influence of temperature and salt concentration on the growth of a facultatively halophilic "*Micrococcus*" species. *Can. J. Microbiol.* 21:107–110.

Novitsky, T.J. and D.J. Kushner. 1976. *Planococcus halophilus* sp. nov.; a facultatively halophilic coccus, *Int. J. Syst. Bacteriol.* 26:53–57.

Oesterhelt, D. and W. Marwan, W. 1993. Signal transduction in halobacteria, in *The Biochemistry of Archaea (Archaebacteria),* M. Kates, D.J. Kushner, and A.T. Matheson, Eds. New Comprehensive Biochemistry, Vol. 26. Elsevier, Amsterdam. 173–188.

Peerbaye, Y. and D.J. Kushner. 1993. Effects of penicillin on a moderately halophilic bacterium, *Vibrio costicola. Curr. Microbiol.* 26:229–232.

Sanchez, M.E., P. Londei, and R. Amils. 1996. Total reconstitution of active small ribosomal subunits of the extreme halophilic archaeon, *Haloferax mediterranei. Biochim. Biophys. Acta* 1292:140–144.

Skulachev, V.P. 1993. Bioenergetics of extreme halophiles, in *The Biochemistry of Archaea (Archaebacteria),* M. Kates, D.J. Kushner, and A.T. Matheson, Eds. New Comprehensive Biochemistry, Vol. 26. Elsevier, Amsterdam. 25–40.

Zillig, W., P. Palm, H.-P. Klenk, D. Langer, U. Hüdepohl, J. Hain, M. Lanzendörfer, and I. Holz. 1993. Transcription in Archaea, in *The Biochemistry of Archaea (Archaebacteria),* M. Kates, D.J. Kushner, and A.T. Matheson, Eds. New Comprehensive Biochemistry, Vol. 26. Elsevier, Amsterdam. 367–392.

18 Molecular Interactions in Extreme Halophiles — The Solvation-Stabilization Hypothesis for Halophilic Proteins

Christine Ebel, Pierre Faou, Bruno Franzetti, Blandine Kernel, Dominique Madern, Mihaela Pascu, Claude Pfister, Stéphane Richard, and Giuseppe Zaccai

CONTENTS

18.1 Introduction 227
    18.1.1 Solvent Interactions in Protein Stabilization...................................227
    18.1.2 Salt Effects on Protein Stability 228
18.2 Malate Dehydrogenase from *Haloarcula marismortui*................................229
    18.2.1 Thermal Inactivation and Solvent Interactions:
          the Complementary Approaches 229
    18.2.2 The Solvation-Stabilization Model..231
    18.2.3 Mutants 233
    18.2.4 Crystallographic Studies..233
18.3 Rnase E from *H. marismortui* as a Model for the Study of
    Nucleic Acid–Protein Interactions..235
18.4 Summary 235
References..236

18.1 INTRODUCTION

18.1.1 SOLVENT INTERACTIONS IN PROTEIN STABILIZATION

In a hypothetical world in which folded proteins are stable in vacuum, the thermodynamics of unfolding is very simply described in terms of a two-state model. Unfolding leads to an increase in enthalpy ($DH_{unf} > 0$) because of the breaking of

H-bonds and other bonds that maintained the structure. The entropy also increases upon unfolding ($DS_{unf} > 0$) because of the much greater number of configurations that can be adopted by the unfolded chain compared with the tertiary structure. There is no reason that these values should depend on temperature. The free energy of unfolding as a function of temperature is expected to decrease linearly and to pass through zero at a melting temperature T_m. Careful calorimetric measurements, however, have shown that this is not the case. The experimental free energy of unfolding of a protein follows a bell-shaped curve as a function of temperature. It passes through zero at low and at high temperatures, with a maximum value in between (Makhatdze and Privalov, 1995). The reason is that a protein is never in a vacuum (Figure 18.1). In solution, there are protein-solvent interactions that occur in both the folded and unfolded structures. Bonds broken by unfolding are compensated by other bonds formed between the unfolded chain and solvent components. The increased entropy of the unfolded chain can be compensated by a decrease in configurational freedom of water molecules around apolar groups. Furthermore, because of solvent interactions, DS_{unf} and DH_{unf} are temperature-dependent. Protein-solvent interactions therefore have a crucial contribution to the energetics of protein folding.

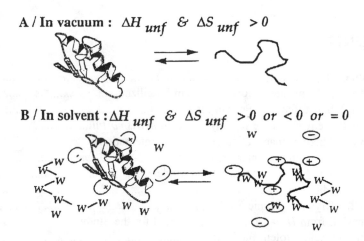

FIGURE 18.1 The thermodynamics of unfolding. A folded protein, symbolized by α-helices, β-sheets and loops unfolds into a polypeptide extended chain: (A), in vacuum, (B), in a solvent symbolized by water molecules (w) and ions (+ or –). Water molecules and ions can be in the bulk or bound to the protein of the unfolded polypeptide chain.

18.1.2 Salt Effects on Protein Stability

Salts are known to affect protein stability, and salt anion and cation effects have been found to be additive in the multimolar concentration range (Von Hippel and Schleich, 1969). Ions such as sulfate and phosphate decrease protein solubility and increase protein stability (salting-out), while ions such as bromide, magnesium or lithium increase protein solubility and reduce protein stability (salting-in). Arakawa et al. (1990a, 1990b) have shown that proteins in solution in the presence of salting-out salts are generally preferentially hydrated; the folded structure is favored, as is precipitation, to minimize the surface accessible to the solvent. On the other hand, salting-in salts are preferentially bound to the polypeptide chain, favoring unfolding.

18.2 MALATE DEHYDROGENASE FROM *HALOARCULA MARISMORTUI*

Malate dehydrogenase from *Haloarcula marismortui* (halophilic MalDH) was first purified by Mevarech et al. (1977) and characterized as a halophilic enzyme (Mevarech and Neumann, 1977). It has become the paradigm of a halophilic protein, following extensive biophysical and biochemical experiments (Bonneté et al., 1993; Eisenberg et al., 1992). This tetrameric enzyme can now be obtained in large amounts, since the protein and its mutants can easily be purified following overexpression in *Escherichia coli* (Cendrin et al., 1993; Madern et al., 1995). Its crystal structure has been solved to 3.2 Å resolution by Dym et al. (1995).

18.2.1 THERMAL INACTIVATION AND SOLVENT INTERACTIONS: THE COMPLEMENTARY APPROACHES

All macromolecules in solution associate with solvent components — ions and water molecules in the case of salt solutions — that participate in the thermodynamics of stabilization of the macromolecular tertiary and quaternary structure. These interactions play a particularly interesting role in the case of halophilic proteins because of the extreme solvent environment. To understand this role, a complementary approach has been developed, correlating protein stability with solvent interactions under similar conditions.

The stability of halophilic MalDH as a function of various parameters — salt type and concentration, temperature, pH, various mutations — was investigated by measuring its intrinsic fluorescence, circular dichroism, and activity under standard conditions (Bonneté et al., 1994; Zaccai et al., 1989). It deactivates with first-order kinetics. Figure 18.2A shows selected Arrhenius plot profiles for the inactivation rate constants k of halophilic MalDH in ammonium sulfate and NaCl (data derived from Bonneté et al., 1994). From these data, the inactivation thermodynamic parameters $\Delta G^{\#0}$, $\Delta H^{\#0}$, and $\Delta S^{\#0}$ are calculated from the expressions:

$$d(\ln k)/d(1/T) = -(\Delta H^{\#0} + RT)/R$$

$$\Delta G^{\#0} = -RT\ln(hk/k_BT)$$

$$\Delta G^{\#0} = \Delta H^{\#0} - T\Delta S^{\#0}$$

in which R is the gas constant, h is Planck's constant, and k_B Bolzmann's constant. Straight lines of negative slope (e.g., as in the NaCl conditions) indicate that unfolding is enthalpy-dominated. Positive or zero slopes indicate entropy terms are dominant

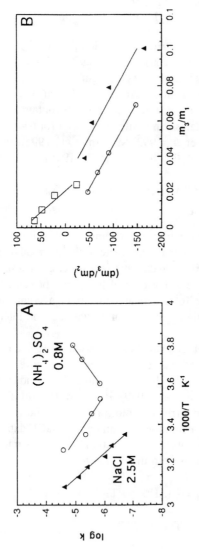

FIGURE 18.2 Characterization of the stability and solvation of halophilic MalDH. (A) Arrhenius plots for the kinetic constant (k) of denaturation and derived activation thermodynamic parameters (see text). The data points for the two examples given (2.5 M NaCl — closed triangles, 0.8 M $(NH_4)_2SO_4$ — open circles) are taken from Bonneté et al., 1994. Solid lines correspond to linear fits on data points. (B) Determination of the solvation of halophilic MalDH. The preferential interaction parameters $(\partial m_3/\partial m_2)_\mu$ are plotted as a function of the molar ratio between salt and water in the solvent m_3/m_1. The data were obtained from density and neutron-scattering experiments at 4°C, pH 8.2 in NaCl (closed triangles), ammonium sulfate (open circles) and magnesium chloride (open squares) (experimental details to be published). Lines correspond to linear fits on data points. The hydration (slope) is two times higher in magnesium chloride as compared with the two other salts. Salt binding is observed in the case of sodium chloride and magnesium chloride (positive intercept); salt exclusion is observed in the case of ammonium sulfate (negative intercept).

TABLE 18.1
Thermodynamic Activation Parameters for Halophilic Malate Dehydrogenase in Different Salt Solutions

Parameter (kJ mol^{-1})[a]	2.5 M NaCl, 25°C	0.8 M (NH$_4$)$_2$SO$_4$	
		−5°C	25°C
$\Delta G^{\#}$	112	92	102
$\Delta H^{\#}$	132	−99	87
$\Delta S^{\#}$	20	−191	−15

[a] Values were calculated from data presented in Figure 18.2A.

(e.g., the data in ammonium sulfate). The experimental data corresponding to the chosen examples are given in Table 18.1.

Solvent-macromolecule interactions are very difficult to measure, and new procedures have been developed by using complementary approaches on halophilic MalDH: high-precision densimetry, analytical centrifugation, and neutron scattering (Ebel, 1995; Eisenberg, 1976, 1981; Zaccai et al., 1986), which allow the protein to be weighed, as well as the associated solvent components, water and salt. Figure 18.2B gives selected examples obtained with halophilic MalDH (Ebel et al., to be published). From the experiments, density increments or neutron scattering-length density increments are calculated, from which values for the thermodynamical preferential interaction parameter $(\partial m_3/\partial m_2)\mu$ can be extracted. This has units of number of moles of salt (component 3 in the usual nomenclature) per mol of protein (component 2), considered in the present treatment as the tetramer plus its 156 counterions. Component 1 is water. If $(\partial m_3/\partial m_2)\mu$ is measured in various salt concentrations, and if the plot of $(\partial m_3/\partial m_2)\mu$ as a function of the molar ratio m_3/m_1 between salt and water in the solvent is a straight line, the solvation of the protein can be considered as invariant. Its values can be determined from the value of the slope and intercept of this line, the value of the slope yielding bound water, and the value of the intercept yielding bound salt. A positive value for the intercept (e.g., for sodium chloride or magnesium chloride in Figure 18.2B) corresponds to salt binding, a negative value for the intercept (e.g., for ammonium sulfate in Figure 18.2B) corresponds to salt exclusion. It must be kept in mind that in the present description, salt binding or exclusion are relative to the electroneutral protein (component 2, as defined above).

18.2.2 The Solvation-Stabilization Model

It is clear from Figure 18.2 that the thermal inactivation and solvation data for halophilic MalDH in concentrated sodium chloride and ammonium sulfate are profoundly different. Depending on the nature of the salt, stability is dominated by different energy terms. In a model described by Zaccai et al. (1989) the stabilization of halophilic proteins has been related to protein–solvent interactions.

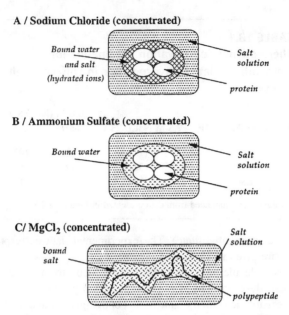

FIGURE 18.3 Salt ion effects (at molar concentrations) on halophilic MalDH stability, solubility, and solvation. (A) At high concentrations of NaCl (above ca. 2 M) the stabilization occurs via hydrated ion binding to the folded tetramer. The same situation holds for high concentrations of KCl, or in $MgCl_2$ concentrations of 0.5–1.3 M. (B) At high concentrations of $(NH_4)_2SO_4$ (above ca. 1 M) the classical salting-out model is valid. (C) At high concentrations of $MgCl_2$ (above 2 M) the protein unfolds, and the classical salting-in model may be valid.

The dominant mechanisms depend on the nature of the salts in solution (Figure 18.3). There are two main mechanisms, apparent in sodium chloride and ammonium sulfate, respectively.

Inactivation kinetics in all NaCl conditions show a positive enthalpy term $\Delta H^{\#}$ that dominates the free activation energy upon denaturation (Bonneté et al., 1993, see also Figure 18.2A). The solvation shell surrounding the protein is composed of water and salt, with a high concentration of sodium chloride (Bonneté et al., 1993, see also Figure 18.2B). A quite high salt content in the solvation shell was also found for two other halophilic proteins, elongation factor Tu and glyceraldehyde-3-phosphate dehydrogenase (Ebel et al., 1992, 1995). This is in contrast to nonhalophilic proteins that do not bind salt (or about 10 times less). In the hypothesis presented by Zaccai et al. (1989), the stabilization of the folded structure involves a network of hydrated ions associated with the acidic residues localized in patches at the surface of the halophilic protein (Figure 18.3A).

In ammonium sulfate, the enthalpy term $\Delta H^{\#}$ changes sign between 10 and 20°C, but the free energy of activation for the inactivation process always corresponds to a negative activation entropy $T\Delta S^{\#}$ (Bonneté et al., 1993, see also Figure 18.2A). This suggests that in this salt, hydrophobic interactions are important for protein stability. Moreover, the solvation shell surrounding the protein can be described as composed essentially of water (see Figure 18.2B). This behavior (Figure 18.3B) is similar to that observed for nonhalophilic proteins.

Some salts destabilize halophilic MalDH at high concentration. One example is $MgCl_2$ (Madern and Zaccai, 1997; Zaccai et al., 1989). We measured the solvation of the folded form, which contains a high amount of both salt and water (Figure 18.2B). We have found that the relative capability of the ions to unfold halophilic MalDH at high salt is that expected from the Hofmeister series (Ebel et al., to be published). We have also observed that the ability of the cations to denature the protein at high salt is related to their capacity to maintain the folded halophilic MalDH structure when decreasing the salt concentration. Keeping in mind that salting-in ions interact with the polypeptide chain (Figure 18.3C), and that an increase of their concentration leads to protein unfolding, our observations support the importance of solvation by ions for the stabilization of the folded structure.

18.2.3 Mutants

A better understanding of the mode of structural adaptation of halophilic enzymes to function at high salt concentrations can be obtained by combining biochemical characterization of these enzymes with structural analysis and molecular genetic methodologies. The amino acid compositions of halophilic proteins show a significant increase in acidic nature, which is supposed to play a role in the binding of the solvation network. The consequences for structure–stability relationships of changes from acidic to non-acidic residues were probed through a site-directed mutagenesis study by using two mutants E243R and E101K of halophilic MalDH. They were found to require a higher NaCl concentration or a lower temperature for equivalent stability. We estimated the solvation of both mutant enzymes in NaCl solutions by small angle neutron-scattering experiments. The results were not significantly different from those obtained with the wild-type enzyme, suggesting that changes in stability do not result from a massive change in water or salt binding, but more likely are due to subtle changes in the organization of the network (Madern et al., 1995).

Comparative structural analysis of halophilic MalDH with its nonhalophilic homologues revealed the existence of a significant number of complex salt bridges in the halophilic enzyme. It has been suggested that some of these are the only interactions in the inter-dimer-dimer interface that serve to stabilize the tetramer (Dym et al., 1995). To test this hypothesis, arginine 205 and arginine 291, which are involved in complex salt bridges located at the monomer 1–monomer 2 interface, were mutated into serines. The sedimentation velocity experiments (Figure 18.4) suggest that the resulting species at a concentration of about 1 mg ml^{-1} is a tetramer, which dissociates into low molecular weight species at lower protein concentration. Other parameters are currently under investigation (Madern et al., to be published).

18.2.4 Crystallographic Studies

The three-dimensional structure of halophilic MalDH was solved by Dym et al. (1995) in the presence of cofactor (NADH). So far, however, the 3.2 Å resolution

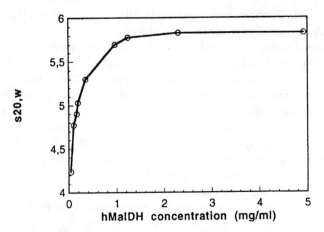

FIGURE 18.4 The stabilization of the tetrameric structure of halophilic MalDH by salt bridges: a sedimentation velocity study. A mutant has been expressed in which two arginines located at the monomers interface have been mutated into serines. The figure presents the corrected sedimentation coefficient $s_{20,w}$ measured at ca. 250,000 g in 4 M NaCl at 20°C as a function of protein concentration. At high concentrations, the value of the sedimentation coefficient approaches that of the wild-type tetramer. The drop in sedimentation coefficient indicates the dissociation of the tetramer.

of the structure is insufficient to allow the observation of the interactions between the protein and the water or the ions involved in its stabilization. We have developed a rational approach for the crystallization of halophilic proteins, using the MPD-NaCl-H_2O phase diagram (Richard et al., 1995), and extended it to the MPD-KCl-H_2O system. Currently, we are engaged in the study of solvent effects in halophilic MalDH, combining site-directed mutagenesis with high-resolution crystallography, using synchrotron radiation under cryo-conditions (T = 100°K). The crystals of halophilic MalDH in the presence of 4 M NaCl or KCl are bi-pyramidal, from 300 nm to 1 mm in their longest dimension; the space group is C2221 from 21°C to 6°C, with a phase transition to C2 in cryo-conditions — the flash-cooling procedure always produces a shift of the β angle from 90° to near approximately 92°. The structure with NADH is now under refinement at 2.2 Å resolution in cryo-conditions. The unambiguous assignment of electron density to water or ions in these structures is very difficult, but essential to validate the solvation–stabilization model involving cooperative solvent–protein interactions. The two halophilic MalDH mutants for which acidic amino acids at the surface have been exchanged have been successfully crystallized, providing good crystallographic data. The structure of the E243R mutant has just been solved in absence of NADH at 2.6 Å resolution (Rf = 19% and Rfree = 27% at T = 6°C). Since we are interested here in the solvent structure, neither non-crystallographic symmetry nor solvent-flattening techniques were applied, and solvent interpretation is in progress.

18.3 RNASE E FROM *H. MARISMORTUI* AS A MODEL FOR THE STUDY OF NUCLEIC ACID–PROTEIN INTERACTIONS

The molecular and biophysical basis of RNA recognition is still poorly understood because of the complexity of RNA folding in solution and the instability of the RNA-protein complexes. In this context, halophilic proteins that bind nucleic acids might represent attractive systems to understand protein–nucleic acid interactions; how can a protein bind RNA in an environment where ribonucleoprotein complexes normally dissociate? In these extreme salt concentrations, medium-range non-specific electrostatic interactions are virtually impossible. Structural studies of these complexes might therefore facilitate the elucidation of the specificity determinants.

RNase E is an enzyme that appears to control the rate-limiting step that mediates the degradation of many mRNA species in bacteria. Our data on halophilic RNase E activity indicate that this enzyme can specifically recognize and cleave mRNA from *E. coli* in an extreme salt environment (3 M KCl) (Franzetti et al., 1997). Having recently been discovered in mammalian cells (Wennborg et al., 1995), RNase E-like activity is now identified in all three evolutionary domains: Archaea, Bacteria, and Eukarya. This strongly suggests that mRNA decay mechanisms are highly conserved, despite enormously different environmental conditions. RNase E activity, followed by *in vitro* RNA processing assays, shows that — contrary to the *E. coli* enzyme — the halophilic enzyme requires a high salt concentration for stability and mRNA cleavage specificity. The alteration of the processing pattern observed at low salt could be due to a partial denaturation of the enzyme, leading to an accumulation of intermediate processing products or to an alteration of the enzymatic cleavage specificity. Since such complex processing patterns are also observed with highly purified RNase E from *E. coli*, it is more likely that halophilic RNase E exhibits a better cleavage specificity when working in extreme salt conditions. In this respect, halophilic RNase E would represent a good system to study RNA recognition processes. Further studies on the halophilic RNase E processing enzyme are therefore under way to address the biophysical mechanisms of RNA recognition.

18.4 SUMMARY

The biochemical machinery of extreme halophiles is adapted to conditions that are usually very unfavorable for protein function in general and for stability, solubility and protein–nucleic acid interactions, in particular. Protein–solvent interactions, which play essential roles in all these processes, are the focus of studies on a number of halophilic proteins. An approach was developed using a variety of complementary experimental methods and applied to malate dehydrogenase from *H. marismortui* and its mutants. Biochemical and spectroscopic characterizations of function and stability in various salt conditions are correlated with neutron scattering, analytical centrifugation and densimetry measurements of solution structure and interactions. Results are summarized in a solvation-stabilization model. Similarly to non-halophilic proteins, dominant stabilization mechanisms in the halophilic protein depend on the nature of the solvent salt and its concentration. However, the halophilic protein in salt environments close to physiological conditions has adapted to bind hydrated ions cooperatively via a network of acidic groups on its surface. This keeps it stable and soluble in conditions where nonhalophilic proteins would precipitate. The solvation–stabilization model is being tested and refined by high-resolution X-ray crystallography on the protein and surface residue mutants. How proteins and nucleic acids interact in high salt remains a mystery and, in a search for suitable candidates to study transient protein–nucleic acid interactions in extreme halophiles, an RNase E activity was discovered in Archaea, showing this enzyme to occur in all three domains of life.

REFERENCES

Arakawa, T., R. Bhat, and S.N. Timasheff. 1990a. Preferential interactions determine protein solubility in three-component solutions: the $MgCl_2$ system. *Biochemistry* 29:1914–1923.

Arakawa, T., R. Bhat, and S.N. Timasheff. 1990b. Why preferential hydration does not always stabilize the native structure of globular proteins. *Biochemistry* 29:1924–1931.

Bonneté F., C. Ebel, G. Zaccai, and H. Eisenberg. 1993. Biophysical study of halophilic malate dehydrogenase in solution: revised subunit structure and solvent interactions of native and recombinant enzyme. *J. Chem. Soc. Faraday Trans.* 89:2659–2666.

Bonneté, F., D. Madern, and G. Zaccai. 1994. Stability against denaturation mechanisms in halophilic malate dehydrogenase "adapt" to solvent conditions. *J. Mol. Biol.* 244:436–447.

Cendrin, F., J. Chroboczek, G. Zaccai, H. Eisenberg, and M. Mevarech. 1993. Cloning, sequencing and expression in *Escherichia coli* of the gene coding for malate dehydrogenase of the extremely halophilic archaebacterium *Haloarcula marismortui*. *Biochemistry* 32:4308–4313.

Dym, O., M. Mevarech, and J.L. Sussman. 1995. Structural features stabilizing halophilic malate dehydrogenase from an archaebacterium. *Science* 267:1344–1346.

Ebel, C. 1995. Characterisation of the solution structure of halophilic proteins. Analytical centrifugation among complementary techniques (light, neutron and X-ray scattering, density measurements). *Progr. Colloid Polym. Sci.* 99:17–23.

Ebel, C., F. Guinet, J. Langowski, C. Urbanke, J. Gagnon, and G. Zaccai. 1992. Solution studies of the elongation factor Tu from the extreme halophile *Halobacterium marismortui*. *J. Mol. Biol.* 223:361–371.

Ebel, C., W. Altekar, J. Langowski, C. Urbanke, E. Forest, and G. Zaccai. 1995. Solution structure of glyceraldehyde-3-phosphate dehydrogenase from *Haloarcula vallismortis*. *Biophys. Chem.* 54:219–227.

Eisenberg, H. 1976. *Biological Macromolecules and Polyelectrolytes in Solution*. Clarendon, Oxford, UK. 272 pp.

Eisenberg, H. 1981. Forward scattering of light, X-rays and neutrons. *Q. Rev. Biophys.* 14:141–172.

Eisenberg, H., M. Mevarech, and G. Zaccai. 1992. Biochemical, structural, and molecular genetic aspects of halophilism. *Adv. Prot. Chem.* 43:1–61.

Franzetti, B., B. Sohlberg, G. Zaccai, and A. von Gabain. 1997. Biochemical and serological evidence for a RNase E-like activity in halophilic Archaea. *J. Bacteriol.* 179:1180–1185.

Madern. D. and G. Zaccai. 1997. Stabilisation of halophilic malate dehydrogenase from *Haloarcula marismortui* by divalent cations. *Eur. J. Biochem.* 249:607–611.
Madern, D., C. Pfister, and G. Zaccai. 1995. Mutation at a single acidic amino acid enhances the halophilic behaviour of malate dehydrogenase from *Haloarcula marismortui* in physiological salts. *Eur. J. Biochem.* 230:1088–1095.
Makhatadze G.I. and P.L. Privalov. 1995. Energetics of protein structure. *Adv. Prot. Chem.* 47:307–425.
Mevarech, M. and E. Neumann. 1977. Malate dehydrogenase isolated from extremely halophilic bacteria of the Dead Sea. 2. Effect of salt on catalytic activity and structure. *Biochemistry* 16:3786–3792.
Mevarech, M., H. Eisenberg, and E. Neumann. 1977. Malate dehydrogenase isolated from extremely halophilic bacteria of the Dead Sea. 1. Purification and molecular characterization. *Biochemistry* 16:3781–3785.
Richard, S., F. Bonneté, O. Dym, and G. Zaccai. 1995. The MPD-NaCl-H_2O system for the crystallization of halophilic proteins, in *Archaea: a Laboratory Manual*, F. T. Robb, A. R. Place, K. S. Sowers, H. J. Schreier, S. DasSarma, and E. M. Fleischmann, Eds. Cold Spring Harbor Laboratory, New York. 149–154.
Von Hippel, P. and T. Schleich. 1969. The effects of neutral salts on the structure and conformational stability of macromolecules in solution, in *Structure of Biological Macromolecules*, S. N. Timasheff and G. D. Fasman, Eds. Marcel Dekker Inc., New York. 417–575.
Wennborg, A., B. Sohlberg, D. Angerer, G. Klein, and A. von Gabain. 1995. A human RNAse E-like activity that cleaves RNA sequences involved in mRNA stability control. *Proc. Natl. Acad. Sci. USA* 92:7322–7326.
Zaccai, G., E. Wachtel, and H. Eisenberg. 1986. Solution structure of halophilic malate dehydrogenase from small-angle neutron and X-ray scattering and ultracentrifugation. *J. Mol. Biol.* 190:97–106.
Zaccai, G., F. Cendrin, Y. Haik, N. Borochov, and H. Eisenberg. 1989. Stabilization of halophilic malate dehydrogenase. *J. Mol. Biol.* 208:491–500.

19 New Insights into the Molecular Enzymology of Pyruvate Metabolism in the Halophilic Archaea

*Michael J. Danson, Keith A. Jolley,
Deborah G. Maddocks,
Michael L. Dyall-Smith, and David W. Hough*

CONTENTS

19.1 Introduction 239
19.2 Dihydrolipoamide Dehydrogenase in the Halophilic Archaea 241
19.3 Homologous Expression of Dihydrolipoamide Dehydrogenase in
 Haloferax volcanii 241
19.4 Dihydrolipoamide Dehydrogenase as a Model for Studying
 Protein Halophilicity 242
  19.4.1 A Molecular Model of the Halophilic Dihydrolipoamide
   Dehydrogenase 242
  19.4.2 Site-Directed Mutagenesis 243
  19.4.3 The Structural Basis of Protein Halophilicity 245
19.5 A PDHC Operon in the Halophilic Archaea? 246
19.6 Conclusion 247
Acknowledgments 247
References 247

19.1 INTRODUCTION

The conversion of pyruvate to acetyl-CoA provides the connection between the catabolism of sugars and pathways such as fat metabolism and the citric acid cycle. In the Eukarya and most aerobic Bacteria, pyruvate is converted to acetyl-CoA via the pyruvate dehydrogenase multienzyme complex (PDHC) (Mattevi et al., 1992; Perham, 1991). This complex is a member of the 2-oxoacid dehydrogenase complex

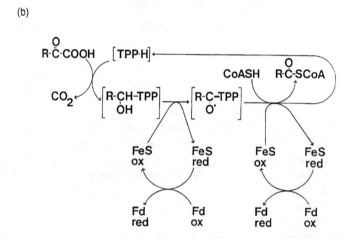

FIGURE 19.1 The enzymatic reaction mechanisms for the conversion of pyruvate to acetyl-CoA. (a) The pyruvate dehydrogenase multienzyme complex of Bacteria and Eukarya. E1 (pyruvate decarboxylase); E2 (dihydrolipoyl acetyltransferase); E3 (dihydrolipoamide dehydrogenase, DHLipDH). (b) The pyruvate: ferredoxin oxidoreductase of the halophilic Archaea. Symbols: B (a histidine base on DHLipDH); R (CH_3^-); Fd (ferredoxin); FeS (an enzyme-bound iron-sulfur cluster); Lip (enzyme-bound lipoic acid); TPP-H (thiamine pyrophosphate).

family, and is a three-component system consisting of multiple copies of enzymes E1 (pyruvate decarboxylase), E2 (lipoate acetyl-transferase) and E3 (dihydrolipoamide dehydrogenase). Central to the catalytic mechanism (Figure 19.1a) is the acyl-carrying cofactor, lipoic acid, which is covalently attached to E2 and serves to connect the three active sites and channel substrate through the complex. In addition to forming the catalytic core of PDHC, the E2 component also composes the structural core. In Gram-negative Bacteria there are 24 E2 polypeptides per PDHC molecule, whereas Gram-positive Bacteria and eukaryal PDHCs have 60 E2 chains in each complex. The E1 and E3 components are noncovalently bound to this E2 core.

In contrast, the Archaea catalyze this reaction via a simpler pyruvate oxidoreductase (Danson, 1993). In the halophilic Archaea, this enzyme has an $\alpha_2\beta_2$ structure (Plaga et al., 1992) and the catalytic mechanism does not involve the participation of a lipoic acid moiety (Figure 19.1b). Instead, the hydroxyethyl-group is transferred directly from TPP to Coenzyme-A and the reducing equivalents are passed via an iron-sulfur center to ferredoxin (Kerscher and Oesterhelt, 1981).

Until now no PDHC activity has been found in the Archaea. The existence of the pyruvate oxidoreductase throughout this domain and the anaerobic Bacteria has therefore led to the suggestion that this may be the ancestral enzyme system for the conversion of pyruvate to acetyl-CoA, and that PDHC emerged after the development of oxidative phosphorylation (Kerscher and Oesterhelt, 1982). However, the data reviewed in this chapter indicate that the situation is not as clear cut as first thought; we have evidence that the genes encoding the PDHC components may be present in the halophilic Archaea, and, from at least one of them, an active component enzyme is produced.

19.2 DIHYDROLIPOAMIDE DEHYDROGENASE IN THE HALOPHILIC ARCHAEA

The only known function of dihydrolipoamide dehydrogenase (DHLipDH) is as the third enzyme component of PDHC and the other members of the 2-oxoacid dehydrogenase family (2-oxoglutarate dehydrogenase and the branched-chain 2-oxoacid dehydrogenases), along with the glycine cleavage system (Danson, 1988, 1993). Given that none of these complex activities have been detected in the Archaea, it was a surprise to discover the presence of DHLipDH and its substrate, lipoic acid, in the halophilic Archaea (Danson et al., 1984; Pratt et al., 1989). As indicated in Figure 19.1A, DHLipDH catalyzes:

$$\text{dihydrolipoamide} + NAD^+ \longleftrightarrow \text{lipoamide} + NADH + H^+$$

but the metabolic significance of this function in the Archaea is still a mystery.

The experimental approaches reviewed in this chapter describe the ways in which we are currently probing the possible functions of DHLipDH from the halophilic Archaea, and how we are using this enzyme to explore the structural basis of protein halophilicity.

19.3 HOMOLOGOUS EXPRESSION OF DIHYDROLIPOAMIDE DEHYDROGENASE IN *HALOFERAX VOLCANII*

The gene encoding the DHLipDH from the halophilic Archaeon *Haloferax volcanii* has been cloned and sequenced (Vettakkorumakankav and Stevenson, 1992), and from sequence alignments it is clearly related to the DHLipDHs from bacterial and eukaryal 2-oxoacid dehydrogenase complexes. We have recently developed a homologous overexpression system for the cloned gene, based on an *Escherichia coli–Hf. volcanii* shuttle vector containing a strong promoter from the rRNA operon of *Halobacterium salinarum* (formerly named *Halobacterium cutirubrum*) (Jolley et al., 1996). The expressed DHLipDH has been purified to homogeneity and shown to retain its dimeric nature and to have catalytic properties comparable to the non-recombinant enzyme. The enzyme can now be produced in the quantities required for structural studies.

A DHLipDH-minus mutant of *Hf. volcanii* has been created by homologous recombination with the subcloned gene after insertion of the mevinolin resistance determinant into the protein-coding region (Jolley et al., 1996). Growth experiments with this strain and the parent cells on minimal media containing a range of amino acids and sugars as sole carbon sources revealed no difference between the two organisms. At this stage, therefore, the metabolic function of DHLipDH and lipoic acid in the halophilic Archaea remains unknown.

19.4 DIHYDROLIPOAMIDE DEHYDROGENASE AS A MODEL FOR STUDYING PROTEIN HALOPHILICITY

The sequence identity of the *Hf. volcanii* DHLipDH to its nonarchaeal counterparts is high (up to 50%), implying very similar 3D-structures. Therefore, we believe that the archaeal enzyme is a good model protein in which to explore the structural basis of protein halophilicity, as detailed comparisons can be made with well-characterized, nonhalophilic DHLipDHs. To date, this study has taken the form of a structurally based investigation using site-directed mutagenesis (Jolley et al., 1997). Our data are summarized below.

19.4.1 A MOLECULAR MODEL OF THE HALOPHILIC DIHYDROLIPOAMIDE DEHYDROGENASE

Using a structurally based sequence alignment, the primary sequence of the *Hf. volcanii* DHLipDH has been homology-modeled to the crystal structure at 2.5 Å resolution of the *Pseudomonas fluorescens* enzyme, with which it shares 43% identity and 65% similarity. Only a single monomer was modeled, with the dimer generated by applying a twofold rotation to the monomer. The dimer was then subjected to 200 cycles of energy minimization.

The overall shapes of the two molecules bear a close resemblance, with structurally identical active sites. However, characteristic of other halophilic proteins (Frolow et al., 1996), the surface of the halophilic DHLipDH has a high degree of negative charge relative to its nonhalophilic homologue. Each monomer has 31 basic and 86 acidic surface residues, resulting in a net charge of –55 if all side chains are present in their conjugate forms. In contrast, the *P. fluorescens* structure has a surface charge of –7 per monomer. Therefore, in the absence of salt, the molecular surface of the halophilic protein is predominantly negatively charged. However, on calculating the surface electrostatic potential in increasing concentrations of KCl, the charge profile of the *Haloferax* enzyme at 1 M KCl is comparable to nonhalophilic proteins in their native environment. It is at approximately 1 M KCl that *Hf. volcanii* reaches its maximum activity.

19.4.2 SITE-DIRECTED MUTAGENESIS

From an analysis of the energetically favorable binding sites on the modeled structure, we have identified a strong K^+-binding site at the dimer interface of the *Hf. volcanii* DHLipDH. The site consists of four coordinated glutamate residues, two from each monomer (E423 and E426) (Figure 19.2), one of which (E426) is conserved as a glutamate or an aspartate residue throughout all known DHLipDH

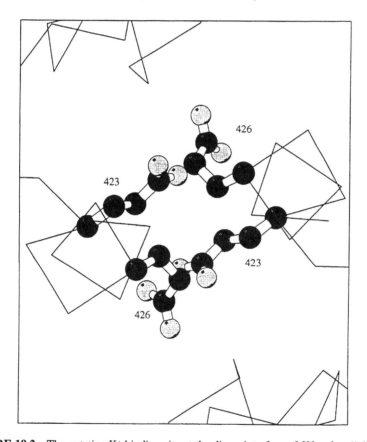

FIGURE 19.2 The putative K^+-binding site at the dimer interface of *Hf. volcanii* dihydrolipoamide dehydrogenase. The site consists of four coordinated glutamate residues, two from each monomer (Glu 423 and Glu 426). The site was identified using the program GRID (Molecular Discovery Ltd.) and is visualized here using MOLSCRIPT (Kraulis, 1991). Carbon atoms are shown as dark spheres and oxygen atoms as light spheres. (Adapted from Jolley et al., 1997, with permission.)

sequences. The other, E423, is unique to the *Hf. volcanii* enzyme and therefore was chosen as a target for site-directed mutagenesis.

Four mutants were created (E423D, E423A, E423S and E423Q), and these and the wild-type DHLipDH were expressed in the DHLipDH-minus strain of *Hf. volcanii* described above. All the recombinant enzymes were purified and characterized with respect to their kinetic properties, salt-dependent activities and thermal stabilities. In summary, the data obtained were as follows:

- The dependence of enzyme activity on the concentration of KCl is shown in Figure 19.3. E423D behaved very similarly to the wild-type DHLipDH, whereas (E423A/S/Q) DHLipDHs are no longer activated by salt. In all cases, activities that were lower in the absence of KCl than in its presence are due to changes in V_{max} values.

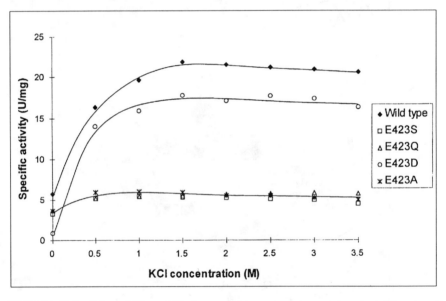

FIGURE 19.3 Salt activity profiles for recombinant wild-type and mutant *Hf. volcanii* dihydrolipoamide dehydrogenases in KCl. The assays were conducted at 30°C in 50 mM potassium phosphate buffer, pH 7.0, with varying concentrations of KCl. All activities refer to equilibrium values. (Reproduced from Jolley et al., 1997, with permission.)

- All the DHLipDHs have a pH optimum of pH 8.5, suggesting no difference in electrostatic potential around the active site.
- The thermal stability properties of the (E423A/S/Q) DHLipDHs were identical to those of the wild-type DHLipDH, although the E423D mutant was inactivated at temperatures around 20°C lower. In the absence of salt, similar relative thermostabilities are observed, but both wild-type and mutant enzymes are inactivated at temperatures around 16–18°C lower than in the presence of salt.

These observations are discussed in detail in Jolley et al. (1997) and are referred to below in the context of the structural basis of protein halophilicity. In essence, they show that the adaptation of catalytic activity to high salt concentrations can depend markedly on the precise spatial arrangement of a small number of charged amino acids. On the other hand, the requirement for high concentrations of salt for conformational stability may depend on the whole surface charge, although spatial considerations will still be important.

19.4.3 The Structural Basis of Protein Halophilicity

As reviewed in Danson and Hough (1997), the surface of contact between protein and water constitutes an interface, and therefore there must be a surface tension (surface energy) at this interface due to the cohesive, hydrogen-bonding nature of the solvent. Consequently, proteins attain a globular shape (minimal surface area to volume ratio) to minimize this surface energy. Solvent additives will perturb the cohesive force of water and hence its surface tension (Timasheff, 1992a, b), which in turn will affect the conformation of the protein.

Most inorganic salts increase the surface tension of water to varying extents that parallel the Hofmeister series; cations and anions tend to act independently, and their effects are additive. The surface area of a denatured protein is greater than that of the folded state, and therefore, this salt-induced increase in the surface tension of water promotes the formation of the compact native protein. As the concentration of the salt is increased, the surface tension of the water increases and the protein aggregates to reduce its solvent-accessible surface area even further, resulting in the salt-precipitation reaction. Therefore, the problem faced by a halophilic protein in extreme salt concentrations is to attract water molecules back to its surface, possibly in the form of hydrated salt ions. At the same time, it might reduce its surface hydrophobicity to counteract the tendency to precipitate at these solute levels.

The structural basis of protein halophilicity (Bonnetté et al., 1993; Danson and Hough, 1997; Eisenberg et al., 1992) is essentially as predicted from the above considerations. As seen in the crystal structures of ferredoxin (Frolow et al., 1996) and malate dehydrogenase (Dym et al., 1995), and the homology-modeled dihydrofolate reductase (Böhm and Jaenicke, 1994) and DHLipDH (Jolley et al., 1997), all from halophilic Archaea, the solvation of the protein in high concentrations of KCl is via hydrated salt ions and is brought about by the high content of surface carboxyl groups of aspartate and glutamate side-chains. Furthermore, the ferredoxin structure shows that the salt is bound in a cooperative manner by the precisely arranged carboxyl groups (Frolow et al., 1996).

The view that the arrangement and interaction of the negatively charged amino acids are as important as the total net charge in determining the adaptation of proteins to high salt concentrations has been noted from mutagenesis studies of malate dehydrogenase from *Haloarcula marismortui* (Madern et al., 1995). Our site-directed mutagenesis experiments with DHLipDH (Jolley et al., 1997) add further weight to these conclusions, and demonstrate the sometimes separate nature of stability to, and activity in, high salt concentrations.

19.5 A PDHC OPERON IN THE HALOPHILIC ARCHAEA?

Our creation of a DHLipDH-minus mutant of *Hf. volcanii* failed to reveal a metabolic function for the enzyme (Jolley et al., 1996). However, homologous expression studies suggest that there is no promoter for the DHLipDH gene in the region upstream of the coding sequence, and expression was only achieved when a rRNA promoter was inserted in the appropriate region (Jolley et al., 1996). Given that the chromosomal gene is normally expressed in wild-type *Hf. volcanii*, there was the distinct possibility that the DHLipDH gene is part of an operon. Therefore, we have sequenced the DNA upstream of the *Hf. volcanii* DHLipDH gene (K.A. Jolley, D.G. Maddocks, M.L. Dyall-Smith, D.W. Hough and M.J. Danson, unpublished observations).

We have thus discovered the presence of three genes upstream of the DHLipDH gene, and preceding these there is a clear halophilic-promoter sequence and a ribosome-binding site. The analysis of these genes is still in progress, but comparison with the sequence data bases suggests that they are most homologous to the genes encoding the E1α, E1β and E2 components of the PDHC from Gram-positive Bacteria. The gene order in the *Hf. volcanii* PDHC-like operon is E1α, E1β, E2, DHLipDH, with -1, 7 and 6 base-pairs between the respective open reading frames.

This is the first report of the genes for a putative 2-oxoacid dehydrogenase complex in the Archaea, and it came as a surprise given that no enzymatic activities for any such complex have been detected to date. A number of questions are raised by our discovery:

- Are the E1 and E2 genes of this operon expressed in *Hf. volcanii*, given that DHLipDH, the fourth gene in the operon, is transcribed and translated into an active enzyme?
- If the genes are expressed, do the components assemble into a multienzyme complex? If not, why not? If they do, why is no PDHC activity detectable in cell extracts?
- By sequence homology, the genes are closest to the PDHC genes of Bacteria; however, could there be an alternative substrate specificity in the extreme halophiles that is not found in non-archaeal organisms?
- If this is a PDHC operon, why does *Hf. volcanii* possess two enzyme systems for the conversion of pyruvate to acetyl-CoA?
- Is this operon present in other halophilic Archaea, given that they do possess an active DHLipDH (Danson et al., 1984)? Are the genes present in the nonhalophilic Archaea?

We are currently engaged in experiments to answer these questions, and the increasing availability of archaeal genome sequences will greatly help in some of the aspects to be explored.

19.6 CONCLUSION

The work reviewed in this chapter had its origins in the surprising discovery that halophilic Archaea contain an active DHLipDH but seemingly no active 2-oxoacid dehydrogenase complex of which it would be expected to be an integral component. While DHLipDH is proving to be a good model protein with which to study the structural basis of protein halophilicity, its function in the Archaea remains a mystery. Our discovery that DHLipDH is the fourth component of a PDHC-like operon has raised as many questions as it has potential answers. In terms of the evolution of metabolism, the origin of this operon in the Archaea, whether it is ancient or has been acquired more recently by lateral gene transfer, is an important consideration. If it is ancient, then it adds to the growing body of evidence that the majority of central metabolic pathways were established before the separation of Archaea, Bacteria and Eukarya (Danson et al., 1998). If its origins in the Archaea lie in a gene-transfer event, even this must be ancient, as DHLipDH has been found throughout the extreme halophiles (Danson et al., 1984) and the enzyme is both functionally and structurally adapted to high salt concentrations.

Clearly, the molecular and metabolic enzymology of the halophilic Archaea still warrant considerable investigation.

ACKNOWLEDGMENTS

We thank the Biotechnology and Biological Sciences Research Council, UK (Research Grant to M.J.D., D.W.H. and G.L. Taylor, and Earmarked Studentships to K.A.J. and D.G.M.) and The Royal Society and The British Council and NATO (travel grants to M.J.D. and D.W.H.) for financial support. We are also indebted to Drs. G.L. Taylor and R.J.M. Russell (University of Bath, UK) for help with the molecular modeling and for many helpful discussions.

REFERENCES

Böhm G. and R. Jaenicke. 1994. A structure-based model for the halophilic adaptation of dihydrofolate reductase from *Halobacterium volcanii*. *Prot. Eng.* 7:213–220.

Bonneté, F., C. Ebel, G. Zaccai, and H. Eisenberg. 1993. Biophysical study of halophilic malate dehydrogenase in solution: revised subunit structure and solvent interactions of native and recombinant enzyme. *J. Chem. Soc. Faraday Trans.* 89:2659–2666.

Danson, M.J. 1988. Dihydrolipoamide dehydrogenase: a "new" function for an old enzyme? *Biochem. Soc. Trans.* 16:87–89.

Danson, M. J. 1993. Central metabolism of the Archaea. *New Compr. Biochem.* 26:1–24.

Danson, M. J. and D. W. Hough. 1997. The structural basis of protein halophilicity. *Comp. Biochem. Physiol.* 117A:307–312.

Danson, M.J., R. Eisenthal, S. Hall, S.R. Kessell, and D.L. Williams. 1984. Dihydrolipoamide dehydrogenase from halophilic archaebacteria. *Biochem. J.* 218:811–818.

Danson, M.J., R.J.M. Russell, D.W. Hough, and G.L. Taylor. 1998. Comparative enzymology as an aid to understanding evolution, in *Thermophiles: The Keys to Molecular Evolution and the Origin Of Life?* M.W.W. Adams and J. Wiegel, Eds. Taylor & Francis, London, in press.

Dym, O., M. Mevarech, and J.L. Sussman. 1995. Structural features that stabilize halophilic malate dehydrogenase from an archaebacterium. *Science* 267:1344–1346.

Eisenberg H., M. Mevarech, and G. Zaccai. 1992. Biochemical, structural and molecular genetic aspects of halophilism. *Adv. Prot. Chem.* 43:1–62.

Frolow, F., M. Hare, J.L. Sussman, M. Mevarech, and M. Shoham. 1996. Insights into protein adaptation to a saturated salt environment from the crystal structure of a halophilic 2Fe-2S ferredoxin. *Nature Struct. Biol.* 3:452–457.

Jolley, K.A., E. Rapaport, D.W. Hough, M.J. Danson, W.G. Woods, and M.L. Dyall-Smith. 1996. Dihydrolipoamide dehydrogenase from the halophilic Archaeon *Haloferax volcanii*–homologous overexpression of the cloned gene. *J. Bacteriol.* 178:3044–3048.

Jolley, K.A., R.J.M. Russell, D.W. Hough, and M.J. Danson. 1997. Site-directed mutagenesis and halophilicity of dihydrolipoamide dehydrogenase from the halophilic Archaeon, *Haloferax volcanii. Eur. J. Biochem.* 248:362–368.

Kerscher, L. and D. Oesterhelt. 1981. The catalytic mechanism of 2-oxoacid: ferredoxin oxidoreductases from *Halobacterium halobium. Eur. J. Biochem.* 116:595–600.

Kerscher, L. and D. Oesterhelt. 1982. Pyruvate: ferredoxin oxidoreductase—new findings on an ancient enzyme. *Trends Biochem. Sci.* 7:371–374.

Kraulis, P. 1991. MOLSCRIPT: a program to produce both detailed and schematic plots of proteins. *J. Appl. Crystallogr.* 24:946–950.

Madern D., C. Pfister, and G. Zaccai. 1995. Mutation at a single amino acid enhances the halophilic behaviour of malate dehydrogenase from *Haloarcula marismortui* in physiological salts. *Eur. J. Biochem.* 230:1088–1095.

Mattevi, A., A. de Kok, and R.N. Perham. 1992. The pyruvate dehydrogenase multienzyme complex. *Curr. Opin. Struct. Biol.* 2:877–887.

Perham, R.N. 1991. Domains, motifs and linkers in 2-oxo acid dehydrogenase multienzyme complexes: a paradigm in the design of a multifunctional protein. *Biochemistry* 30:8501–8512.

Plaga, W., F. Lottspeich, and D. Oesterhelt. 1992. Improved purification, crystallization and primary structure of pyruvate: ferredoxin oxidoreductase from *Halobacterium halobium. Eur. J. Biochem.* 205:391–397.

Pratt, K.J., C. Carles, T.J. Carne, M.J. Danson, and K.J. Stevenson. 1989. Detection of bacterial lipoic acid: a modified gas chromatographic–mass spectrometric procedure. *Biochem. J.* 258:749–754.

Timasheff, S.N. 1992a. Solvent effects on protein stability. *Curr. Opin. Struct. Biol.* 2:35–39.

Timasheff, S.N. 1992b. Stabilization of protein structure by solvent additives, in *Stability of Protein Pharmaceuticals*. Part B, T.J. Ahern and M.C. Manning, Eds. Plenum, New York. 265–285.

Vettakkorumakankav, N.N. and K.J. Stevenson. 1992. Dihydrolipoamide dehydrogenase from *Haloferax volcanii*: gene cloning, complete primary sequence and comparison to other dihydrolipoamide dehydrogenases. *Biochem. Cell Biol.* 70:656–663.

20 Regulation of Gene Expression in *Halobacterium salinarum*: The *arcrACB* Gene Cluster and the TATA Box-Binding Protein

*Jörg Soppa, Petra Vatter, Andreas Ruepp,
Alexandra zur Mühlen, and Thomas A. Link*

CONTENTS

20.1 Introduction 250
    20.1.1 The General Transcription Apparatus of Archaea 250
    20.1.2 Regulation of Gene Expression 250
    20.1.3 Haloarchaeal Model Systems to Study
          Gene-Specific Regulation 251
20.2 The Arginine Deiminase Pathway 252
    20.2.1 The Arginine Deiminase Pathway in Bacteria and Archaea 252
    20.2.2 Genes and Gene Products of the Arginine Deiminase
          Gene Cluster from *Halobacterium salinarum* 252
    20.2.3 Analysis of Transcripts from the *arc* Gene Cluster and
          Characterization of Transcriptional Regulation 253
    20.2.4 Comparison of a Bacterial and an Archaeal ADI Pathway 256
20.3 Elements of the General Transcription Apparatus 256
    20.3.1 The Haloarchaeal TATA Box 256
    20.3.2 Archaeal and Eukaryal TATA Box-Binding Proteins 257
    20.3.3 Heterologous Production of the Haloarchaeal TBP and its
          Folding into a Native Conformation 258

20.3.4 Binding of the Haloarchaeal TBP to TATA Box-Containing
  DNA Fragments..259
20.4 Conclusion 260
Acknowledgments..261
References..261

20.1 INTRODUCTION

20.1.1 THE GENERAL TRANSCRIPTION APPARATUS OF ARCHAEA

On the highest taxonomic level, three groups of organisms, called domains, have been recognized: Archaea, Bacteria, and Eukarya (Woese et al., 1990). It was surprising to find that the Archaea, one of the two "prokaryotic" domains, are phylogenetically closer to the Eukarya than to the Bacteria (Iwabe et al., 1989).

Recently, the genome sequence of the archaeon *Methanococcus jannaschii* became available (Bult et al., 1996). Comparison of the deduced proteins from all open reading frames revealed that elements involved in transcription, translation, and replication have a higher degree of similarity to eukaryal than to bacterial proteins. The general transcription apparatus is an especially striking example, showing the high degree of similarity between Archaea and Eukarya, while the transcription apparatus of the Bacteria is fundamentally different.

It was discovered about 15 years ago that the archaeal RNA polymerase is related to the eukaryal enzyme (Huet et al., 1983). In the years since, excellent work, especially in the group of Zillig in Martinsried, has shown that this holds true for the subunit structure, immunological cross-reactivity, transcription-factor dependence, and, ultimately, sequences of large and small subunits (Langer et al., 1995).

About 10 years ago it was first proposed that the major promoter element involved in transcription initiation was similar to the TATA box of Eukarya, and not to bacterial promoter elements (Reiter et al., 1988; Thomm and Wich, 1988). The similarity of archaeal and eukaryal transcription initiation was further underscored by the recent discovery that Archaea harbor homologs to the two transcription factors involved in the earliest steps of promoter recognition in Eukarya, i.e. the TATA box-binding protein (TBP) and TFIIB. The current knowledge about archaeal transcription initiation has recently been reviewed (Keeling and Doolittle, 1995; Thomm, 1996).

In this chapter, our group's characterization of the haloarchaeal TATA box and TBP will be summarized, and a comparison with other archaeal and eukaryal systems will be given.

20.1.2 REGULATION OF GENE EXPRESSION

Knowledge about the general archaeal transcription machinery has increased significantly in recent years, but detailed knowledge about mechanisms of gene-specific regulation is still sparse. In Archaea, a variety of energy-transducing pathways with similarity to bacterial pathways have been detected, e.g. anaerobic respirations with different electron acceptors, or arginine fermentation. Analysis of the genome sequence of *M. jannaschii* has shown that, in contrast to proteins involved in transcription, translation, and replication, many proteins involved in central metabolism are more similar to their bacterial counterparts than to eukaryal proteins. It will be interesting to unravel the principles of archaeal gene regulation, and especially to elucidate the regulation of "bacteria-like" genes in organisms with a "eukaryal" general transcription apparatus. A variety of studies have shown that archaeal gene expression is highly regulated on the transcriptional level. It is generally assumed that observed increases in steady-state levels of transcripts are due to the induction of transcription. However, an alternative explanation, i.e. the repression of transcript degradation, cannot be excluded.

A general method to measure transcript stability is not yet available, and in only one case have transcript stabilities been reported (Hennigan and Reeve, 1994). It has become evident that, besides transcriptional regulation, archaeal gene expression can be regulated on different levels. Examples are the involvement of anti-sense RNA in gene expression (Stolt and Zillig, 1993; Pfeifer et al., 1997), or the discoveries of examples that the levels of transcript and protein are not correlated, and thus the translational efficiencies of some transcripts are not constant (Danner and Soppa, 1996; Cheung et al., 1997; Pfeifer et al., 1997). Furthermore, it was shown that the haloarchaeal genome includes sequences with Z-DNA conformation, which could be involved in regulation (Kim and DasSarma, 1996). Therefore, a variety of different regulatory mechanisms appear to exist, and for a detailed understanding of archaeal regulation of gene expression, it seems desirable to investigate several systems on the molecular level.

20.1.3 HALOARCHAEAL MODEL SYSTEMS TO STUDY GENE-SPECIFIC REGULATION

For a detailed understanding of regulatory processes, the ability to perform studies *in vivo* is very important. As many studies on eukaryal transcription factors have shown, it is not always clear whether protein interactions discovered *in vitro* are of physiological significance. Halophilic Archaea were the first archaeal group of organisms for which a transformation system became available (Cline et al., 1989), and archaeal molecular genetics is still most advanced for the haloarchaeal group. Methods available for haloarchaea include the selection of point mutants of desired phenotypes (Soppa and Oesterhelt, 1989), and gene replacement to either delete genes of interest (to deduce their possible functions from the null phenotype) or to introduce specifically modified versions of haloarchaeal genes or DNA elements to study structure/function relationships. Translational (Danner and Soppa, 1996) as well as a transcriptional (Palmer and Daniels, 1995) reporter systems have been used to study transcription initiation, and very recently a haloarchaeal β-galactosidase gene became available, which will facilitate the characterization of cis-acting DNA elements involved in transcriptional regulation (Holmes et al., 1997).

Several model systems have been developed to study gene-specific regulation in halophilic Archaea. These include: (1) the bacterio-opsin gene expression (Sumper and Herrmann, 1976), which is thought to be regulated by light, oxygen concentration, and DNA conformation (Gropp et al., 1995; Yang et al., 1996), (2) genes of the gas vesicle gene clusters of several species (Pfeifer et al., Chapter 23; Pfeifer et al., 1997), (3) a halocin gene expressed exclusively during the transition phase from exponential to stationary phase (Cheung et al., 1997; Shand et al., Chapter 24), and (4) genes regulated by the external salt concentration (Ferrer et al., 1996; Pfeifer et al., 1997). In our group, analysis of the arginine deiminase pathway from *Halobacterium salinarum* was chosen for the investigation of gene-specific regulation.

20.2 THE ARGININE DEIMINASE PATHWAY

20.2.1 THE ARGININE DEIMINASE PATHWAY IN BACTERIA AND ARCHAEA

Fermentative arginine degradation via the arginine deiminase (ADI) pathway has been found in a variety of bacterial species of diverse phylogenetic groups, e.g., different species of *Pseudomonas, Mycobacterium*, and lactic acid bacteria. Within the Archaea, the ADI pathway has thus far been detected only in several haloarchaeal species (Hartmann et al., 1980; Ruepp et al., 1995), e.g. *Hb. salinarum*. The reactions of the ADI pathway are catalyzed by the three enzymes arginine deiminase, catabolic ornithine transcarbamylase (cOTCase), carbamate kinase (CK), and a membrane-bound arginine/ornithine antiporter (Figure 20.1). The antiport of the substrate arginine and the product ornithine is electroneutral. Thus, it is driven by the concentration gradients of these substances. The degradation of arginine to ornithine, CO_2, and ammonia leads to the equimolar generation of ATP by substrate level phosphorylation. It was shown that anaerobic utilization of arginine by *Hb. salinarum* enhances the intracellular ATP level and leads to the accumulation of the product ornithine in the medium (Hartmann et al., 1980). Up to arginine concentrations of about 2%, growth yields of fermentatively grown *Hb. salinarum* cultures are directly dependent on the arginine concentration (data not shown).

20.2.2 GENES AND GENE PRODUCTS OF THE ARGININE DEIMINASE GENE CLUSTER FROM *HALOBACTERIUM SALINARUM*

The genes of the arginine deiminase gene cluster (designated *arc* genes, for *arginine catabolism*) have been cloned and sequenced from *Hb. salinarum* (Ruepp and Soppa,

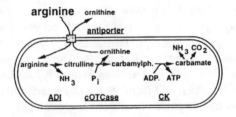

FIGURE 20.1 Fermentative arginine degradation via the arginine deiminase pathway. Substrate, intermediates, products, enzymes and their cellular localization are shown schematically. The abbreviations of the three cytoplasmic enzymes and a membrane-bound transport protein are underlined. ADI–arginine deiminase, cOTCase catabolic ornithine transcarbamylase, CK–carbamate kinase, antiporter–arginine/ornithine antiporter.

1996; Ruepp et al., 1995). The gene order is *arcRACB*; *arcR* codes for a putative regulatory protein (see below), *arcA* codes for the ADI, *arcC* for the CK, and *arcB* for the cOTCase. The gene cluster is followed by a T-rich region thought to be involved in transcriptional termination. No further open reading frames with haloarchaeal codon usage have been found 1.3 kbp upstream and 500 bp downstream of the gene cluster. Therefore, it appears that additional members of the *arc* regulon, e.g. the gene for the arginine-ornithine antiporter and genes involved in regulation of expression (see below), are situated at different locations in the chromosome.

The coding capacity of the three structural genes was verified experimentally. The ADI and the cOTCase were isolated from fermentatively grown *Hb. salinarum* cultures, and N-terminal sequence determination of several peptides confirmed that they are the products of the *arcA* and *arcB* genes, respectively. The *arcC* gene was expressed heterologously in *Haloferax volcanii*, which does not have the ADI pathway. After transformation, a CK activity, which was absent from untransformed control cultures, could be measured. Very recently, the *arcR* gene could be expressed in *Escherichia coli*, but no functional assay is yet available (Herrmann and Soppa, data not shown). The deduced protein sequence of ArcR is similar to several bacterial repressors of catabolic pathways, i.e. KdgR (from *Erwinia chrysanthemi*), GylR (from *Streptomyces griseus* and *S. coelicolor*), and IclR (from *E. coli* and *Salmonella typhimurium*). This sequence similarity, which is evenly distributed over the entire length of ArcR, was a first indication that ArcR might be involved in regulation of expression (see below). Surprisingly, ArcR is considerably shorter than the bacterial regulators. The N-terminal 60-100 amino acids of the bacterial proteins, which include helix-turn-helix motifs responsible for DNA-binding, are lacking in ArcR. Therefore, either ArcR has a different DNA-binding motif not yet identified, or the regulatory function of ArcR does not require DNA binding, but is mediated in a "eukaryal" way solely by protein-protein interactions.

20.2.3 ANALYSIS OF TRANSCRIPTS FROM THE *ARC* GENE CLUSTER AND CHARACTERIZATION OF TRANSCRIPTIONAL REGULATION

The transcripts from the *arc* gene cluster were characterized by northern blot analysis, and the results are summarized schematically in Figure 20.2. The major transcripts are monocistronic transcripts from the three structural genes, and a bicistronic *arcCB* transcript. Only minor amounts of longer transcripts were found. The *arcR* transcript level is very low and does not seem to be inducible, in agreement with the proposed regulatory role of ArcR. In contrast, the transcripts from the three structural genes are highly regulated. Following a shift from aerobic to fermentative growth conditions, transcript levels increase more than 20-fold. The increase of transcript levels can be detected about 10 minutes after induction, and transcript levels reach a maximum a few hours later. As an example, a northern blot analysis with an *arcB* specific probe is shown in Figure 20.3; induction of the *arcB* (lower band) and *arcCB* transcripts (upper band) can be seen. The presence of the substrate arginine is the major signal for transcriptional induction; arginine is sufficient for

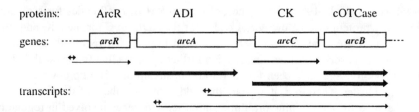

FIGURE 20.2 Overview on genes, gene products, and transcripts of the arginine deiminase gene cluster of *Hb. salinarum*. The genes, denoted *arc* for *ar*ginine *c*atabolism, are shown as rectangles. The nomenclature of the *Pseudomonas aeruginosa* genes is used. Above the genes, abbreviations for the names of the gene products are given (compare Figure 20.1). ArcR–putative regulatory protein. Below the genes, major and minor transcripts are shown schematically. Small double arrows indicate that the 5′-ends of these transcripts have not been determined, and therefore their location could vary from the location shown here to some extent.

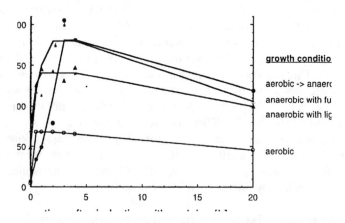

FIGURE 20.3 Induction of *arcA* gene transcription under different conditions. Four *Hb. salinarum* cultures were grown at different conditions: (1) aerobically, (2) phototrophically, (3) by fumarate respiration, and (4) pregrown aerobically and shifted to anaerobic conditions at the start of the experiment. Arc gene expression was induced by the addition of arginine (1% final concentration) at time zero. At the indicated times, aliquots were removed, and total RNA was isolated. The *arcA* transcript was visualized by northern blot hybridization using an *arcA*-specific probe. Densitometry was applied to quantitate the (relative) amounts of *arcA* transcript (arbitrary units).

induction under all conditions tested. This is illustrated in Figure 20.4, which summarizes a series of induction experiments.

Hb. salinarum cultures were grown under four different conditions, aerobically, phototrophically, via fumarate fermentation, and pregrown aerobically and shifted to anaerobic conditions, respectively. Arginine was added, the amount of *arcA* transcript was determined by northern blot hybridization, and quantitated by densitometry. Induction of transcript levels can clearly be seen in all cases. A

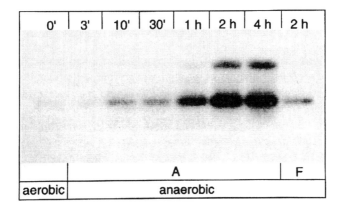

FIGURE 20.4 Kinetics of induction of *arcB* gene transcription. A *Hb. salinarum* culture was pregrown aerobically. At time zero, the culture was shifted to anaerobic conditions, and arginine was added (1% final concentration). At the indicated times, aliquots were removed, and total RNA was isolated. Northern blot analysis with an *arcB*-specific probe was used to visualize specific transcripts. The monocistronic *arcB*-transcript (lower band) and the bicistronic *arcCB*-transcript (upper band) can be seen. The lane on the right shows one time point of a parallel experiment in which 1% fumarate, the substrate for anaerobic fumarate respiration, was added instead of arginine.

second signal, which may be the availability of oxygen, influences the transcript level, which is about twice as high at anaerobic than at aerobic conditions. Furthermore, there seems to be cross-talk between the different anaerobic energy-yielding pathways, e.g. a low level of induction of *arc* gene transcripts can be seen with fumarate (Figure 20.3, right lane), and prior to arginine induction, the *arc* transcript levels are higher in anaerobically grown than in aerobically grown cultures (Figure 20.4, time zero). Recently, it was discovered that citrulline is also an inducer of *arc* gene transcription, albeit not as efficient as arginine. It was confirmed that *Hb. salinarum* can indeed grow anaerobically with citrulline. The growth rate is slower and the cell yield is lower than with arginine, possibly because citrulline transport is inefficient and energy-dependent.

Although all three structural genes are inducible, there are differences in the transcription of *arcA* in comparison with *arcB/arcC*. At the end of the exponential growth phase, *arcA* is induced in contrast to *arcC* and *arcB* (Ruepp and Soppa, 1996). This observation might be correlated to the differences in the regions upstream of the three genes. Upstream of the *arcB* and *arcC* TATA boxes, a conserved sequence element was detected (>60% identity in 40 bp), which is totally absent upstream of *arcA*. This differential transcription of *arcA* vs. *arcC/arcB* will be characterized further. It could indicate flexibility in the usage of this pathway, i.e. the usage of citrulline instead of arginine, or mainly the usage of the arginine deiminase reaction under nitrogen-limited conditions.

20.2.4 COMPARISON OF A BACTERIAL AND AN ARCHAEAL ADI PATHWAY

The ADI pathway had been most thoroughly studied in *Pseudomonas aeruginosa*. The four proteins have been isolated and characterized, and the *arcDABC* operon, including the *arcD* gene coding for the arginine/ornithine antiporter, has been cloned and sequenced. Transcription is initiated from a single promoter upstream of *arcD*, and the primary transcript is subsequently processed to smaller transcripts of different stabilities. The processing is dependent on RNaseE. Transcription is controlled by the ANR protein, the major signal for induction is the absence of oxygen (Gamper and Haas, 1993; Gamper et al., 1991, and references therein).

Comparison of the only two well-characterized ADI pathways from the Bacterium *P. aeruginosa* and the Archaeon *Hb. salinarum* has revealed fundamental differences in a variety of points: (1) the genes included in the bacterial *arcDABC* operon and the archaeal *arcRACB* gene cluster as well as the gene order are different, (2) the bacterial operon is transcribed from a single promoter, whereas the archaeal *arc* genes are under the control of individual promoters, (3) the major signals for the induction of expression are anaerobiosis in *P. aeruginosa* and, in contrast, arginine in *Hb. salinarum*, (4) ANR regulates the bacterial operon. ANR is a member of the FNR protein family and therefore, probably functions via interaction with the bacterial sigma factor, which is not present in Archaea. Accordingly, no indication for the presence of ANR or ANR-binding sites in *Hb. salinarum* have been detected. Instead, TATA boxes are present upstream of the three structural genes, and it can be presumed that ArcR is involved in regulation, (5) the cOTCases of both organisms exhibit major differences in quaternary structure, way of enzyme binding, allosteric effectors, and primary sequence (Ruepp et al., 1995).

Taken together, these data indicate that the ADI pathways evolved independently in these two organisms. However, detailed understanding of ADI pathways of further organisms would be required to be able to decide whether these differences can be generalized and the two pathways can be regarded as prototypic for Bacteria and Archaea.

20.3 ELEMENTS OF THE GENERAL TRANSCRIPTION APPARATUS

20.3.1 THE HALOARCHAEAL TATA BOX

Two approaches revealed that a sequence element around 25 bp upstream of the transcriptional start site was important for archaeal transcription initiation: (1) comparison of sequences upstream of archaeal genes, and (2) functional analysis of promoter-sequence *in vitro* transcription systems from *Sulfolobus* and *Methanococcus* (Hain et al., 1992; Hausner et al., 1991). This sequence element was termed "box A" or "distal promoter element," and will be called "TATA box" in this chapter. Based on these results, a characterization of the haloarchaeal TATA box was performed (Danner and Soppa, 1996) with the aim of extending the existing studies in the following points: (1) to ensure that this element is also essential for haloarchaeal transcription initiation, (2) to characterize the TATA box functionally *in vivo*, (3) to characterize the TATA box of a protein coding gene instead of a gene for a stable RNA (rRNA, tRNA), and (4) to construct a TATA box *de novo* from random sequences.

The positions −19 to −32 upstream of a reporter gene were randomized to generate a plasmid library of about 10^8 different entries, including all possible sequence combinations in the randomized region. *Hf. volcanii* was transformed with this plasmid library, and all plasmids with sequences acting as functional TATA boxes *in vivo* were selected. Selection was performed making use of the enzyme encoded by the reporter gene. The mixture of plasmids obtained after selection, comprising the subset of the original random library with functional TATA boxes, was sequenced directly, and, furthermore, randomly chosen plasmids after selection were sequenced, and a consensus sequence was derived. In parallel to this study, Palmer and Daniels (1995) characterized a haloarchaeal TATA box *in vivo* using site-specific mutants of a tRNA promoter. A region of 11 nucleotides was chosen, and all possible single base pair mutations were constructed. The wild-type TATA box and the 33 mutant sequences were cloned upstream of a transcriptional reporter gene, and the amount of reporter transcript generated *in vivo* was quantitated (normalized to an internal control transcript).

The results of both approaches are in excellent agreement, and the following conclusions can be drawn: (1) As in other archaeal groups, in haloarchaea transcription initiation also depends on a TATA box. A more than 100-fold difference in transcription efficiency could be seen, dependent on the sequence of the TATA box. (2) The optimal spacing of the TATA box center to the transcriptional start site is about 27 nucleotides, but a different spacing by one or two nucleotides is compatible with high TATA box activity. (3) The consensus sequences derived by both studies were TWWWWRAC and TTATAAAC, respectively (W = A or T, R = A or G; only the positions −30 to −23 are shown). (4) Both approaches revealed that the sequence specificity is astonishingly low, i.e. lower than found in bacterial promoter elements. At most positions two, or at some positions even three, different nucleotides lead to similar transcript levels; however, other nucleotides at the same positions are not compatible with a high TATA box activity. For example, the positions −29 to −26 must be AT-rich, but many different combinations of A and T generate good TATA boxes.

20.3.2 Archaeal and Eukaryal TATA Box-Binding Proteins

Recently, it became clear that the transcription factor binding to the TATA box, TBP, is conserved in Eukarya and Archaea, and TBP sequences of several archaeal species became available. On the basis of highly conserved regions of eukaryal TBPs, a *tbp* gene from *Hb. salinarum* was cloned and sequenced (Soppa and Link, 1997). A phylogenetic tree with four archaeal and four eukaryal species was constructed, based on the first two codon positions of the respective *tbp* genes. The *tbp* tree was in agreement with the phylogeny of organisms, e.g. archaeal and eukaryal *tbp*s formed monophyletic groups. The protein sequence identity within the Archaea is 40 to 45%, about the same degree of conservation found for TBPs from higher and lower Eukarya. The similarity between archaeal and eukaryal TBPs is somewhat lower (30 and 40% identity). However, several features distinguish archaeal and eukaryal TBPs, showing that they form two distinct protein subfamilies: (1) Archaeal TBPs are about 180 amino acids long and consist of two repeats of about 90 amino acids. Eukaryal TBPs are longer, in addition to the conserved repeat domains, located at their C-terminus, they possess N-terminal domains, which are not conserved in length or sequence. (2) Conservation of the repeat domains in Archaea and Eukarya indicates that a gene duplication of an ancient *tbp* gene in the common ancestor led to the recent *tbp* genes.

The internal symmetry, i.e. the similarity of the first TBP repeat to the second, is dramatically different in Archaea and Eukarya. In Archaea, the identity of the two repeats is around 40%, whereas it is only 26% for eukaryal TBPs. The lower degree of internal conservation in eukaryal TBPs could be explained by the fact that, in contrast to archaeal TBPs, eukaryal TBPs must interact with transcription initiation complexes of three different polymerases. Furthermore, the differences in internal conservation might indicate that archaeal TBPs must interact with a lower number of different transcription factors than their eukaryal counterparts. In accordance with this view, the "internally conserved" amino acids are located only on the DNA-binding surface of yeast TBP, whereas they are found also on the opposite, transcription-factor-binding side of the haloarchaeal TBP (data not shown; the yeast TBP structure was taken as a template). Furthermore, only homologs to TBP and TFIIB have been detected in the genome of *M. jannaschii*, also indicating that the number of archaeal general transcription factors is limited. (3) Archaeal TBPs have an excess of acidic over basic amino acids, in contrast to eukaryal TBPs, where basic amino acids predominate. The halophilic TBP is an extreme case with a value of 12% for the excess of acidic residues (other Archaea, about 3%; Eukarya, around 6% excess of basic residues). The high excess of negatively charged residues is typical for haloarchaeal proteins and the basis for their solubility at high salt concentrations (Eisenberg et al., 1992; Eisenberg, 1995; Frolow et al., 1996).

It was discovered that *Hb. salinarum* and *Hf. volcanii* both contain more than one *tbp* gene in their genomes (Daniels, personal communication). This is probably not typical for Archaea, e.g. in the genome sequence of *M. jannaschii* only one *tbp* gene could be found. Gene duplications in haloarchaea have been discussed as a means to adapt to rapidly changing salt concentrations (Dennis and Shimmin, 1997).

20.3.3 Heterologous Production of the Haloarchaeal TBP and its Folding into a Native Conformation

The *tbp* gene from *Hb. salinarum* was cloned into the *E. coli* expression vector pQE30, resulting in the production of a fusion protein with an N-terminal hexahistidine tag, 6H-TBP, in a transgenic producer strain (Soppa and Link, 1997). As the haloarchaeal TBP is adapted to the high internal salt concentrations of *Hb. salinarum*, production in *E. coli* yielded a denatured protein. After affinity purification in 8 M urea to ensure complete unfolding, 6H-TBP was dialyzed against buffers of different salt concentrations. CD spectra were recorded to monitor secondary structure for-

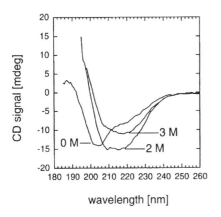 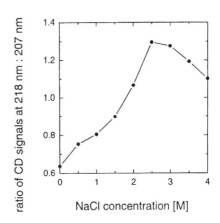

FIGURE 20.5 Secondary structure formation of 6H-TBP at different salt concentrations. A fusion protein of the TATA box-binding protein from *Hb. salinarum* and an N-terminal hexahistidine tag, 6H-TBP, was isolated in the presence of 8 M urea after heterologous production in *E. coli*. Aliquots were dialyzed against buffers containing different salt concentrations, and CD spectra were recorded. (left) Three selected CD spectra are shown, and the salt concentrations are indicated. In the absence of NaCl, a negative CD band around 202 nm is indicative for random coil conformation. Secondary structure formation in the presence of NaCl is characterized by a negative CD band around 218 nm. (right) The ratios of CD signal intensities at 218 nm and 207 nm were plotted against the NaCl concentration of the 6H-TBP sample. The wavelengths were chosen, because the highest negative CD signal around 218 nm is indicative for the folded state, and at 207 nm an isosbestic point was observed for CD spectra taken at concentrations of up to about 2 M NaCl.

mation. As shown in Figure 20.5, increased salt concentration resulted in elevated secondary structure formation, indicating folding of 6H-TBP.

At salt concentrations approaching 4 M sodium chloride, however, a different process predominates. In contrast to typical haloarchaeal cytoplasmatic proteins, 6H-TBP is not soluble at very high salt concentrations, but instead aggregates. A possible explanation is that a high negative-charge density at the DNA binding side of the protein would interfere with its function. In the absence of DNA, TBP would be less soluble in high salt than other haloarchaeal proteins, because its hydrophobic DNA binding surface is exposed, and mediates aggregation of the protein. This situation would not occur *in vivo*, because probably TBP is permanently attached to DNA, and finds its way to TATA boxes by one-dimensional diffusion.

20.3.4 BINDING OF THE HALOARCHAEAL TBP TO TATA BOX-CONTAINING DNA FRAGMENTS

Until now, nothing has been known about the mechanism of specific interactions of proteins and DNA in the presence of high salt. Characterization of the haloarchaeal TATA box and folding of 6H-TBP into a native conformation after heterologous overproduction open the opportunity to establish methods for studying protein DNA interactions under these extreme conditions. Presently, we are optimizing a binding assay that makes use of the hexahistidine affinity tag of the heterologously produced 6H-TBP. The assay consists of several steps: (1) 6H-TBP is incubated together with a radioactively labeled TATA box-containing PCR fragment, allowing complex formation in solution. (2) Nickel-chelating sepharose (NCS) is added, to which 6H-TBP and binary 6H-TBP/DNA complexes bind via the hexahistidine tag. (3) Centrifugation allows binary NCS/6H-TBP and ternary NCS/6H-TBP/DNA fragments to separate from unbound DNA fragments, which remain in solution. Quantitation of the radioactivity in the pellet and in the supernatant thus allows direct determination of the fractions of bound DNA and free DNA. As an example, Figure 20.6 shows the influence of the magnesium sulfate concentration on complex formation. About 40 mM magnesium sulfate are required for binding; in the absence of magnesium, almost no complex formation was observed. Although characterization of this binding assay is far from complete, it is evident that it will become a useful tool for the investigation of protein DNA interaction at high salt.

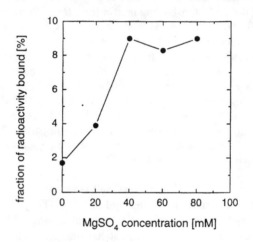

FIGURE 20.6 Optimization of 6H-TBP binding to a TATA box-containing DNA fragment. The assay is described in the text. In principle, 6H-TBP was incubated with a radioactively labeled DNA fragment, 6H-TBP/DNA complexes were separated from free DNA, and the amounts of bound and unbound DNA were determined by liquid scintillation counting. The fraction of the total amount of DNA found in a complex with 6H-TBP (% bound) was plotted against the MgSO$_4$ concentration of the incubation buffer.

20.4 CONCLUSION

The results summarized in this chapter have shown that the ADI pathway of *Hb. salinarum* is an excellent model system to study regulation of gene expression in halophilic Archaea. Three structural genes and a putative regulatory gene have been characterized. Transcription of the structural genes is highly regulated. Differential regulation of *arcA* vs. *arcC/arcB* has been observed, and two signals for induction of expression have been identified. The elements controlling *arc* gene-specific regulation must interact with the general transcription machinery to allow regulated

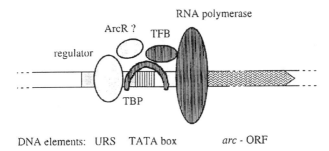

FIGURE 20.7 Schematic representation of the proteins and DNA elements of a minimal preinitiation complex necessary for regulated *arc* gene transcription. The elements specific for *arc* gene expression are shaded, the components of the general transcription apparatus are indicated by stripes. URS–upstream regulatory sequence; ORF–open reading frame; TFB –transcription factor homologous to the eukaryal factor TFIIB.

transcription to occur. Figure 20.7 shows a schematic "working model" of a minimal preinitiation complex (PIC) that can be envisaged to form upstream of *arc* open reading frames.

This working model should help to illustrate some of the problems on which future work will focus: (1) Characterization of TBP binding to model TATA boxes and comparison with its binding at *arc* TATA boxes. (2) Functional characterization of the conserved upstream regulatory sequence (URS) preceding the *arcC/arcB* TATA boxes, and identification of an *arcA*-URS. (3) Identification of a URS-binding protein, and characterization of the role of ArcR and its mode of action. (4) Characterization of the interactions of gene-specific factors, e.g. ArcR, and the general transcription apparatus, e.g. TBP. Hopefully, the *arc* genes, together with other genetic systems mentioned in the introduction, will help to elucidate the mechanisms of haloarchaeal gene regulation on a molecular level.

ACKNOWLEDGMENTS

The work in the group of J.S. summarized here was supported by the Deutsche Forschungsgemeinschaft (DFG) through grants So 264/1, So 264/2, and So 264/5. J.S. is supported by the DFG through a "Heisenberg-Stipendium" grant So 264/4. The CD spectropolarimeter was funded by the DFG through grant Li 474/4-3 to T.A. L.J.S. thanks Dieter Oesterhelt for continuous support during his years at the Max-Planck-Institut für Biochemie in Martinsried.

REFERENCES

Baumann, P., S.A. Qureshi, and S.P. Jackson. 1995. Transcription: New insights from studies on archaea. *Trends Genet.* 11:279–283.

Bult, C.J., O. White, G.J. Olsen, L. Zhou, R.D. Fleischmann, G.G. Sutton, J. Blake, L.M. FitzGerald, R.A. Clayton, J.D. Gocayne, A.R. Kerlavage, B.A. Dougherty, J.-F Tomb, M.D. Adams, C.I. Reich, R. Overbeek, E.F. Kirkness, K. Weinstock, J.M. Merrick, A. Glodek, J.L. Scott, N.S. Geophagen, J.F. Weidman, J.L. Fuhrmann, D. Nguyen, T.R. Utterback, J.M. Kelley, J.D. Peterson, P.W. Sadow, M.C. Hanna, M.D. Cotton, K.M. Roberts, M.A. Hurst, B.P. Kaine, M. Borodovsky, H.-P Klenk, C.M. Fraser, H.O. Smith, C.R. Woese, and J.C. Venter. 1996. Complete genome sequence of the methanogenic archaeon, *Methanococcus jannaschii*. *Science* 273:1058–1073.

Cheung, J., K.J. Danna, E.M. O'Connor, L.B. Price, and R.F. Shand. 1997. Isolation, sequence, and expression of the gene encoding halocin H4, a bacteriocin from the halophilic archaeon *Haloferax mediterranei* R4. *J. Bacteriol.* 179:548–551.

Cline, S.W., W.L. Lam, R.L. Charlebois, L.C. Schalkwyk, and W.F. Doolittle. 1989. Transformation methods for halophilic archaebacteria. *Can. J. Microbiol.* 35:148–152.

Danner, S. and J. Soppa. 1996. Characterization of the distal promoter element of halobacteria *in vivo* using saturation mutagenesis and selection. *Mol. Microbiol.* 19:1265–1276.

Dennis, P.P. and L. Shimmin. 1997. Evolutionary divergence and salinity-mediated selection in halophilic archaea. *Microbiol. Mol. Biol. Rev.* 61:90–104.

Eisenberg, H. 1995. Life in unusual environments–progress in understanding the structure and function of enzymes from extreme halophilic bacteria. *Arch. Biochem. Biophys.* 318:1–5.

Eisenberg, H., M. Mevarech, and G. Zaccai. 1992. Biochemical, structural, and moleculargenetic aspects of halophilism. *Adv. Protein Chem.* 43:1–62.

Ferrer, C., F.J.M. Mojica, G. Juez, and F. Rodríguez-Valera. 1996. Differentially transcribed regions of *Haloferax volcanii* genome depending on the medium salinity. *J. Bacteriol.* 178:309–313.

Frolow, F., M. Harel, J.L. Sussman, M. Mevarech, and M. Shoham. 1996. Insights into protein adaptation to a saturated salt environment from the crystal structure of a halophilic 2Fe–2S ferredoxin. *Nature Struct. Biol.* 3:452–458.

Gamper, M. and D. Haas. 1993. Processing of the *Pseudomonas* arcDABC mRNA requires functional RNase E in *Escherichia coli*. *Gene* 129:119–122.

Gamper, M., A. Zimmermann, and D. Haas. 1991. Anaerobic regulation of transcription initiation in the *arcDABC* operon of *Pseudomonas aeruginosa*. *J. Bacteriol.* 173:4742–4750.

Gropp, F., R. Gropp, and M.C. Betlach. 1995. Effects of upstream deletions on light-regulated and oxygen-regulated bacterioopsin gene expression in *Halobacterium halobium*. *Mol. Microbiol.* 16:357–364.

Hain, J., W.-D Reiter, U. Hüdepohl, and W. Zillig. 1992. Elements of an archaeal promoter defined by mutational analysis. *Nucl. Acids Res.* 20:5423-5428.

Hartmann, R., H. Sickinger, and D. Oesterhelt. 1980. Anaerobic growth of halobacteria. *Proc. Natl. Acad. Sci. USA* 77:3821–3825.

Hausner, W., G. Frey, and M. Thomm. 1991. Control regions of an archaeal gene: a TATA box and an initiator element promote cell-free transcription of the rRNA^{val} gene of *Methanococcus vannielii*. *J. Mol. Biol.* 222:495–508.

Hennigan, A. and J. Reeve. 1994. mRNAs in the methanogenic archaeon *Methanococcus vannielii*: numbers, half-lives, and processing. *Mol. Microbiol.* 11:665–670.

Holmes, M.L., R.K. Scopes, R.L. Moritz, R.J. Simpson, C. Englert, F. Pfeifer, and M.L. Dyall-Smith. 1997. Purification and analysis of an extremely halophilic β-galactosidase from *Haloferax alicantei*. *Biochim. Biophys. Acta* 1337:276–286.

Huet, J., R. Schnabel, A. Sentenac, and W. Zillig. 1983. Archaebacteria and eukaryotes possess DNA-dependent RNA polymerases of a common type. *EMBO J.* 2:1291–1294.

Iwabe, N., K. Kuma, M. Hasegawa, S. Osawa, and T. Miyata. 1989. Evolutionary relationship of archaebacteria, eubacteria, and eukaryotes inferred from phylogenetic trees of duplicated genes. *Proc. Natl. Acad. Sci. USA* 86:9355–9359.

Keeling, P.J. and W.F. Doolittle. 1995. Archaea: narrowing the gap between prokaryotes and eukaryotes. *Proc. Natl. Acad. Sci. USA* 92:5761–5764.

Kim, J. and S. DasSarma. 1996. Isolation and chromosomal distribution of natural z-DNA-forming sequences in *Halobacterium halobium*. *J. Biol. Chem.* 271:19724–19731.

Langer, D., J. Hain, P. Thuriaux, and W. Zillig. 1995. Transcription in Archaea: similarity to that in Eucarya. *Proc. Natl. Acad. Sci. USA* 92:5768–5772.

Palmer, J.R. and C.J. Daniels. 1995. In vivo definition of an archaeal promoter. *J. Bacteriol.* 177:1844–1849.

Pfeifer, F., K. Krüger, R. Röder, A. Mayr, S. Ziesche, and S. Offner. 1997. Gas vesicle formation in halophilic Archaea. *Arch. Microbiol.* 167:259–268.

Reiter, W.-D., P. Palm, and W. Zillig. 1988. Analysis of transcription in the archaebacterium *Sulfolobus* indicates that archaebacterial promoters are homologous to eucaryotic pol II promoters. *Nucl. Acids Res.* 16:1–19.

Ruepp, A. and J. Soppa. 1996. Fermentative arginine degradation in *Halobacterium salinarium* (formerly *Halobacterium halobium*): Genes, gene products, and transcripts of the *arcRACB* gene cluster. *J. Bacteriol.* 178:4942–4947.

Ruepp, A., H.N. Muller, F. Lottspeich, and J. Soppa. 1995. Catabolic ornithine transcarbamylase of *Halobacterium halobium* (*salinarium*): purification, characterization, sequence determination, and evolution. *J. Bacteriol.* 177:1129–1136.

Soppa, J. and T.A. Link. 1997. The TATA-box binding protein (TBP) of *Halobacterium salinarum*: cloning of the tbp gene, heterologous production of TBP and folding of TBP into a native conformation. *Eur. J. Biochem.* 249:318–324.

Soppa, J. and D. Oesterhelt. 1989. Bacteriorhodopsin mutants of *Halobacterium* sp grb.1. the 5-bromo-2'-deoxyuridine selection as a method to isolate point mutants in halobacteria. *J. Biol. Chem.* 264:13043–13048.

Stolt, P. and W. Zillig. 1993. Antisense RNA mediates transcriptional processing in an archaebacterium, indicating a novel kind of RNase activity. *Mol. Microbiol.* 7:875–882.

Sumper, M. and G. Herrmann. 1976. Biogenesis of purple membrane: regulation of bacterio-opsin synthesis. *FEBS Lett.* 69:149–152.

Thomm, M. 1996. Archaeal transcription factors and their role in transcription initiation. *FEMS Microbiol. Rev.* 18:159–171.

Thomm, M. and G. Wich. 1988. An archaebacterial promoter element for stable RNA genes with homology to the TATA box of higher eukaryotes. *Nucl. Acids Res.* 16:151–163.

Woese, C.R., O. Kandler, and M.L. Wheelis. 1990. Towards a natural system of organisms — proposal for the domains archaea, bacteria, and eucarya. *Proc. Natl. Acad. Sci. USA* 87:4576–4579.

Yang, C.F., J.M. Kim, E. Molinari, and S. DasSarma. 1996. Genetic and topological analyses of the bop promoter of *Halobacterium halobium*: stimulation by DNA supercoiling and non-B-DNA structure. *J. Bacteriol.* 178:840–845.

21 Cloning, Sequence and Heterologous Expression of *bgaH*, a Beta-Galactosidase Gene of "*Haloferax alicantei*"

Melissa L. Holmes and Mike L. Dyall-Smith

CONTENTS

21.1 Introduction..265
21.2 *Haloferax alicantei* and Mutant SB1 266
21.3 Cloning the ß-Galactosidase Gene 267
    21.3.1 The Sequence of *bgaH* 267
    21.3.2 ORFS Surrounding *bgaH* 268
21.4 Transformation of *H. volcanii* 269
21.5 Conclusion 269
Acknowledgments...269
References..269

21.1 INTRODUCTION

The extremely halophilic Archaea (or halobacteria) are not only extremophiles, living at saturating salt concentrations, but are experimentally the most convenient representatives of the archaeal domain for genetic study (Doolittle et al., 1992; Dyall-Smith, 1997; Keeling and Doolittle, 1995). They are mesophilic aerobes, able to grow on numerous substrates and in defined media.

    Recently, a number of genetic tools and methods have been developed for halobacteria. There is an efficient DNA transformation method (Cline and Doolittle, 1992), plasmid cloning and expression vectors are available, and gene knockouts and conversions can be performed (Jolley et al., 1996). These improvements have allowed increasingly detailed examinations of halobacterial genes and their regulation (e.g. Offner and Pfeifer, 1995). Halobacterial transcriptional signals have become clearer over this period, and show convincing homology to the eukaryotic system (Hausner et al., 1991; Palmer and Daniels, 1995; Reiter et al., 1990), but the analysis of translational signals has been less tractable, and little has been directly proven by experiment. For example, there remains no direct proof that archaeal mRNAs possess ribosome-binding sites (or Shine-Dalgarno sequences) for positioning the start codon on the ribosome. Some genes show very clear motifs, 3–6 bp upstream of the start codon, that are complementary to the 3' end of the 16S rRNA, e.g. ribosomal protein L1e (Shimmin and Dennis, 1989), but many do not. The question is even more intriguing in halobacteria, which show a preponderance of mRNAs that lack any significant 5' leader sequence, precluding a ribosome-binding site upstream of the start codon (e.g. *bop*, *hmg*) (Lam and Doolittle, 1992; Xu et al., 1995). In such cases, it has been postulated that secondary structures downstream of the start codon might act as ribosome-binding sites but, despite a recent attempt to demonstrate this using the *bop* gene (Xu et al., 1995), the situation remains unclear. Perhaps the so-called "downstream box" found in leaderless bacterial mRNAs (e.g. Winzeler and Shapiro, 1997) plays a role in Archaea.

The study of halobacterial gene expression could be greatly improved if a convenient reporter gene were available. The *Escherichia coli lacZ* gene, coding for ß-galactosidase, is widely used in bacterial and eukaryal systems, and provides a rapid, simple and sensitive means of measuring gene activity (Beale et al., 1992; Slauch and Silhavy, 1991). Unfortunately, the extremely high salt concentrations found in the halobacterial cytoplasm — up to 5 M KCl (Masui and Wada, 1973) — prevent the *lacZ* gene from expressing an active enzyme (M. Dyall-Smith, unpublished observation). Currently, the measurement of gene activity in halobacteria requires technically demanding methods such as quantitative northern blot hybridization (Palmer and Daniels, 1995), spectroscopic measurement of (colored) protein (Xu et al., 1995), or western blot assays (Englert and Pfeifer, 1993).

In light of current difficulties, we searched for a halobacterial enzyme with catalytic activity toward typical *lacZ* substrates, such as ONPG (*o*-nitrophenyl-ß-D-galactopyranoside) and X-Gal (5-bromo-4-chloro-3-indolyl-ß-D-galactopyranoside). Many natural halobacterial isolates possess ß-galactosidase(s) able to hydrolyze ONPG (Torreblanca et al., 1986), and we have studied one such isolate, designated *Haloferax alicantei*, for several years (Holmes and Dyall-Smith, 1990; Holmes et al., 1991). The ß-galactosidase from this strain was purified, its general enzymatic properties examined, and part of its protein sequence determined (Holmes et al., 1997). More recently, the gene coding for this enzyme was isolated, sequenced and expressed in *Haloferax volcanii*. In this chapter, we summarize the work so far concerning the enzyme and its gene, and discuss its potential as a general reporter for gene expression in halobacteria.

21.2 "HALOFERAX ALICANTEI" AND MUTANT SB1

The strain studied, "*Haloferax alicantei*," previously designated phenon K, strain Aa2.2 (Torreblanca et al. 1986), does not grow on lactose as its sole carbon source, but can hydrolyze the synthetic ß-galactoside substrates ONPG and X-Gal. After UV mutagenesis and screening for deep blue colonies on X-Gal-containing plates, a mutant (SB1) was obtained with an enzyme level 25 times higher than the parent's. The enzyme was halophilic and irreversibly lost all activity at low salt (< 0.5 M). Fortunately, 30% (w/v) sorbitol was found to stabilize enzyme activity in the absence of salt. This allowed conventional purification methods (e.g. ion-exchange chromatography) to be utilized. A three-step protocol (i.e. gel filtration, Q-sepharose and immobilized metal ion affinity) produced a 140-fold purification of the enzyme, but SDS-polyacrylamide gel electrophoresis still showed contaminating protein species. Pure enzyme was obtained by preparative electrophoresis on non-denaturing polyacrylamide gels (8%, Tris-glycine buffer) containing 30% sorbitol, and detecting the position of the enzyme band by activity staining with ONPG or X-Gal (Holmes et al., 1997).

Enzyme activity was optimal at 4 M NaCl and was greatly reduced in the absence of reducing agents (dithiothreitol or 2-mercaptoethanol). The size of the native enzyme was estimated to be 180 ± 20 kDa, whereas SDS-PAGE showed a single band of about 78 ± 3 kDa, indicating the enzyme is dimeric. Lactose was not hydrolyzed but the enzyme cleaved several other ß-galactosides, such as ONPG, X-Gal, and lactulose. It showed weaker ß-fucosidase activity and no ß-glucosidase, ß-arabinosidase or ß-xylosidase activity.

21.3 CLONING THE ß-GALACTOSIDASE GENE

An N-terminal amino acid sequence (25 amino acids) and the sequence of some internal peptides were obtained using the purified enzyme, and these data were used to design DNA oligonucleotide primers for PCR amplification of the gene. One of the longer PCR products (1.2 kb) was sequenced to confirm that it coded for a ß-galactosidase, and then used as a specific probe in Southern blot hybridizations to isolate the entire gene from *Kpn*I and *Sau*3A digests of genomic DNA. The final clone, approximately 5.4 kb in length, was sequenced (Genbank accession U70664), and found to contain the entire ß-galactosidase gene, surrounded by other genes involved in carbohydrate metabolism.

21.3.1 THE SEQUENCE OF *BGAH*

The *bgaH* ORF has a coding potential of 665 amino acids, and the predicted size of the protein (74.6 kDa) was close to previous molecular weight estimates (78 ± 3 kDa) of the purified enzyme (Holmes et al., 1997). The sequence contained all the peptide sequences previously determined, confirming the identity of the gene. While the N-terminal protein sequence showed no initial methionine, the nucleotide sequence showed an ATG start codon immediately adjacent to the first amino acid of the protein sequence (threonine), indicating that the initiating methionine is removed post-translationally.

The ß-galactosidase sequence most similar to the halobacterial enzyme was found to be BgaB of the Gram-positive bacterium *Bacillus stearothermophilus* (32% amino acid identity). The high level of similarity between the halobacterial and

Bacillus enzymes was remarkable, given the phylogenetic distance of the hosts and the differing salt requirements of the enzymes. Glycosyl hydrolases have been grouped according to sequence similarity (Henrissat and Bairoch, 1993, 1996), and the *H. alicantei* enzyme clearly belongs to family 42, a group that currently contains three ß-D-galactosidases; *B. stearothermophilus* BgaB, *B. circulans* BgaB, and *Arthrobacter* sp. B7 LacG (Henrissat and Bairoch, 1993; Henrissat, personal communication). Following the nomenclature of the other members of this family, the halobacterial ß-galactosidase gene was designated *bgaH* (H signifying the genus name *Haloferax*).

Family 42 members are dimeric enzymes with component monomers of about 70–80 kDa. Compared with *E. coli lacZ* they differ significantly, the latter enzyme being tetrameric and consisting of much larger monomers (116 kDa). Within family 42, only the *Arthrobacter* enzyme has been tested against a range of substrates (Gutshall et al., 1995), and was also found to be specific for β-D-galactosides, including lactose. Hypersaline lakes are not known to contain sources of lactose, so the natural function of the halobacterial enzyme is unclear. The *Sulfolobus* LacS enzyme (Cubellis et al., 1990) shares only limited amino-acid similarity with BgaH.

21.3.2 ORFs Surrounding *BGAH*

Three long ORFs were found immediately upstream of *bgaH* (Figure 22.1). Database searches revealed that they share strong amino-acid similarity to 2-keto-3-deoxygluconate kinase (KDG kinase), 2-dehydro-3-deoxyphosphogluconate aldolase (KDPG-aldolase) and glucose-fructose oxidoreductase (GFOR). These are all enzymes involved in carbohydrate metabolism. KDG kinase phosphorylates 2-keto-3-deoxygluconate, and the product is cleaved by KDPG aldolase, producing glyceraldehyde-3-phosphate and pyruvate. KDPG aldolase is a member of the modified Entner-Doudoroff pathway found in halobacteria (reviewed by Danson, 1989). Interestingly, the KDP kinase and KDPG-aldolase genes are in opposite orientation and overlap by 312 nt, which probably indicates some regulatory feature (Figure 22.1).

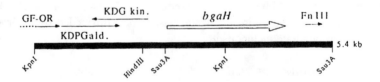

FIGURE 21.1 Schematic representation of the cloned 5.4 kb region of "*H. alicantei*" SB1 genomic DNA. Orientations and lengths of identified ORFs are shown by arrows. GFOR, glucose-fructose oxidoreductase (incomplete as indicated by the dotted line); KDG kinase, 2-keto-3-deoxygluconate kinase; KDPG aldolase, 2-dehydro-3-deoxyphosphogluconate aldolase; *bgaH*, ß-galactosidase gene; FnIII, ORF containing a fibronectin type III motif.

GFOR converts glucose and fructose to gluconolactone and sorbitol (Leigh et al., 1984). In *Zymomonas*, GFOR produces sorbitol (Zachariou and Scopes, 1986) and gluconolactone, the latter compound being subsequently converted to gluconate, which is an intermediate of the modified Entner-Doudoroff pathway of halobacteria.

In summary, it appears that the enzymes upstream of *bgaH* are part of, or peripheral to, the Entner-Doudoroff pathway of this organism.

A surprising feature (discovered by database searches) downstream of the *bgaH* gene is the first archaeal example of a fibronectin type III motif. In animal protein, these are relatively common, being usually about 90 amino acids in length (Bork and Doolittle, 1993). More recently, these motifs have been discovered in bacterial enzymes (Little et al., 1994), and it has been proposed that this motif has been laterally transferred. The functions of this motif have not been clearly established.

21.4 TRANSFORMATION OF *H. VOLCANII*

The entire *bgaH* gene was cloned as a 3.4 kb *Hin*dIII-*Sau*3A fragment into the shuttle plasmid pMDS20 (Holmes et al., 1994), and the resulting construct, pMLH32, was introduced into *H. volcanii* cells. When novobiocin-resistant transformants were grown on selective plates and the colonies sprayed with X-Gal, they turned dark blue after several hours' incubation at 37°C. X-Gal was inhibitory to the growth of cells, most likely due to the insoluble blue product accumulating inside cells, so we do not include this compound in solid media. Enzyme activities of wild-type and transformant cells (using ONPG as a substrate) showed that the transformant expressed high levels of enzyme activity, approximately double that of the SB1 strain of *"H. alicantei"* (data not shown). The higher activity is probably a reflection of the plasmid copy number.

21.5 CONCLUSION

The *bgaH* gene can be expressed when introduced into *H. volcanii* cells, and colonies can be rapidly screened for BgaH activity by spraying plates with X-Gal. ONPG can also be used as a substrate, allowing quantitative assays. *H. volcanii* is a convenient host, as it lacks endogenous ß-galactosidase activity, making possible the detection of even low levels of BgaH. It is hoped that the *bgaH* gene will be as useful in halobacteria as *lacZ* has been in *E. coli*.

ACKNOWLEDGMENTS

We thank M. Danson, R. Scopes, B. Henrissat and R. Doolittle for helpful discussions throughout this work. The project was funded by the Australian Research Council.

REFERENCES

Beale, E.G., E.A. Deeb, R.S. Handley, H. Akhaven-Tafti, and A.P. Schaap. 1992. A rapid and simple chemiluminescent assay for *Escherichia coli* ß-galactosidase. *Biotechniques* 12:320–324.

Bork, P. and R.F. Doolittle. 1993. Fibronectin type III modules in the receptor phosphatase CD45 and tapeworm antigens. *Protein Sci.* 2:1185–1187.

Cline, S.W. and W.F. Doolittle. 1992. Transformation of members of the genus *Haloarcula* with shuttle vectors based on *Halobacterium halobium* and *Haloferax volcanii* plasmid replicons. *J. Bacteriol.* 174:1076–1080.
Cubellis, M.V., C. Rozzo, P. Montecucchi, and M. Rossi. 1990. Isolation and sequencing of a new beta-galactosidase-encoding archaebacterial gene. *Gene* 94:89–94.
Danson, M.J. 1989. Central metabolism of the archaebacteria: an overview. *Can. J. Microbiol.* 35:58–64.
Doolittle, W.F., W.L. Lam, L.C. Schalkwyk, R.L. Charlebois, S.W. Cline, and A. Cohen. 1992. Progress in developing the genetics of the halobacteria. *Biochem. Soc. Symp.* 58:73–78.
Dyall-Smith, M.L. 1998. Molecular genetics of halobacteria, in *Biotechnology Handbook Series: Archaebacteria*, D.A. Cowan, Ed. Plenum, London.
Englert, C. and F. Pfeifer. 1993. Analysis of gas vesicle gene expression in *Haloferax mediterranei* reveals that GvpA and GvpC are both gas vesicle structural proteins. *J. Biol. Chem.* 268:9329–9336.
Gutshall, K.R., D.E. Trimbur, J.J. Kasmir, and J.E. Brenchley. 1995. Analysis of a novel gene and β-galactosidase isozyme from a psychrotrophic *Arthrobacter* isolate. *J. Bacteriol.* 177:1981–1988.
Hausner, W., G. Frey, and M. Thomm. 1991. Control regions of an archaeal gene — a TATA box and an initiator element promote cell-free transcription of the tRNA^{val} gene of *Methanococcus vannielii*. *J. Mol. Biol.* 222:495–508.
Henrissat, B. and A. Bairoch. 1993. New families in the classification of glycosyl hydrolases based on amino acid sequence similarities. *Biochem. J.* 293:781–788.
Henrissat, B. and A. Bairoch. 1996. Updating the sequence-based classification of glycosyl hydrolases. *Biochem. J.* 316:695–696.
Holmes, M.L. and M.L. Dyall-Smith. 1990. A plasmid vector with a selectable marker for halophilic archaebacteria. *J. Bacteriol.* 172:756–761.
Holmes, M.L., S.D. Nuttall, and M.L. Dyall-Smith. 1991. Construction and use of halobacterial shuttle vectors and further studies on *Haloferax* DNA gyrase. *J. Bacteriol.* 173:3807–3813.
Holmes, M., F. Pfeifer, and M. Dyall-Smith. 1994. Improved shuttle vectors for *Haloferax volcanii* including a dual-resistance plasmid. *Gene* 146:117–121.
Holmes, M.L., R.K. Scopes, R.L. Moritz, R.J. Simpson, C. Englert, F. Pfeifer, and M.L. Dyall-Smith. 1997. Purification and analysis of an extremely halophilic ß-galactosidase from *Haloferax alicantei*. *Biochim. Biophys. Acta* 1337:276–286.
Jolley, K.A., E. Rapaport, D.W. Hough, M.J. Danson, W.G. Woods, and M.L. Dyall-Smith. 1996. Dihydrolipoamide dehydrogenase from the halophilic archaeon *Haloferax volcanii*: homologous overexpression of the cloned gene. *J. Bacteriol.* 178:3044–3048.
Keeling, P.J. and W.F. Doolittle. 1995. Archaea: narrowing the gap between prokaryotes and eukaryotes. *Proc. Natl. Acad. Sci. USA* 92:5761–5764.
Lam, W.L. and W.F. Doolittle. 1992. Mevinolin-resistant mutations identify a promoter and the gene for a eukaryote-like 3-hydroxy-3-methylglutaryl-coenzyme A reductase in the archaebacterium *Haloferax volcanii*. *J. Biol. Chem.* 267:5829–5834.
Leigh, D., R.K. Scopes, and P.L. Rogers. 1984. A proposed pathway for sorbitol production in *Zymomonas mobilis*. *Appl. Microbiol. Biotechnol.* 20:413–415.
Little, E., P. Bork, and R.F. Doolittle. 1994. Tracing the spread of fibronectin type III domains in bacterial glycohydrolases. *J. Mol. Evol.* 39:631–643.
Masui, M. and S. Wada. 1973. Intracellular concentrations of Na$^+$, K$^+$ and Cl$^-$ of a moderately halophilic bacterium. *Can. J. Microbiol.* 19:1181–1186.

Offner, S. and F. Pfeifer. 1995. Complementation studies with the gas vesicle-encoding p-vac region of *Halobacterium salinarium* pHH1 reveal a regulatory role for the p-*gvpDE* genes. *Mol. Microbiol.* 16:9–19.

Palmer, J.R. and C.J. Daniels. 1995. In vivo definition of an archaeal promoter. *J. Bacteriol.* 177:1844–1849.

Reiter, W.D., U. Hudepohl, and W. Zillig. 1990. Mutational analysis of an archaebacterial promoter: essential role of a TATA box for transcription efficiency and start-site selection in vitro. *Proc. Natl. Acad. Sci. USA* 87:9509–9513.

Shimmin, L.C. and P.P. Dennis. 1989. Characterization of the L11, L1, L10 and L12 equivalent ribosomal protein gene cluster of the halophilic archaebacterium *Halobacterium cutirubrum*. *EMBO J.* 8:1225–1235.

Slauch, J.M. and T.J. Silhavy. 1991. Genetic fusions as experimental tools. *Meth. Enzymol.* 204:213–248.

Torreblanca, M., F. Rodriguez-Valera, G. Juez, A. Ventosa, M. Kamekura, and M. Kates. 1986. Classification of non-alkaliphilic halobacteria based on numerical taxonomy and polar lipid composition, and description of *Haloarcula* gen. nov. and *Haloferax* gen. nov. *Syst. Appl. Microbiol.* 8:89–99.

Winzeler E. and L. Shapiro. 1997. Translation of the leaderless *Caulobacter dnaX* mRNA. *J. Bacteriol.* 179:3981–3988.

Xu, Z.J., D.B. Moffett, T.R. Peters, L.D. Smith, B.P. Perry, J. Whitmer, S.A. Stokke, and M. Teintze. 1995. The role of the leader sequence coding region in expression and assembly of bacteriorhodopsin. *J. Biol. Chem.* 270:24858–24863.

Zachariou, M. and R.K. Scopes. 1986. Glucose-fructose oxidoreductase, a new enzyme isolated from *Zymomonas mobilis* that is responsible for sorbitol production. *J. Bacteriol.* 167:863–869.

22 What Do the Extreme Halophiles Tell Us About the Evolution of the Proton-Translocating ATPases?

Lawrence I. Hochstein and Roberto Bogomolni

CONTENTS

22.1 Introduction 273
22.2 ATP Synthesis in Vesicles 274
    22.2.1 Effect of Inhibitors on ATP Synthesis 274
    22.2.2 Effect of Azide 276
22.3 Conclusion 278
Acknowledgments 278
References 278

22.1 INTRODUCTION

The proton-translocating ATPases are a super-family of enzymes that appear to have evolved from a common ancestor (Gogarten et al., 1989). They can be distinguished on the basis of their sensitivity to certain inhibitors as well as their subunit structure (for reviews of the F_0F_1-ATP synthases or F-ATPases, and the vacuolar or V-ATPases, see Downie et al. 1979; and Forgac, 1989, respectively). The V-ATPases are unaffected by azide but are inhibited by N-ethylmaleimide (NEM) and nitrate, as well as by the macrolide antibiotics bafilomycin A_1 (Bowman et al., 1988) and concamycin A (Muroy et al., 1993). The opposite pattern is obtained in the case of the F-ATPases. The F and V-ATPases also differ with respect to the molecular masses and function of their subunits.

A third class of ATPases (the archaeal or A-ATPases) is found in several extreme halophiles (Dane et al., 1992; Kristjansson et al., 1986; Nanba and Mukohata, 1987). The enzyme from *Halorubrum saccharovorum* is unaffected by azide (Kristjansson and Hochstein, 1985), but is inhibited by NEM and nitrate (Stan-Lotter et al., 1991). As is the case with the V-ATPases, the catalytic site appears to be associated with the largest of the subunits (subunit I) (Bonet and Schobert, 1992; Sulzner et al., 1992). In addition, antiserum against subunit A of the *Neurospora crassa* V-ATPase (which is the catalytic subunit) reacts with subunit I from the *H. saccharovorum* ATPase (Stan-Lotter et al., 1991). However, not all of the halobacterial A-ATPases have the same properties (Dane et al., 1992; Stan-Lotter and Hochstein, 1989). A similar situation exists with respect to the A-ATPases from other Archaea, and these may represent ATPases whose functions have not yet been established (Hochstein and Stan-Lotter, 1992).

In spite of these differences, the amino-acid sequences of the two largest F, V and A-ATPases subunits indicate that they arose from a common ancestor in a process that included gene duplication of the catalytic subunit, segregation into unique functions, and the loss of ATP synthase activity in the case of the V-ATPases (Gogarten et al., 1989). Implicit in this proposal is that A-ATPases retained their ability to synthesize ATP. We recently reported that ATP synthesis in cells of several extreme halophiles is not affected by NEM or nitrate but is inhibited by azide, and we suggested that an F-type ATPase was involved in the process (Hochstein, 1992; Hochstein and Lawson, 1993). Here we extend these observations and describe experiments demonstrating that the proton-dependent synthesis of ATP in "inside-in" vesicle preparations is affected in the same manner as in cells. In addition, we will present evidence that azide inhibition is not a consequence of its potential to act as a protonophore.

22.2 ATP SYNTHESIS IN VESICLES

Figure 22.1 describes the kinetics of the proton-dependent synthesis of ATP by *H. saccharovorum* vesicles. The addition of citrate buffer ($\Delta pH = 4$) resulted in the production of ATP, which occurred at an initial rate of 590 pmol ATP min^{-1} (mg vesicle protein^{-1}). An apparent steady state was reached after about 5 min, when approximately 840 pmol of ATP (mg vesicle protein)$^{-1}$ had been synthesized. Thereafter, the amount of ATP slowly decreased (not shown).

22.2.1 Effect of Inhibitors on ATP Synthesis

Table 22.1 describes the effect of various agents that would affect the utilization of ΔpH for the synthesis of ATP. There was no synthesis of ATP in the presence of 0.1% Triton X-100 (v/v), a concentration that lysed the vesicles. The protonophore carbonylcyanide m-chlorophenylhydrazone (CCCP) inhibited synthesis, as did dicyclohexylcarbodiimide (DCCD). The inhibition at any DCCD concentration increased with the time that the vesicles were exposed to the inhibitor at pH 8 (data not shown). DCCD inhibits the synthesis and hydrolysis of ATP by F-ATPases when the reaction is carried out in a slightly alkaline medium (Hoppe and Sebald, 1984). DCCD also inhibits ATP hydrolysis when conditions are acidic by reacting with a highly conserved glutamate residue in the β subunit (Vignais and Lunardi, 1985). DCCD inhibits the A-ATPase from *H. saccharovorum*, but only when conditions are acidic

Evolution of the Proton-Translocation ATPases

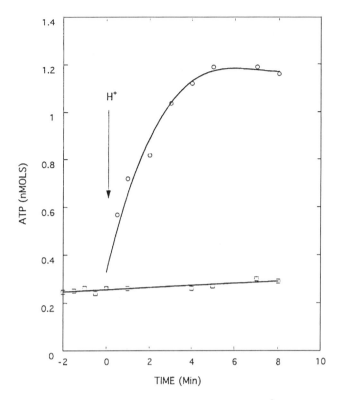

FIGURE 22.1 ATP synthesis in *H. saccharovorum* vesicles. *H. saccharovorum* (ATCC 29252) grown as previously described (Hochstein, 1992), was suspended in 50 mM Tris/4 M NaCl/10 mM $MgCl_2$, pH 8.0 buffer, washed twice, and suspended in the same buffer. The cells were incubated overnight on a rotary shaker at 37°C, centrifuged at 5,900 * g for 30 min at 4°C, and subsequently suspended in 20 mM Tris/4 M NaCl/10 mM $MgCl_2$/10 mM Na_2HPO_4/5 mM ADP, pH 8.0 buffer. The suspension was passed through a French pressure cell operated at 3,000–4,000 psi and incubated with DNAase I for 20 min at 37°C. After centrifuging at 5,900 * g for 30 min at 4°C, the supernatant was decanted and centrifuged at 163,000 * g for 30 min at 4°C. The pellet (vesicle fraction) was suspended in 20 mM Tris/4 M NaCl/10 mM $MgCl_2$, pH 8.0 buffer and incubated overnight at room temperature. The orientation of the vesicles was >90% "inside-in," using menadione reductase activity as the marker (Lanyi and MacDonald, 1979). Vesicles (1.1 mg protein) were suspended in 20 mM Tris/4 M NaCl/10 mM $MgCl_2$, pH 8.0 buffer, and the reaction was initiated by the addition of 20 µl of 250 mM citrate buffer (arrow), resulting in a ΔpH of 4. The open squares represent vesicles maintained at pH 8.

(Kristjansson and Hochstein, 1985). The dependence of DCCD inhibition on preincubation at pH 8 suggests that inhibition of ATP synthesis was not due to DCCD reacting with subunit II of the A-ATPase or the β subunit of the F-ATPase.

Table 22.1 also describes the effect of various agents whose inhibitory actions are used to differentiate the V, A, and F-ATPases. ATP synthesis was not inhibited by the V-ATPase inhibitors nitrate or NEM; in fact, the amount of ATP produced in the presence of NEM was slightly higher than in its absence, a situation previously

TABLE 22.1
Effect of ATPase Inhibitors on ATP Synthesis by *H. saccharovorum* Vesicles[a]

Inhibitor	Concentration	Activity (% of Control)
Triton X-100	0.1% (v/v)	0
CCCP	5 µM	0
DCCD	5 µM	27
Na-nitrate	40 mM	95
NEM	1 mM	116
Na-acetate	10 mM	90
Na-azide	100 µM	5
NBM-Cl	160 µM	57

[a] Vesicles (1.1 mg protein) suspended in 20 mM Tris/4 M NaCl/10 mM $MgCl_2$/10 mM Na_2HPO_4/5 mM ADP, pH 8.0 buffer were incubated in the presence of the indicated inhibitors for 1 h at room temperature before initiating ATP synthesis by the addition of 20 µl of 250 mM pH 3 citrate buffer. After 1 min, aliquots were diluted 1:10 with 0.2% Triton X-100, and ATP was determined as previously described (Hochstein, 1992). The control (100%) synthesized 690 pmol of ATP min^{-1}. None of the inhibitors affected luminescence when added at the concentrations present in the Triton-treated reaction mixtures. NaCl quenched luminescence, which was compensated by preparing ATP standards in 100 µM Tris/20 mM NaCl/50 µM $MgCl_2$, pH 8.0 buffer, being the final concentration of the buffer constituents in the Triton-diluted reaction mixture.

observed with cells (Hochstein and Lawson, 1993). This stimulation is consistent with the existence of two ATPases, one a synthase (F-ATPase) and the other a hydrolytic enzyme (A-ATPase) whose activity was inhibited by the thiol reagent.

22.2.2 Effect of azide

ATP synthesis was inhibited by relatively low concentrations of azide (Table 22.1). Azide inhibits a variety of F-ATPases (Downie et al. 1979), but does not inhibit V-ATPases (Forgac, 1989). Azide is a weak acid, and it can uncouple ATP synthesis (Kobayashi et al. 1977). Two observations suggest that this is not the case with respect to ATP synthesis in *H. saccharovorum*. First, acetate, which has a pK_a approximating that of azide, did not inhibit ATP synthesis (Table 22.1). The second

Evolution of the Proton-Translocation ATPases

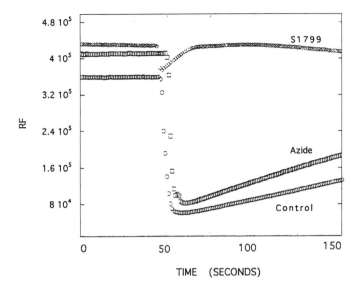

FIGURE 22.2 The effect of azide on the pH-dependent distribution of acridine orange in *H. saccharovorum* vesicles. Vesicles were prepared as previously described except that the cells were suspended in 20 mM MES/4 M NaCl/10 mM $MgCl_2$/10 mM Na_2HPO_4/5 mM ADP, pH 6 buffer prior to being disrupted. Following centrifugation at 163 000 * g, the pellet was suspended in the same buffer and stored at room temperature. In this experiment, vesicles (320 µg protein) were suspended in 2 ml of 20 mM MES/4 M NaCl/10 mM $MgCl_2$/10 mM Na_2HPO_4/5 mM ADP pH 6 buffer that was 4 µM with respect to AO, and stirred in the dark for 5 min. Fluorescence was measured using 450 nm as the excitation wavelength and 530 nm as the emission wavelength. Where indicated, 10 nmol of S1799 or 3.1 µmol of azide were present. The data were acquired at a rate of two data points per second and reported as relative fluorescence (RF) units.

evidence came from the effect of azide on the pH-dependent distribution of acridine orange (AO). AO distributes between compartments as a function of ΔpH (Palmgren, 1991). If azide collapsed ΔpH, then azide should abolish the pH-dependent accumulation of AO within vesicles. Figure 22.2 summarizes the results of a typical experiment to test this hypothesis. *H. saccharovorum* vesicles (inside pH of 6.0) were equilibrated with AO at that pH and then the pH of the bulk medium was changed from 6 to 8. This resulted in a rapid decrease in fluorescence (ca. 80%), reflecting the net influx of AO into the vesicles. When the same experiment was carried out in the presence of the uncoupler S1799, a small change in fluorescence (10–15%) was observed, followed by a rapid recovery to the initial fluorescence value. When the same experiment was carried out in the presence of 1.5 mM azide (which inhibits ATP synthesis for more than 95%), the change in fluorescence was essentially the same as observed in the control with a slightly greater rate of recovery. So, if azide acts as a protonophore, it does not do so at the concentrations and conditions employed.

22.3 CONCLUSION

Our results differ from those observed with *H. salinarum* (Mukohata and Yoshida, 1987a) and *H. volcanii* (Steinert et al., 1997). We have no definitive explanation for these discrepancies at this time. The three vesicle preparations were prepared differently. In the case of *H. salinarum*, "inside-in" vesicles were prepared by sonication in pH 6 buffer that was 1 M and 100 mM with respect to NaCl and $MgCl_2$, respectively. The *H. volcanii* vesicles were also prepared by sonication but these vesicles were inverted. In the case of *H. saccharovorum* and *H. salinarum*, ATP synthesis was brought about by an inward flow of protons, whereas the opposite was the case with *H. volcanii*. In addition, the effect of inhibitors of ATP synthesis with *H. volcanii* vesicles was unlike that observed with *H. salinarum* and *H. saccharovorum*. Synthesis with *H. volcanii* vesicles was not affected by 7-chloro-4-nitrobenzo-2-oxa-1,3-diazole (NBD-Cl), which inhibited ATP synthesis in *H. saccharovorum* (Table 22.1) and *H. salinarum* (Mukohata and Yoshida, 1987b). DCCD, which inhibits ATP synthesis in *H. saccharovorum* at relatively low concentrations, was ineffective in the case of *H. volcanii*, unless the vesicles were incubated for at least 24 h in the presence of extremely high concentrations of DCCD. In addition, a relatively high concentration of p-trifluoromethoxyphenylhydrazone was required to uncouple ATP synthesis.

Earlier, we found that ATP synthesis in cells of *Haloferax mediterranei, H. volcanii*, and *Haloarcula hispanica* mimics what occurs in *H. saccharovorum* and *H. salinarum* (Hochstein and Lawson, 1993). We take this to mean that ATP is also synthesized by F-ATPases in these extreme halophiles. The presence of F-ATPases in the Archaea is not restricted to the extreme halophiles. DNA coding for the β subunit of an F-type ATPase is present in *Methanosarcina barkeri* (Sumi et al., 1992) and functional F-ATPases occur in *M. mazei* (Becher and Müller, 1994) and *Methanobacterium thermoautotrophicum* (Simgan et al., 1995). To what extent they occur and are responsible for ATP synthesis in the other Archaea is an interesting question that remains to be answered. Finally, if F-ATPases synthesize ATP in the extreme halophiles, then the function of the A-ATPases remains to be defined.

ACKNOWLEDGMENTS

We thank Darion Lawson for expert technical assistance. The research was supported by the NASA program in the early evolution of life.

REFERENCES

Becher, B. and V. Müller. 1994. ΔμNa⁺ drives the synthesis of ATP via an ΔμNa⁺-translocating F_1F_0-ATP synthase in membrane vesicles of the archaeon *Methanosarcina mazei* Göl. *J. Bacteriol.* 176:2543–2550.

Bonet, M.L. and B. Schobert. 1992. The catalytic cycle is located on subunit I of the ATPase from *Halobacterium saccharovorum*. *Eur. J. Biochem.* 207:369–376.

Bowman, E.J., A. Siebers, and K. Altendorf. 1988. Bafilomycins: a class of inhibitors of membrane ATPases from microorganisms, animal cells, and plant cells. *Proc. Natl. Acad. Sci. USA* 85:7972–7976.

Dane, M., K. Steinert, K. Esser, S. Bickel-Sandkötter, and F. Rodriguez-Valera. 1992. Properties of the plasma membrane ATPases of the halophilic archaebacteria *Haloferax mediterranei* and *Haloferax volcanii*. *Z. Naturforsch.* 47c:835–844.

Downie, J.A., F. Gibson, and G.B. Cox. 1979. Membrane adenosine triphosphatases of prokaryotic cells. *Ann. Rev. Biochem.* 48:103–131.

Forgac, M. 1989. Structure and function of vacuolar classes if ATP-driven proton pumps. *Physiol. Rev.* 69:765–796.

Gogarten, J.P., H. Kibak, P. Dittrich, L. Taiz, E.J. Bowman, B.J. Bowman, M.F. Manolson, R.J. Poole, T. Date, T. Oshima, J. Konishi, K. Denda, and M. Yoshida. 1989. Evolution of the vacuolar H$^+$-ATPase: implication for the origin of eukaryotes. *Proc. Natl. Acad. Sci. USA* 86:6661–6665.

Hochstein, L.I. 1992. ATP synthesis in *Halobacterium saccharovorum*: evidence that synthesis may be catalyzed by an F_0F_1-ATP synthase. *FEMS Microbiol. Lett.* 97:155–160.

Hochstein, L.I. and D. Lawson. 1993. Is ATP synthesized by a vacuolar-ATPase in the extremely halophilic bacteria? *Experientia* 49:1059–1063.

Hochstein, L.I. and H. Stan-Lotter. 1992. Purification and properties of an ATPase from *Sulfolobus solfataricus*. *Arch. Biochem. Biophys.* 295:153–160.

Hoppe, J. and W. Sebald. 1984. The proton conducting F_0 part of bacterial ATP synthase. *Biochim. Biophys. Acta* 768:1–27.

Kobayashi, H., M. Maeda, and Y. Anraku. 1977. Membrane-bound adenosine triphosphatase of *Escherichia coli*. III. Effects of sodium azide on the enzyme functions. *J. Biochem.* 81:1071–1077.

Kristjansson, H. and L.I. Hochstein. 1985. Dicyclohexylcarbodiimide-sensitive ATPase in *Halobacterium saccharovorum*. *Arch. Biochem. Biophys.* 241:590–595.

Kristjansson, H., M. Sadler, and L.I. Hochstein. 1986. Halobacterial adenosine triphosphatases and the adenosine triphosphatase from *Halobacterium saccharovorum*. *FEMS Microbiol. Rev.* 39:151–157.

Lanyi, J. and R.E. MacDonald. 1979. Light-induced transport in *Halobacterium halobium*, in *Methods in Enzymology*, Vol. 56. S. Fleischer and L. Packer, Eds. Academic, New York. 398–347.

Mukohata, Y. and M. Yoshida. 1987a. Activation and inhibition of ATP synthesis in cell envelope vesicles of *Halobacterium halobium*. *J. Biochem.* 101:311–318.

Mukohata, Y. and M. Yoshida. 1987b. The H$^+$-translocating ATP synthase in *Halobacterium halobium* differs from F_0F_1-ATPase/synthase. *J. Biochem.* 102:797–802.

Muroi, M., N. Shiragami, K. Nagao, M. Yamasaki, and A. Takatsuki. 1993. Folimycin (concanamycin A), a specific inhibitor of V-ATPase, blocks intracellular translocation of the glycoprotein of vesicular stomatitis virus before arrival to the Golgi apparatus. *Cell Struct. Funct.* 18:139–149.

Nanba, T. and Y. Mukohata. 1987. A membrane-bound ATPase from *Halobacterium halobium*: purification and characterization. *J. Biochem.* 102:591–598.

Palmgren, M.G. 1991. Acridine orange as a probe for measuring pH gradients across membranes: mechanism and limitations. *Anal. Biochem.* 192:316–321.

Smigan, P., A. Majernik, P. Polak, I. Hapala, and M. Greksak. 1995. The presence of H$^+$ and Na$^+$ translocating ATPases in *Methanobacterium thermoautotrophicum* and their possible function under alkaline conditions. *FEBS Lett.* 371:119–122.

Stan-Lotter, H., E.J. Bowman, and L.I. Hochstein. 1991. Relationship of the membrane ATPase from *Halobacterium saccharovorum* to vacuolar ATPases. *Arch. Biochem. Biophys.* 284:116–119.

Stan-Lotter, H. and L.I. Hochstein. 1989. A comparison of an ATPase from the archaebacterium *Halobacterium saccharovorum* with the F_1 moiety from *Escherichia coli* ATP synthase. *Eur. J. Biochem.* 179:155–160.

Steinert, K., V. Wagner, P.G. Kroth-Panic, and S. Bickel-Sandkötter. 1997. Characterization and subunit structure of the ATP synthase of the halophilic archaeon *Haloferax volcanii* and organization of the ATP synthase genes. *J. Biol. Chem.* 272:6261–6269.

Sulzner, M., H. Stan-Lotter, and L.I. Hochstein. 1992. Nucleotide-protectable labeling of sulfhydryl groups in subunit I of the ATPase from *Halobacterium saccharovorum*. *Arch. Biochem. Biophys.* 296:347–349.

Sumi, M., M. Sato, K. Denda, T. Date, and M. Yoshida. 1992. A DNA fragment homologous to F_1-ATPase β subunit was amplified from genomic DNA of *Methanosarcina barkeri*. *FEBS Lett.* 314:207–210.

Vignais, P.V. and J. Lunardi. 1985. Chemical probes of the mitochondrial ATTP synthesis and translocation. *Ann. Rev. Biochem.* 54:977–1014.

23 Comparative Analysis of the Halobacterial Gas Vesicle Gene Clusters

Felicitas Pfeifer, Andrea Mayr, Sonja Offner, and Richard Röder

CONTENTS

23.1 Introduction 281
23.2 The Gas Vesicle Wall Consists of Mainly One Protein, GvpA 282
23.3 Gas Vesicle Formation in *Haloferax mediterranei* Involves the
     14 *gvp* Genes of the mc-vac Region 283
23.4 The p-vac and c-vac Region of *Halobacterium salinarum* 284
23.5 The nv-vac Region of *Natronobacterium vacuolatum* 285
23.6 Expression of the Various vac Regions at the RNA Level 286
23.7 Possible Functions of Gvp Proteins During Gas Vesicle Formation 288
23.8 The Two Regulatory Proteins GvpD and GvpE 288
23.9 Conclusion 290
Acknowledgments..290
References...291

23.1 INTRODUCTION

Enrichment cultures from samples derived from salt ponds usually contain a certain percentage of cells harboring gas vesicles. Rod-shaped, triangular, or square halobacteria contain these gas-filled, proteinaceous structures that are easily visible, using a phase-contrast microscope, as light refractile bodies inside the cell. Gas vesicles are not only limited to halophilic Archaea, many cyanobacteria (such as *Calothrix* spp. or *Anabaena flos-aquae*), and other bacteria contain gas vesicles (Walsby, 1994). A surprisingly high number of gas-vesiculated cells is also found among the bacteria isolated from the sea ice of Antarctica (Staley et al., 1989). In the cyanobacteria, gas vesicles enable the cells to move without a large energy expense to regions of high light intensity and oxygen supply (Walsby, 1994). Besides providing buoyancy, the possession of gas vesicles has an additional effect: the dense filling of the cell interior with gas vesicles raises the surface area-to-volume ratio of the cytoplasm, and the resulting shorter diffusion times could be advantageous for the cells, especially for those growing at low temperatures (Staley et al., 1989).

Gas-vesiculated halophiles include the extremely halophilic purple membrane producing *Halobacterium salinarum* (formerly *Hb. salinarium* and *Hb. halobium* strains; Ventosa and Oren, 1996), and other rod-shaped isolates such as *Halobacterium* spp. GN101, GRA, GRB and SB3 (Ebert et al., 1984). All these organisms are facultative anaerobes and contain retinal proteins as light-driven proton and chloride pumps, together with sensory rhodopsins. In contrast, the disc-shaped *Haloferax mediterranei* (Rodriguez-Valera et al., 1983) and the haloalkaliphilic species *Natronobacterium vacuolatum* (Mwatha and Grant, 1993), recently renamed *Halorubrum vacuolatum* (M. Kamekura et al., *Int. J. Syst. Bacteriol.* U7:853–857, 1997), are moderate halophiles and form gas vesicles depending on the salt concentration of the medium — *Hf. mediterranei* grown in 15% salt medium, and *N. vacuolatum* grown in 13% salt medium are gas-vesicle-free, whereas cells cultivated in media containing higher salt concentrations synthesize gas vesicles (Englert et al., 1990; Mayr and Pfeifer, 1997).

23.2 THE GAS VESICLE WALL CONSISTS OF MAINLY ONE PROTEIN, GvpA

Gas vesicles of halophilic archaea can be isolated by lysing the cells with water, and sampling the floating gas vesicles from the surface of the lysate after a few days. Flotation can be enhanced by the application of low-speed centrifugation (60 * g for 12 h). Gas vesicles sustain a critical maximal pressure before collapsing, and since halobacterial gas vesicles are synthesized without any turgor pressure, they collapse rather rapidly. Their critical pressure is 0.1 MPa (1 bar), whereas cyanobacterial gas vesicles withstand pressures between 0.4 and 3.5 MPa (4-35 bar), depending on the strain (Walsby, 1994).

Electron microscopic inspection of gas vesicles indicates different shapes: the *Hb. salinarum* wild-type strains PHH1 and NRC-1 contain predominantly spindle-shaped gas vesicles, whereas all other halobacteria form cylinder-shaped gas vesicles with sizes varying from 0.4 µm up to 1.5 µm in length, and 200 nm width in the cylinder section (Englert et al., 1990; Walsby, 1994). A striking feature is that 4.5-nm-wide ribs are arranged perpendicular to the length of the gas vesicles. The dissection of gas vesicles and separation of the protein subunits by SDS-PAGE is, however, hampered by their extreme resistance to solubilization in aqueous solutions. Almost all of the proteins contained in a gas vesicle preparation remain aggregated near the top of the separating gel (Englert and Pfeifer, 1993; Simon, 1981; Surek et al., 1988). N-terminal amino acid sequence determination of a gas vesicle preparation reveals the sequence of the 7–8 kDa GvpA, which is the major gas vesicle structural protein, and is highly conserved among cyanobacteria and halobacteria (Figure 23.1) (Englert et al., 1990; Surek et al., 1988; Walsby, 1994). Immunological studies indicate a second structural protein, GvpC, that is present in minor amounts in gas vesicle preparations of both cyanobacteria and halobacteria (Englert and Pfeifer, 1993; Halladay et al., 1993; Hayes et al., 1992). In cyanobacteria, GvpC is located

```
c-GvpA      MAQPDSSSLAEVLDRVLDKGVVVDVWARISLVGIEILTVEARVVAASVDT
p-GvpA      MAQPDSSGLAEVLDRVLDKGVVVDVWARVSLVGIEILTVEARVVAASVDT
mc-GvpA     MVQPDSSSLAEVLDRVLDKGVVVDVWARISLVGIEILTVEARVVAASVDT
nv-GvpA     MAQPDSSSLAEVLDRVLDKGVVVDVYARLSLVGIEILTVEARVVAASVDT

c-GvpA      FLHYAEEIAKIEQAELTAGAEAPEPAP...EA      79 aa
p-GvpA      FLHYAEEIAKIEQAELTAGAEA...AP...EA      76 aa
mc-GvpA     FLHYAEEIAKIEQAELTAGAEA...APTPEA       78 aa
nv-GvpA     FLHYAEEIAKIEQAELTAGAEA...APTPEA       78 aa
```

FIGURE 23.1 Comparison of the amino acid sequences of GvpA deduced from the various *gvpA* genes in halophilic Archaea. The sequences derive from: c-GvpA: *Hb. salinarum* c-vac region; p-GvpA: *Hb. salinarum* p-vac region; mc-GvpA: *Hf. mediterranei* mc-vac region; nv-GvpA: *N. vacuolatum* nv-vac region.

on the outside of the gas vesicle wall and stabilizes the ribbed structure formed by GvpA by binding several ribs like a clamp (Hayes et al., 1992; Kinsman et al., 1995; Walsby, 1994). A particular structural feature is important for this function: the amino acid sequence of GvpC exhibits four to seven repeated sequences of 25–38 amino acids. Each of these repeats is long enough to span one rib, and it has been demonstrated for cyanobacterial gas vesicles that the strengthening is increased with GvpC proteins containing at least two of these repeats (Kinsman et al., 1995). The genes encoding the two structural proteins GvpA and GvpC have been isolated from the cyanobacteria *Calothrix* and *Anabaena flos-aquae* (Damerval et al., 1987; Hayes and Powell, 1995), as well as from *Hb. salinarum* and *Hf. mediterranei* (DasSarma et al., 1987; Englert et al., 1990; Horne and Pfeifer, 1989).

23.3 GAS VESICLE FORMATION IN *HALOFERAX MEDITERRANEI* INVOLVES THE 14 *gvp* GENES OF THE mc-vac REGION

Only in the case of halobacteria has gas vesicle formation been tested by transformation experiments using fragments containing these genes. Halobacterial vector plasmids that carry compatible origins of replication and a selectable marker conferring resistance either against mevinolin or novobiocin are available for complementation studies (Blaseio and Pfeifer, 1990; Holmes and Dyall-Smith, 1990; Holmes et al., 1991; Lam and Doolittle, 1989). The resistance against mevinolin (or lovastatin) is due to an enhanced production of the 3-hydroxy-3-methylglutaryl-CoA-reductase caused by a promoter-up mutation in the respective gene (Lam and Doolittle, 1992), whereas the novobiocin resistance marker isolated from "*Haloferax alicantei*" encodes a mutated gyrase B subunit (Holmes and Dyall-Smith, 1991). The gas vesicle-negative strain *Haloferax volcanii* is a perfect recipient strain for such transformation experiments.

It appeared to be a fortunate coincidence that the mc-*gvpA* gene of *Hf. mediterranei* was originally isolated as a relatively large *Pst* I fragment of 12 kb (Englert et al., 1990), and that the vector plasmid pUBP2 had a single *Pst* I cloning site. Transformation of *Hf. volcanii* using this 12 kb fragment revealed Vac⁺ transformants, and this experiment enabled the dissection of the genes involved in gas vesicle formation (Blaseio and Pfeifer, 1990). The Vac⁺ phenotype of the transformants proved that all the genes required for gas vesicle formation are contained in this fragment. By introduction of various deletions and subsequent transformations of *Hf. volcanii*, it was demonstrated that a minimal DNA region of 9.4 kb is required for gas vesicle synthesis. This DNA region was termed mc-vac region since it derived from the *Hf. mediterranei* chromosome (Englert et al., 1992b). DNA sequence determination of the mc-vac region revealed the presence of 14 mc-*gvp* genes that are arranged as two units (Englert et al., 1992a) — the mc-*gvpA* gene is followed by the three genes *gvpC*, *gvpN*, and *gvpO*. The 10 genes mc-*gvpDEFGHIJKLM* are located upstream of mc-*gvpA* and oriented in the opposite direction. All of these reading frames are tightly packed with close or even overlapping start and stop codons (Figure 23.2).

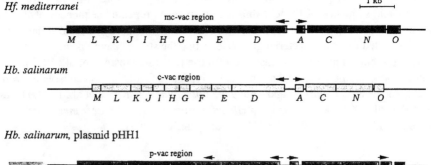

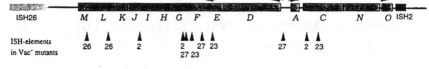

FIGURE 23.2 Genetic map of the gas vesicle gene clusters found in *Hf. mediterranei* (mc-vac region) and *Hb. salinarum* PHH1 (p-vac and c-vac region). Each *gvp* gene is represented by a box labeled A and C through O. Arrowheads above each map indicate the start sites and directions of vac transcripts. The p-vac region of plasmid pHH1 is flanked by the halobacterial insertion elements ISH2 and ISH26. The arrowheads arranged vertically below the p-vac region depict the integration sites of the various ISH-elements (ISH2, ISH23, ISH26, or ISH27) found in Vac⁻ mutants.

23.4 THE p-vac AND c-vac REGION OF *HALOBACTERIUM SALINARUM*

Similar *gvp* genes present in an identical arrangement are found in *Hb. salinarum* PHH1 (Figure 23.2). In contrast to *Hf. mediterranei*, this species contains two related, but different, vac regions. One, the so-called p-vac region (p for plasmid) is located on the 150 kb plasmid pHH1 (which is part of the 58% G+C-rich fraction FII-DNA), whereas the second vac region (c-vac) is present in the 68% G+C-rich FI-DNA of the chromosome (Horne and Pfeifer, 1989; Horne et al., 1991). Transformation experiments were also performed with the p-vac region, and *Hf. volcanii* transformants were Vac+, indicating that the p-vac region by itself is sufficient for gas vesicle formation, and thus does not depend on products of the c-vac region (Englert et al., 1992a).

The p-vac region is often the target for the integration of an insertion element: Vac− mutants of *Hb. salinarum* appear with a frequency of 10^{-2}, and each of these mutants contains alterations in pHH1 affecting the p-vac region (Horne et al., 1991; Pfeifer et al., 1981). The halobacterial IS-elements ISH2, ISH23, ISH26, and ISH27 are found in the p-vac region of Vac− mutants (Horne et al., 1991), which is also flanked by a copy of ISH26 and ISH2 (Pfeifer and Ghahraman 1993) (Figure 23.2). Other plasmids of Vac− mutants incurred deletions encompassing the p-vac region: The 35 kb plasmid pHH4 of *Hb. salinarum* PHH4 is such a deletion derivative, although it does not originate directly from pHH1 (Pfeifer and Blaseio, 1989). An almost identical vac region is located on plasmid pNRC100 of *Hb. salinarum* NRC-1, and a similar genetic instability due to ISH-elements has also been described for this *gvp* gene cluster (DasSarma et al., 1994; Jones et al., 1989, 1991).

In the *Hb. salinarum* PHH1 wild-type, the c-*gvpA* gene is usually not expressed, but *Hb. salinarum* PHH4 (lacking the p-vac region) synthesizes gas vesicles by expressing the c-vac region (Horne and Pfeifer, 1989). In contrast to the continuous synthesis of the spindle-shaped gas vesicles in *Hb. salinarum* PHH1, gas vesicles of *Hb. salinarum* PHH4 are cylinder shaped and appear during the stationary growth phase only. Similarly, the *Halobacterium* spp. SB3, GN101 and GRB each contain the c-vac region only, and the expression in stationary growth leads to cylinder-shaped gas vesicles (Horne and Pfeifer, 1989). The mutation rate for the c-vac region is much lower compared with the p-vac region: Vac− mutants of *Hb. salinarum* PHH4 occur with a frequency of 10^{-8} (the normal point mutation rate). None of these mutants incurred an ISH-element, supporting the idea that the G+C-rich FI-DNA fraction of the *Hb. salinarum* genome is much more stable compared with the FII-DNA sequences (Pfeifer and Betlach, 1985).

23.5 THE nv-vac REGION OF *NATRONOBACTERIUM VACUOLATUM*

Recently, the gas vesicles of the haloalkaliphilic species *N. vacuolatum* and the respective nv-*gvpA* gene encoding the major gas vesicle protein have been characterized (Mayr and Pfeifer, 1997). The nv-GvpA protein resembles mc-GvpA except for three amino acid exchanges (positions 2, 26, 29; see Figure 23.1). The DNA-sequence determination of sequences adjacent to nv-*gvpA* uncovered additional *gvp* genes that are all arranged consecutively as the nv-*gvpACNOFGHIJKLM* operon (Figure 23.3) (Mayr and Pfeifer, 1997; Mayr and Pfeifer, unpublished results). Neither a *gvpD* nor a *gvpE* homologue has been detected so far. This arrangement

N. vacuolatum

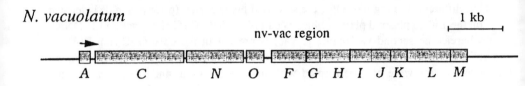

Anabaena flos-aquae

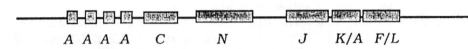

FIGURE 23.3 Genetic map of the nv-vac region of *N. vacuolatum* and the arrangement of *gvp* genes in *Anabaena flos-aquae* as described by Kinsman and Hayes (1997). Each *gvp* gene is represented by a box labeled A, C, and F through O. The arrowhead above the nv-vac map designates the single start site of the various nv-vac transcripts (Mayr and Pfeifer, 1997).

of the nv-*gvp* genes resembles somewhat the consecutive arrangement of *gvp* genes in the cyanobacterium *Anabaena flos-aquae*, where several *gvpA* genes are followed by *gvpC* and *gvpN*, and 650 nt downstream of *gvpN* three genes homologous to *gvpJ*, *gvpK/A*, and *gvpF/L* have been detected (Figure 23.3) (Kinsman and Hayes, 1997). Additional cyanobacterial *gvp* genes have not been described so far, and the sufficiency of the *gvpACNJK/AF/L* gene cluster for gas vesicle synthesis has not been investigated. The similarity of each nv-*gvp* gene to the respective halobacterial counterpart is, however, much higher (approximately 20% higher in each case) compared with the similarity of the nv-*gvp* genes to the respective cyanobacterial *gvp* genes (unpublished results).

23.6 EXPRESSION OF THE VARIOUS vac REGIONS AT THE RNA LEVEL

Since the gas vesicles appear at different time points during growth cycle, RNA analyses have been performed to determine the size and time points of transcript appearance, and to locate the respective promoter sequences. The nv-vac region of *N. vacuolatum* contains a single promoter located upstream of nv-*gvpA* that drives the expression of the entire nv-*gvp* gene cluster (Mayr and Pfeifer, 1997). The nv-*gvpA* transcript and longer cotranscripts up to 9.6 kb are detectable throughout the growth, starting at a common start site three nucleotides upstream of the nv-*gvpA* reading frame. Except for nv-*gvpA*, all other nv-*gvp* genes are expressed as part of a cotranscript, up to the nv-*gvpACNOFGHIJKLM* mRNA. The promoter boxA element, centered around position −28 relative to the transcript start site, resembles the archaeal consensus promoter sequence (Hain et al., 1992; Mayr and Pfeifer, 1997).

The p-vac region of *Hb. salinarum* PHH1 is also expressed throughout the growth cycle; however, four promoters located upstream of p-*gvpA*, p-*gvpO*, p-*gvpD*, and p-*gvpF* drive the transcription. A high level of p-*gvpA* mRNA, and minor amounts of p-*gvpACNO* transcripts are observed during growth, both starting 20 nt upstream of the p-*gvpA* reading frame (DasSarma et al., 1987; Horne and Pfeifer, 1989). The additional promoter located upstream of p-*gvpO* is mainly active during exponential growth (Offner and Pfeifer, 1995). The divergently oriented p-*gvpD-M* gene cluster is transcribed as 4 kb p-*gvpFGHIJKLM* mRNA during exponential growth, whereas the 2.4 kb p-*gvpDE* transcript, starting 69 nt upstream of p-*gvpD*, occurs during stationary growth only (Jones et al., 1989; Offner and Pfeifer, 1995). The consecutive appearance of the p-*gvpF-M* and p-*gvpDE* transcripts indicates a growth-phase-dependent regulation of p-vac gene expression at the transcriptional level. All p-vac promoter sequences resemble the archaeal consensus promoter (Pfeifer et al., 1997).

In contrast, the c-vac and mc-vac regions contain only two promoters located in front of the respective *gvpA* and *gvpD* genes (Englert et al., 1990, 1992a; Horne and Pfeifer, 1989). The transcription of the *gvpACNO* gene cluster results in similar mRNAs: in both cases no transcripts are detectable during exponential growth, whereas an abundant *gvpA* mRNA occurs in the early stationary growth phase, together with minor amounts of the *gvpACNO* transcript. The transcription of the *gvpD-M* gene cluster is, however, somewhat different (Krüger and Pfeifer, 1996; Röder and Pfeifer, 1996). The appearance of minor amounts of a relatively unstable mRNA encompassing the entire *gvpD-M* gene cluster is common to both vac-regions. This transcript occurs prior to the *gvpACNO* mRNAs and before gas vesicles are detectable. Its early presence during growth implies that some of the products encoded here are important for initial stages of the gas vesicle formation. In the case of the c-vac region, an additional 1.7 kb c-*gvpD* mRNA occurs during the stationary growth phase, whereas high levels of various, often truncated, mc-*gvpD* transcripts are found in *Hf. mediterranei*. The function of the latter RNAs is still unknown.

The salt-dependent synthesis of gas vesicles in *N. vacuolatum* and *Hf. mediterranei* is of special interest, and the expression of both vac-regions has been investigated in Vac⁻ and Vac⁺ cells at the transcript level. In all cases, the respective vac region is transcribed: *N. vacuolatum* grown in 13% salt medium (Vac⁻ cells) contains the nv-*gvpA* mRNA and minor amounts of the longer nv-vac transcripts (up to the nv-*gvpACNOFGHIJKLM* mRNA), but gas vesicles are not formed (Mayr and Pfeifer, 1997). Also in *Hf. mediterranei*, both mc-vac promoters are active when cells are grown in 15% salt medium (Vac⁻ cells); however, the respective mRNAs occur in lower amounts, and later, during growth (Röder and Pfeifer, 1996). Despite this transcript formation, the cells are unable to produce gas vesicles. The GvpA protein is even detectable in Vac⁻ cells in both cases, but the assembly of gas vesicles is somehow hampered — possibly due to the lack of accessory proteins, or to the improper formation of an assembly machinery. Additional experiments are required to unravel the reason(s) for the lack of gas vesicle formation in cells grown in media with lower salt concentrations.

23.7 POSSIBLE FUNCTIONS OF Gvp PROTEINS DURING GAS VESICLE FORMATION

The functions of the various *gvp* genes during gas vesicle formation have been investigated by transformation experiments using *Hb. salinarum* (DasSarma et al., 1994), or *Hf. volcanii* as recipient strain (Offner and Pfeifer, 1995; Offner et al., 1996). Two methods (insertion of foreign DNA or deletion of a single *gvp* gene) have been employed to mutate specific *gvp* genes; the results obtained were, however, often different (e.g., in six out of 14 cases). The review by Pfeifer et al. (1997) discusses the difficulties encountered by both methods in more detail; the results described here are those obtained by the deletion analyses.

The two gas vesicle structural proteins, GvpA and GvpC, are encoded by the p-*gvpACNO* operon, but only the deletion of the p-*gvpA* gene (ΔA transformant, containing the p-vac region lacking p-*gvpA*) results in a Vac$^-$ phenotype (Offner et al., 1996). The lack of the p-*gvpC* gene in ΔC transformants leads to large numbers of irregularly shaped gas vesicles, with no distinct cylindrical part and varying diameters throughout a single vesicle (Offner et al., 1996). Complementation of ΔC transformants with the p-*gvpC* gene present on a second vector construct restores the wild-type shape of the gas vesicles (i.e. the spindle shape), suggesting that GvpC is involved in the formation of the correct gas vesicle shape. ΔN transformants still produce minor amounts of gas vesicles, whereas the lack of p-*gvpO* in ΔO transformants reveals a Vac$^-$ phenotype (Offner et al., 1996). The products of p-*gvpD* and p-*gvpE* are not required for gas vesicle formation and appear to be involved in the regulation of gas vesicle synthesis (see below). ΔF, ΔG, ΔJ, ΔK, and ΔM transformants are Vac$^-$, indicating that each of these gene products is essential for gas vesicle formation. In contrast, ΔH and ΔI transformants are Vac$^+$ (Offner, 1996).

GvpJ and GvpM exhibit more then 60% sequence similarity to the GvpA protein, suggesting that they are structural components of the gas vesicle (Jones et al., 1991; Pfeifer and Englert, 1992); however, antisera raised against these proteins are not available so far, and none of these proteins has been detected in gas vesicle preparations. The other Gvp proteins could have chaperone or scaffolding functions, or might be part of a gas vesicle assembly machinery.

23.8 THE TWO REGULATORY PROTEINS GvpD AND GvpE

The evidence that GvpD is involved in the regulation of gas vesicle formation came from a ΔD transformant that overproduced gas vesicles (Englert et al., 1992b). Complementation of the ΔD transformant with the mc-*gvpD* gene present on a second vector construct reduced the amount of gas vesicles to the wild-type level, suggesting that the product of *gvpD* is involved in the repression of gas vesicle formation. However, additional analyses at the transcript level indicated that GvpD is not a repressor per se. The mc-*gvpA* and the mc-*gvpD* promoter both require activation and are not active by themselves, since an A+D transformant neither expresses the mc-*gvpA* nor the mc-*gvpD* gene (Röder and Pfeifer, 1996). Both promoters are active only in the presence of mc-*gvpE* because an A+DE transformant contains large amounts of both mc-vac transcripts throughout the growth. Due to these results mc-GvpE appears to be a transcriptional activator, whereas mc-GvpD has no repressor effect in this system. In particular, GvpD does not repress the two mc-vac promoters during exponential growth, which is the case in a transformant harboring the entire mc-vac region (Röder and Pfeifer, 1996).

Further investigations of the effect of the mc-*gvpD* gene on the gas vesicle formation in the ΔD transformant demonstrated that the reduction depends on the expression of the *gvpD* reading frame: a strong expression of mc-*gvpD* under the ferredoxin promoter control drastically reduces the gas vesicle production, whereas a ΔD transformant containing only the mc-vac promoter region on a second plasmid still exhibits the gas vesicle overproducing phenotype (Röder, 1997). These observations demonstrate that the product of mc-*gvpD* gene is somehow involved in the repression of gas vesicle formation, but cannot repress both mc-vac promoters — at least not by itself (Röder and Pfeifer, 1996). Thus, an additional factor encoded by the mc-vac region is responsible for the repression of both promoters during exponential growth. GvpD might be required for the activity of this additional factor, or this factor modifies GvpD so that it can act as repressor. The amino acid sequence of GvpD contains a P-loop motif indicative for nucleotide binding (Saraste et al., 1990) which could play a role for such a function. The function of GvpD in the repression of gas vesicle formation could be even more indirect.

Also, in the case of the c-vac region of *Hb. salinarum* PHH4, the c-*gvpA* promoter is inactive by itself, but transformants containing the c-*gvpA* gene on one plasmid, and c-*gvpE* expressed under *fdx* promoter control on a different plasmid, produce c-*gvpA* mRNA constitutively throughout the growth (Krüger, 1996). Since the c-*gvpA* expression and formation of gas vesicles in *Hb. salinarum* PHH4 occurs only during the stationary growth phase, the activation by c-GvpE must be prevented during exponential growth. This could be achieved either by an inhibition of the c-GvpE synthesis, or by a modification of the c-GvpE activity during exponential growth. Transcript analysis indicates that the c-*gvpDEF-GHIJKLM* mRNA is present throughout the growth. However, immunological studies demonstrate that the proteins encoded by the first two reading frames are synthesized consecutively (Krüger and Pfeifer, 1996): The c-GvpD protein occurs exclusively during exponential growth, whereas c-GvpE is present only during gas vesicle synthesis in the stationary growth phase. Thus, a modification of c-GvpE by the c-GvpD protein resulting in an active/inactive form (as observed for bacterial two-component regulators) appears to be unlikely. The lack of gas vesicles during exponential growth of *Hb. salinarum* PHH4 is thus due to the lack of c-GvpE, rather than to a modification of the c-GvpE activity.

Small RNAs complementary to the 5'- and 3'-terminal part of c-*gvpD* appear coincidently with the decrease of the c-GvpD protein, and are most likely the reason for the low translation of the mRNA into c-GvpD protein during stationary growth (Krüger and Pfeifer, 1996). The lack of the c-*gvpE*-expression during exponential growth is, however, not accompanied by an antisense RNA formation. A different mechanism (such as mRNA secondary structure to prevent translation, or rapid degradation of GvpE during exponential growth) might be involved.

GvpE is essential for the expression of the c-vac and mc-vac region, whereas such activator functions are not required for the transcription in the p-vac and nv-vac region. The promoter sequences of the p-*gvpA* and nv-*gvpA* genes are highly similar to the archaeal consensus promoter, and are obviously recognized by the halobacterial RNA polymerase without the help of an activator protein. The amino acid sequence of GvpE indicates a cluster of positively charged amino acids (lysine and arginine residues at positions 140–143) followed by an amphiphilic α-helix near the C-terminus (Hermann, Krüger and Pfeifer, unpublished results). These motifs resemble leucine-zipper-type regulatory proteins, and experiments are in progress to prove this hypothesis.

23.9 CONCLUSION

The expression of each halobacterial vac region appears to be characteristic, but common themes are found:

(1) In each case, the *gvpFGHIJKLM* genes are expressed in minor amounts during an early stage of growth and often prior to gas vesicle formation. This mRNA usually disappears during stationary growth because of an early termination of transcription in case of the mc-vac, c-vac, and nv-vac region, or by the repression of the transcription at the p-*gvpF* promoter as seen for the p-vac region.
(2) The expression of the *gvpACNO* cluster containing the two genes encoding the gas vesicle structural proteins leads mainly to high amounts of *gvpA* mRNA. Read-through of a terminator located behind *gvpA* results in a minor amount of cotranscripts encompassing *gvpACNO*, and once this mRNA is synthesized, gas vesicles are usually formed.
(3) GvpE is a transcriptional activator and essential for the expression of the mc-vac and c-vac region. The reason for the late translation of the c-*gvpE* reading frame is not known so far. GvpD is involved in repression, but is not a repressor protein per se. It is possible that GvpD is modified by an additional protein, or modifies an additional protein that is the actual repressor.

The different levels of regulation determined for each vac region reflect the complexity of gene regulation in halobacteria. The gas vesicle formation offers an intriguing model for the investigation of such a regulatory network.

ACKNOWLEDGMENTS

This work was financially supported by grants obtained from the Deutsche Forschungsgemeinschaft (Pf 165/3; Pf 165/4; Pf 165/6).

REFERENCES

Blaseio, U. and F. Pfeifer. 1990. Transformation of *Halobacterium halobium*: Development of vectors and investigation of gas vesicle synthesis. *Proc. Natl. Acad. Sci USA* 87:6772–6776.

Damerval, T., J. Houmard, G. Guglielmi, K. Csisar, and N. Tandeau de Marsac. 1987. A developmentally regulated *gvpABC* operon is involved in the formation of gas vesicles in the cyanobacterium *Calothrix* 7601. *Gene* 54:83–92.

DasSarma, S., T. Damerval, J.G. Jones, and N. Tandeau de Marsac. 1987. A plasmid-encoded gas vesicle protein gene in a halophilic archaebacterium. *Mol. Microbiol.* 1:365–370.

DasSarma, S., P. Arora, F. Lin, E. Molinari, and L. Ru-Siu Yin. 1994. Wild-type gas vesicle formation requires at least ten genes in the *gvp* gene cluster of *Halobacterium halobium* plasmid pNRC100. *J. Bacteriol.* 176:7646–7652.

Ebert, K., W. Goebel and F. Pfeifer. 1984. Homologies between heterogeneous extrachromosomal DNA populations of *Halobacterium halobium* and four new halobacterial isolates. *Mol. Gen. Genet.* 194:91–97.

Englert, C. and F. Pfeifer. 1993. Analysis of gas vesicle gene expression in *Haloferax mediterranei* reveals that GvpA and GvpC are both gas vesicle structural proteins. *J. Biol. Chem.* 268:9329–9336.

Englert, C., M. Horne, and F. Pfeifer. 1990. Expression of the major gas vesicle protein gene in the halophilic archaebacterium *Haloferax mediterranei* is modulated by salt. *Mol. Gen. Genet.* 222:225–232.

Englert, C., K. Krüger, S. Offner, and F. Pfeifer. 1992a. Three different but related gene clusters encoding gas vesicles in halophilic archaea. *J. Mol. Biol.* 227:586–592.

Englert, C., G. Wanner, and F. Pfeifer. 1992b. Functional analysis of the gas vesicle gene cluster of the halophilic archaeon *Haloferax mediterranei* defines the vac-region boundary and suggest a regulatory role for the *gvpD* gene or its products. *Mol. Microbiol.* 6:3543–3550.

Hain, J., W.D. Reiter, U. Hüdepohl, and W. Zillig. 1992. Elements of an archaeal promoter defined by mutational analysis. *Nucleic Acids Res.* 20:5423–5428.

Halladay, J.T., J.G. Jones, F. Lin, A.B. MacDonald, and S. DasSarma. 1993. The rightward gas vesicle operon in *Halobacterium* plasmid pNRC100: identification of the *gvpA* and *gvpC* gene products by use of antibody probes and genetic analysis of the region downstream of *gvpC*. *J. Bacteriol.* 175:684–692.

Hayes, P.K. and R.S. Powell. 1995. The *gvpA/C* cluster of *Anabaena flos-aquae* has multiple copies of a gene encoding GvpA. *Arch. Microbiol.* 164:50–57.

Hayes, P.K., B. Buchholz, and A.E. Walsby. 1992. Gas vesicles are strengthened by the outer-surface protein, GvpC. *Arch. Microbiol.* 157:229–234.

Holmes, M. and M.L. Dyall-Smith. 1990. A plasmid vector with a selectable marker for halophilic archaebacteria. *J. Bacteriol.* 172:756–761.

Holmes, M. and M.L. Dyall-Smith. 1991. Mutations in DNA gyrase result in novobiocin resistance in halophilic archaebacteria. *J. Bacteriol.* 173:642–648.

Holmes, M., S.D. Nuttall, and M.L. Dyall-Smith. 1991. Construction and use of halobacterial shuttle vectors and further studies on *Haloferax* DNA gyrase. *J. Bacteriol.* 173:3807–3813.

Horne, M. and F. Pfeifer. 1989. Expression of two gas vacuole protein genes in *Halobacterium halobium* and other related species. *Mol. Gen. Genet.* 218:437–444.

Horne, M., C. Englert, C. Wimmer, and F. Pfeifer. 1991. A DNA region of 9 kbp contains all genes necessary for gas vesicle synthesis in halophilic archaebacteria. *Mol. Microbiol.* 5:1159–1174.

Jones, J.G., N.R. Hackett., J.T. Halladay, D.J. Scothorn, C.F. Yang, W.L. Ng, and S. DasSarma. 1989. Analysis of insertion mutants reveals two new genes in the pNRC100 gas vesicle gene cluster of *Halobacterium halobium*. *Nucleic Acids Res.* 17:7785–7793.

Jones, J.G., D.C. Young, and S. DasSarma. 1991. Structure and organization of the gas vesicle gene cluster on the *Halobacterium halobium* plasmid pNRC100. *Gene* 102:117–122.

Kinsman, R., A.E. Walsby, and P.K. Hayes. 1995. GvpC's with reduced numbers of repeating sequence elements bind to and strengthen cyanobacterial gas vesicles. *Mol. Microbiol.* 17:147–154.

Kinsman, R. and P.K. Hayes. 1997. Genes encoding proteins homologous to halobacterial Gvps N, J, F, and L are located downstream of *gvpC* in the cyanobacterium *Anabaena flos-aquae*. *DNA Sequence* 7:97–106.

Krüger, K. 1996. *Untersuchungen zur Regulation der Gasvesikelsynthese in dem halophilen Archaeon Halobacterium salinarium PHH4*. Dissertation, LMU München (Shaker Verlag).

Krüger, K. and F. Pfeifer. 1996. Transcript analysis of the c-*vac* region and differential synthesis of the two regulatory gas vesicles proteins GvpD and GvpE in *Halobacterium salinarium* PHH4. *J. Bacteriol.* 178:4012–4019.

Lam, W.L. and W.F. Doolittle. 1989. Shuttle vectors for the archaebacterium *Halobacterium volcanii*. *Proc. Natl. Acad. Sci. USA* 86:5478–5482.

Lam, W.L. and W.F. Doolittle. 1992. Mevinolin-resistant mutations identify a promoter and the gene for a eukaryote-like 3-hydroxy-3-methylglutaryl-coenzyme A reductase in the archaebacterium *Haloferax volcanii*. *J. Biol. Chem.* 267:5829–5834.

Mayr, A. and F. Pfeifer. 1997. The characterization of the nv-*gvpACNOFGH* gene cluster involved in gas vesicle formation in *Natronobacterium vacuolatum*. *Arch. Microbiol.* 168:24–32.

Mwatha, W.E. and W.D. Grant. 1993. *Natronobacterium vacuolata* sp. nov., a haloalkaliphilic archaeon isolated from the Lake Magadi, Kenya. *Int. J. Syst. Bacteriol.* 43:401–404.

Offner, S. 1996. *Gasvesikelsynthese des halophilen Archaeons Halobacterium salinarium PHH1*. Dissertation LMU München (Shaker-Verlag).

Offner, S. and F. Pfeifer. 1995. Complementation studies with the gas vesicle-encoding p-*vac* region of *Halobacterium salinarium* PHH1 reveal a regulatory role for the p-*gvpDE* genes. *Mol. Microbiol.* 16:9–19.

Offner, S., G. Wanner, and F. Pfeifer. 1996. Functional studies of the *gvpACNO* operon of *Halobacterium salinarium* reveal that the GvpC protein shapes gas vesicles. *J. Bacteriol.* 178:2071–2078.

Pfeifer, F. and M. Betlach. 1985. Genome organization in *Halobacterium halobium*: a 70 kb island of more (AT) rich DNA in the chromosome. *Mol. Gen. Genet.* 198:449–455.

Pfeifer, F. and U. Blaseio. 1989. Insertion elements and deletion formation in a halophilic archaebacterium. *J. Bacteriol.* 171:5135–5140.

Pfeifer, F. and C. Englert. 1992. Function and biosynthesis of gas vesicles in halophilic Archaea. *J. Bioenerg. Biomembr.* 24:577–585.

Pfeifer, F. and P. Ghahraman. 1993. Plasmid pHH1 of *Halobacterium salinarium*: characterization of the replicon region, the gas vesicle gene cluster and insertion elements. *Mol. Gen. Genet.* 238:193–200.

Pfeifer, F., G. Weidinger, and W. Goebel. 1981. Genetic variability in *Halobacterium halobium*. *J. Bacteriol.* 145:375–381.

Pfeifer, F., K. Krüger, R. Röder, A. Mayr, S. Ziesche, and S. Offner. 1997. Gas vesicle formation in halophilic archaea. *Arch. Microbiol.* 167:259–268.

Röder, R. 1997. *Untersuchungen zur genetischen Regulation der Gasvesikelbildung bei dem halophilen Archaeon Haloferax mediterranei.* Dissertation, TH Darmstadt.

Röder, R. and F. Pfeifer. 1996. Influence of salt on the transcription of the gas vesicle genes of *Haloferax mediterranei* and identification of the endogenous activator gene. *Microbiology* 142:1715–1723.

Rodriguez-Valera, F., G. Juez, and D. Kushner. 1983. *Halobacterium mediterranei* spec. nov., a new carbohydrate-utilizing extreme halophile. *Syst. Appl. Microbiol.* 4:369–381.

Saraste, M., P.R. Sibbald, and A. Wittighofer. 1990. The p-loop — a common motif in ATP- and GTP-binding proteins. *Trends Biochem. Sci.* 15:430–434.

Simon, R.D. 1981. Morphology and protein composition of gas vesicles from wild type and gas vacuole deficient strains of *Halobacterium salinarium* strain 5. *J. Gen. Microbiol.* 125:103–111.

Staley, J.T., R.L. Irgens, and R.P. Herwig. 1989. Gas vacuolate bacteria from the sea ice of Antarctica. *Appl. Environ. Microbiol.* 42:5–11.

Surek, B., B. Pillay, U. Rdest, K. Beyreuther, and W. Goebel. 1988. Evidence for two different gas vesicle proteins and genes in *Halobacterium halobium*. *J. Bacteriol.* 70:1746–1751.

Ventosa, A. and A. Oren. 1996. *Halobacterium salinarum* nom. corrig., a name to replace *Halobacterium salinarium* (Elazari-Volcani) and to include *Halobacterium halobium* and *Halobacterium cutirubrum*. *Int. J. Syst. Bacteriol.* 46:347.

Walsby, A. E. 1994. Gas vesicles. *Microbiol. Rev.* 58:94–144.

24 Halocins: Protein Antibiotics from Hypersaline Environments

Richard F. Shand, Lance B. Price, and Elizabeth M. O'Connor

CONTENTS

24.1 Bacteriocins and Halocins: Historical Perspective 295
    24.1.1 Bacteriocins 295
    24.1.2 Halocins 296
    24.1.3 Halocin Activity Assays 297
24.2 Specific Halocins 297
    24.2.1 Halocin H4 297
    24.2.2 Halocin H6 299
    24.2.3 Halocin Hal R1 300
    24.2.4 Halocin S8 302
24.3 Conclusion 303
Acknowledgments 304
References 304

24.1 BACTERIOCINS AND HALOCINS: HISTORICAL PERSPECTIVE

24.1.1 BACTERIOCINS

Bacteriocins and bacteriocin-like molecules are a diverse collection of proteinaceous compounds often referred to as "protein antibiotics" or "bactericidal proteins," with molecular weights ranging from ~1 kDa to ~100 kDa (Barefoot et al., 1992; Braun et al., 1994; Hoover, 1992; Jack et al., 1995; James et al., 1992; Vining and Stuttard, 1995). They are ribosomally synthesized, which distinguishes them from "peptide antibiotics" such as bacitracin, gramicidin and valinomycin, which are synthesized nonribosomally by multienzyme complexes or by sequential enzyme reactions (Kleinkauf and von Döhren, 1990).

Bacteriocins have been isolated and characterized from some 30 genera (Barefoot et al., 1992), including halophilic organisms from the domain Archaea, and are often referred to by a common property. Thus, colicins are bacteriocins produced by *Escherichia coli*, klebicins are bacteriocins produced by *Klebsiella* spp., halocins are bacteriocins produced by extremely halophilic members of the domain Archaea, and so on. Their bactericidal modes of action are wide ranging and include inhibition of transcription, translation, DNA replication and peptidoglycan synthesis, DNA and RNA nuclease activity, pore formation, bacteriolysis, and disruption of cellular membranes (Barefoot et al., 1992). Over the years, colicins have been the most carefully studied bacteriocins; consequently, six key colicin characteristics have been used to define a bacteriocin: (1) a proteinaceous compound, (2) exhibits bactericidal action, (3) activity spectrum is restricted to a narrow range of closely related organisms, (4) attaches to a specific cell receptor, (5) structural and immunity genes plasmid-encoded, and (6) production by lethal biosynthesis (i.e., lysis or partial lysis upon induction of colicin synthesis) (Barefoot et al., 1992; Jack et al., 1995). However, this definition has broadened, in part due to characteristics displayed by bacteriocins from Gram-positive bacteria (e.g., lack of a specific receptor for adsorption, relatively low molecular weight, cleavage of leader sequences from preproteins, etc.; Jack et al., 1995). While colicins typically have molecular weights ranging from 29 to 75 kDa (Braun et al., 1994), "microcins" (Mcc) are a separate category of bacteriocins (including those produced by *E. coli*) that are very small (typically ~1 kDa, but include bacteriocins as large as 5–6 kDa; Moreno et al., 1995). In addition to their small size, microcins exhibit stability to heat, are resistant to proteases and are secreted in a noncolicin fashion (Barefoot et al., 1992; Braun et al., 1994). Representative of this class is microcin C7 (McсС7), a heptapeptide of 1,177 Da (by mass spectrometry), that is translated from the smallest known gene (*mccA*) with an open reading frame of only 21 bp (González-Pastor et al., 1994; Guijarro et al., 1995).

24.1.2 Halocins

Halocins were first discovered by Francisco Rodrigùez-Valera and co-workers in 1982. In this first report, 40 extreme halophiles (39 rods and 1 halococcus) were screened against each other for the production of halocins; seven of the 40 were found to produce "inhibitory substances" as judged by zones of inhibition around colonies grown on lawns of the other 39 halobacterial isolates. Five of the seven producers inhibited a large number (19 to 35) of the 40 strains, while the remaining two inhibited only a few (one to three). The presence of a zone of inhibition surrounding a colony can arise from causes other than a bacteriocin (e.g., release of bacteriophages or secretion of metabolic by-products). Thus, experiments must be performed to eliminate these non-bacteriocin-related causes (Jack et al., 1995). For example, in the initial study, culture supernatants containing halocin H4 were analyzed and the following results obtained: halocin H4 (1) was bactericidal and bacteriolytic, (2) lost all activity upon desalting, exposure to a protease, and heating at 80°C for 10 min, and (3) passed through a 100 kDa filter. In addition, inhibition was not due to a bacteriophage (Rodríguez-Valera et al., 1982). Since then, work has focused on the following: halocin H4 (Cheung et al., 1997; Meseguer and

Rodríguez-Valera, 1985, 1986); halocin H6 (Meseguer et al., 1995; Torreblanca et al., 1989, 1990); halocin Hal R1 (Rdest and Sturm, 1987); and antagonism studies involving hundreds of haloarchaea (Meseguer et al., 1986; Torreblanca et al., 1994). These last studies have led to the remarkable conclusion that "production of halocin is a practically universal feature of archaeal halophilic rods" (Torreblanca et al., 1994). In the following sections, halocins H4, H6 and Hal R1 will be discussed in turn, concluding with the description of a new halocin, halocin S8. Note that in the following discussion, *Halobacterium salinarum* (previously *halobium*) nearly always is used as the halocin-sensitive indicator strain.

24.1.3 HALOCIN ACTIVITY ASSAYS

Determining the titer of halocin activity in a culture supernatant or in a chromatography column fraction is most commonly done by performing critical double-serial dilutions to extinction. The dilutions are spotted onto top agar lawns of sensitive cells, or are placed into wells in the agar, and the plate is allowed to incubate. The halocin activity of a sample is then reported as the reciprocal of the first dilution in which activity (i.e., the zone of inhibition) disappears (referred to as the "extinction dilution"). It is often important to determine where activity begins and ends in a series of column fractions. Some column fractions will produce a zone of inhibition at zero dilution (i.e., undiluted), with activity disappearing at a dilution of 1/2, while others will have no activity at zero dilution. Using the extinction dilution as the end point allows one to distinguish between a fraction that shows activity only when undiluted, and a fraction that has no activity. In contrast, using the last dilution that shows activity as the end point gives erroneous information; fractions that have no activity, and fractions that have activity only when spotted undiluted, are both reported as zero. In addition, there are several factors that can influence the sensitivity of the assay: (1) the physiological state of the indicator cells, (2) the density of indicator cells in the top agar, (3) how long and at what temperature lawns have been stored before use, (4) the nature of the diluent, and (5) the amount of agar used in the lawn. Consequently, assay conditions should be optimized for each halocin.

24.2 SPECIFIC HALOCINS

24.2.1 HALOCIN H4

Halocin H4, the first halocin to be characterized, is produced by *Haloferax mediterranei* R4 (ATCC 33500). Early work by Meseguer and Rodriguez-Valera (1985, 1986) suggested that it was a single protein of about 28 kDa (SDS-PAGE). Maximal activity in culture supernatants was present as the culture entered stationary phase and similar levels of activity were reached when the cells were grown in either complex or minimal medium. The production of many colicins is induced by ultraviolet light, but exposure of *Hf. mediterranei* to UV light did not induce halocin production. Similarly, most bacteriocins are plasmidally encoded, but attempts to cure halocin activity by treatment with acridine orange failed. A battery of experiments designed to elucidate the mode of action of halocin H4 showed that killing occurred by single-hit kinetics. Exposure of sensitive cells to halocin H4 (as monitored by electron microscopy) resulted in a swollen, spherical morphology after 5 h, and either lysis or near lysis after 24 h. Although macromolecular synthesis was relatively unaffected by exposure of sensitive cells to halocin H4, two separate experiments involving the transport of 2-amino-[^{14}C]isobutyric acid (AIB, a nonmetabolizable amino acid) showed that AIB uptake was markedly inhibited and AIB efflux markedly enhanced in halocin H4 sensitive cells just 10 min after exposure to the halocin. These and other data suggested that halocin H4 acted at the level of the membrane (see section 24.2.2).

Cheung et al. (1997) cloned and sequenced the structural gene for halocin H4, the *halH4* gene (GenBank accession number U16389). The gene has a 1,080 bp open reading frame that encodes a polypeptide of 359 amino acids. The promoter is typically haloarchaeal, but the start site of transcription is only four bases from the 5' end of the initiator methionine codon, making the *halH4* transcript another example of a "leaderless" mRNA molecule from the haloarchaea. Other examples of haloarchaeal genes that produce leaderless mRNA transcripts include *bop* (DasSarma et al., 1984), *brp* (Betlach et al., 1984), *hop* (Blanck and Oesterhelt, 1987), and *arcA*, *arcB* and *arcC* (Ruepp and Soppa, 1996). The transcriptional terminator is a perfect 11-base inverted repeat separated by a 16-base interval. Separation of the three megaplasmids (490, 320 and 130 kbp) found in *Hf. mediterranei* R4 (López-Garcia et al., 1992) by contour-clamped homogeneous electric field (CHEF) gel electrophoresis and probing by Southern hybridization showed that the *halH4* gene was plasmidally encoded on the 320 kbp plasmid (Cheung et al., 1997).

The halocin H4 protein has a 46 amino acid signal peptide that is cleaved from the preprotein upon secretion. Figure 24.1 schematically compares the Hal H4 signal peptide to a typical signal peptide. This is an unusual feature for any bacteriocin, as colicins and most other bacteriocins do not have leader sequences (Braun et al., 1994). While bacteriocins from Gram-positive organisms do have cleavable leader peptides, they do not conform to typical signal sequences (Jack et al., 1995). The

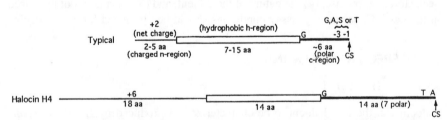

FIGURE 24.1 Schematic comparison (to scale) of a "typical" signal peptide (Izard and Kendall, 1994) to the signal peptide of halocin H4. Note that the halocin H4 n and c regions are unusually long, and the n region carries a high, net positive charge. The amino acids at positions -1 and -3 relative to the cleavage site follow von Heijne's rules (von Heijne, 1984). G, A, S, T, single amino acid codes for glycine, alanine, serine, and threonine, respectively; CS, cleavage site; thin lines, n region; thick lines, c region; open rectangles, hydrophobic h region. Net charges and lengths for the n regions, lengths of the hydrophobic regions, number of polar residues, lengths and initial amino acid of the c regions, and the amino acids at positions −1 and −3 relative to the CS are indicated.

molecular weight of the Hal H4 preprotein before cleavage is 39.6 kDa while the molecular weight of the mature, processed halocin is 34.9 kDa. A second interesting feature is a hydrophobic region of 32 amino acid residues in the middle of the peptide sequence, which may play a role in its mode of action.

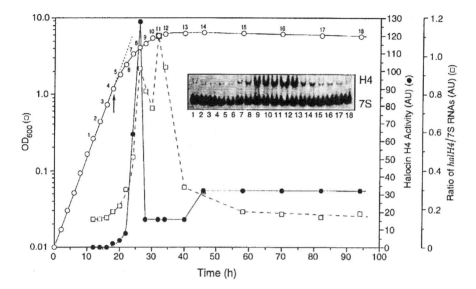

FIGURE 24.2 Expression of the *halH4* gene in *Hf. mediterranei* R4 (Cheung et al., 1997). Growth curve (open circles), halocin activity (closed circles), and ratios of *halH4*/7S transcript levels (open squares) are shown. Arrow, onset of halocin H4 activity; broken line, extrapolation of the exponential growth portion of the curve (g = 2.75 h). Inset, autoradiograph of the northern blot of total RNA probed with ^{32}P-labeled probes from the *halH4* and 7S RNA genes. The 18 samples (lanes 1 to 18) correspond to the 18 time points on the growth curve. (From Cheung et al., *J. Bacteriol.*, 179, 548, 1997. With permission.)

Correlation of growth physiology to both *halH4* expression and halocin activity is shown in Figure 24.2. Halocin H4 activity is first detectable during the transition from exponential phase to stationary phase. The rapid rise in halocin activity is paralleled by an equally rapid rise in *halH4* transcript levels, showing that the *halH4* gene is inducible. However, there are two places where *halH4* transcript levels do not correlate with halocin activity. First, *halH4* transcripts are clearly visible during mid-exponential phase (albeit at low levels; time points 1–3) when no halocin activity is detectable. Second, when *halH4* transcript levels are at their maximum (time points 8–12), halocin activity decreases, then levels out. These data suggest that synthesis of halocin H4 is regulated posttranscriptionally.

24.2.2 HALOCIN H6

Halocin H6, the second halocin to be characterized (Meseguer et al., 1995; Torreblanca et al., 1989, 1990), is produced by *Haloferax gibbonsii* (ATCC 33959). Halocin H6 shares several characteristics with halocin H4: activity is maximal as the culture enters stationary phase; halocin production is not induced by exposure to UV light, nor eliminated when cells are treated with acridine orange; killing is by single-hit kinetics; and the molecular weight is 32 kDa by HPLC using a Spherogel column, but is just below 31 kDa by SDS-PAGE. Treatment of sensitive cells with halocin H6 produced morphological effects similar to those seen with halocin H4, although the effects of H6 occurred more rapidly and with relatively fewer units of activity. Sensitive cells first swelled and then lysed, leaving behind "ghosts" that appeared to be intact cell envelopes. Maximal halocin activity is attained when cultures are grown in either complex or minimal medium, or when grown in salt concentrations ranging from 15 to 25%. In contrast, halocin H6 appears to be more robust than halocin H4, as it is insensitive to desalting, to trypsin (but not pronase), and to heating at 90°C for 10 min.

The mode of action of halocin H6 appears to be the direct inhibition of the Na^+/H^+ antiporter (Meseguer et al., 1995). The approach used to elucidate the primary target of halocin H6 involved two types of experiments: (1) the analysis of whole-cell parameters indirectly related to Na^+/H^+ antiporter activity (effects on intracellular volume, intracellular pH, membrane potential, and proton motive force), and (2) experiments directly related to Na^+/H^+ exchange using membrane vesicles (light-dependent pH change and light-dependent Na^+ efflux). Halocins H4 and H6 were tested in parallel in these experiments, and while halocin H6 clearly targets the Na^+/H^+ antiporter, halocin H4 does not. The specific target of halocin H4 is still unknown, but a possible effect on passive H^+ permeability on the membrane has been suggested (Meseguer et al., 1995).

24.2.3 Halocin Hal R1

Halocin Hal R1 is a microcin, and very different from either halocin H4 or H6. It is produced by a partially characterized extreme halophile called "*Halobacterium* sp. GN101" (GN = Guerrero Negro, Mexico) (Ebert and Goebel, 1985). This isolate contains a small plasmid of 1,765 bp (Hall and Hackett, 1989), and a megaplasmid of ~270 kbp as determined by CHEF gel electrophoresis. Figure 24.3A shows that the physiology of expression of halocin Hal R1 activity is typical, with activity first detected during the transition from exponential to stationary phase. The onset of detectable activity is closely tied to the optical density of the culture (OD_{600} ~0.86) regardless of growth rate. Furthermore, oxygen starvation or amino acid starvation prior to an OD_{600} of 0.86 does not result in halocin production. Unlike halocins H4 (Figure 24.2), H6 (Torreblanca et al., 1989) and S8 (Figure 24.3B), once Hal R1 activity reaches its maximum, it remains at that level (Figure 24.3A).

The initial characterization of this halocin was performed by Rdest and Sturm (1987). They determined that halocin Hal R1 was bacteriostatic, as it did not lyse or impart other morphological changes to sensitive cells. They concluded from their purification scheme that Hal R1 had a molecular weight of 6.2 kDa and that it might be complexed with a larger protein. Our purification data paint a similar picture, except we estimate halocin Hal R1 to be much smaller. Briefly, supernatants from stationary-phase cultures of GN101 were concentrated by passage through a series

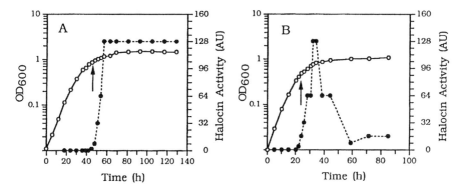

FIGURE 24.3 Growth curves (open circles) and halocin activity (closed circles) in strains GN101 (panel A) and S8a (panel B). Arrow, onset of halocin activity. Halocin activity (in arbitrary units, AU) was determined as described in the text.

of tangential flow filters with progressively smaller nominal molecular weight cutoffs (NMWC). A small fraction of total halocin activity (7.6%) was retained by a 100 kDa NMWC filter, which is in agreement with Rdest and Sturm (1987). The majority of the activity (89.4%) was held back by a 30 kDa NMWC filter. Application of the 30 kDa retentate to a BioRad P60M gel filtration column equilibrated with GN101 basal salts revealed two peaks of activity — one at an apparent molecular weight of 30 kDa, and a second, much smaller peak at < 2.5 kDa. The 30 kDa peak accounted for 75% of the total activity loaded onto the column. Gel filtration column fractions containing halocin activity at 30 kDa were pooled and desalted using tangential flow filters. The desalted material was subjected to reverse-phase HPLC and eluted with acetonitrile in 0.1% trifluoroacetic acid. Activity was associated with a single 280/214 nm peak at 50% ACN, and Figure 24.4 shows the silver-stained SDS-PAGE gel of the reverse phase HPLC fractions from this peak. The very small band (< 2.5 kDa) at the bottom of the gel correlates with activity, while the larger 14 kDa doublet does not. Furthermore, anion exchange chromatography of the 30 kDa retentate has produced fractions that correlate activity with the < 2.5 kDa band in the absence of the 14 kDa band (data not shown). We conclude that halocin Hal R1 is very small (< 2.5 kDa), but is complexed in the culture supernatant with one or more "carrier" proteins that increase its apparent size to about 30 kDa. Hence, it is held back by a 30 kDa NMWC filter and appears as a 30 kDa moiety by gel filtration, but only dissociates from the "carrier" molecule(s) when subjected to SDS-PAGE or reverse phase HPLC.

A remarkably stable protein, halocin Hal R1 is insensitive to desalting. It retains 50% of its activity after boiling for 1 h (Rdest and Sturm, 1987), and 30 kDa retentates of culture supernatants have retained 100% of their activity after storage for ~three years at 4°C. Although sensitive to nonspecific proteases such as proteinase K, pronase P and elastase, it is insensitive to specific proteases like papain, trypsin and thermolysin (Rdest and Sturm, 1987). It is also insensitive to organic solvents like n-propanol, acetonitrile/trifluoroacetic acid and methanol.

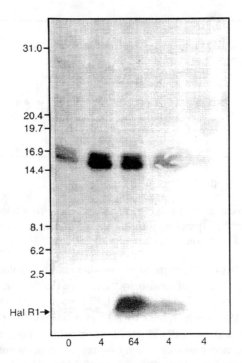

FIGURE 24.4 Silver-stained SDS-PAGE gel of five column fractions from reverse-phase HPLC. The material loaded onto the reverse-phase column was from pooled and desalted fractions from the 30 kDa material eluted from a BioRad P60M gel filtration column. Molecular weight standards in (kDa) and the arrow indicating the band corresponding to halocin Hal R1 are indicated at the left of the figure. Halocin activity (in AU, see text) of each HPLC fraction is indicated at the bottom of the figure.

24.2.4 HALOCIN S8

Halocin S8 is a second example of a halophilic microcin. It is produced by an uncharacterized extremely halophilic rod (strain S8a) isolated from salt crystals from the Great Salt Lake, UT (Penny Amy, University of Las Vegas, NV). Analysis of total genomic DNA by CHEF gel electrophoresis reveals two megaplasmids of ~300 and ~400 kbp. The physiology of expression of halocin activity is shown in Figure 24.3B. Once again, activity is first detected at the transition from exponential phase to stationary phase. However, unlike halocin Hal R1, S8 activity increases rapidly from undetectable to maximum, then decreases and levels off at a lower activity level. Halocin S8 has been purified using a protocol similar to that described for halocin Hal R1. During tangential flow filtration, S8 activity was retained equally by 30- and 10 kDa NMWC filters. The 30 kDa retentate was applied to a BioRad P30M gel filtration column equilibrated with S8a basal salts. A single peak of activity eluted at the lower end of the size exclusion range of the gel support (~2.5 kDa). Taken together, these data suggested that the halocin was small, but could be nonspecifically associated with a variety of larger proteins. Thus, activity was retained by filters with different pore sizes. The fractions from the gel filtration

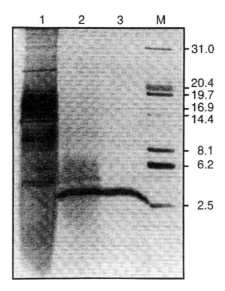

FIGURE 24.5 Silver-stained SDS-PAGE gel of material at various stages of the purification procedure for halocin S8. Lane 1, 30 kDa retentate applied to a BioRad P30M gel filtration column; lane 2, a typical gel filtration column fraction with halocin S8 activity; lane 3, purified halocin S8 after reverse-phase HPLC; M, molecular weight standards in kDa.

column containing activity were pooled, desalted and subjected to reverse-phase HPLC. A single 280/214 nm peak contained all of the activity. Figure 24.5 is a silver-stained SDS-PAGE gel of the purification protocol showing material from the 30 kDa retentate, a typical fraction from the P30M gel filtration column and material from the HPLC peak. The final HPLC step shows a single band that migrates just above the 2.5 kDa marker at about 3–4 kDa, placing halocin S8 in the microcin class.

Halocin S8 also appears robust as it is totally insensitive to desalting, lyophilization, organic solvents like acetonitrile, treatment with trypsin, and boiling for 1 h. However, activity is destroyed by treatment with proteinase K.

24.3 CONCLUSION

The teleological explanation for bacteriocin production is twofold: (1) to reduce competition in a given environmental niche, and (2) to maintain bacteriocin-encoding plasmids containing other factors that contribute to the survival of the cell – loss of the plasmid would result in susceptibility to the bacteriocin produced by competitors (Braun et al., 1994). In the case of halocins, this argument can be taken further to include two observations: the appearance of halocins as cultures enter stationary phase, and the impressive robustness of three of the four halocins so far characterized. As growth slows due to the deterioration of the environment, the cells prepare themselves for stationary-phase survival. Production of a halocin at this time can be advantageous. First, bacteriolytic halocins can lyse sensitive cells, releasing their contents and enriching the environment. Second, bacteriolytic and bacteriostatic halocins reduce competition. When nutrients become abundant again, the halocin-producing organism is positioned to take advantage of this nutrient influx in a less competitive environment. If this is the case, the longer the halocin can remain active in the environment the better, and it is easy to understand why halocins have evolved as environmentally stable protein antibiotics. In addition, since halocin-producing organisms are so prevalent (Torreblanca et al., 1994), they possibly contribute to the overall lack of diversity found in hypersaline environments (Borowitzka, 1981; Oren, 1993).

Halocins are relatively easy to purify (especially if one employs a defined medium) as they are secreted into the environment, separating themselves from the vast majority of cellular proteins. They are a potential model system for a variety of avenues of haloarchaeal research, including stationary-phase gene regulation (Cheung et al., 1997), secretion, and structure/function studies on halophilic proteins. With the exception of halocin H6, their modes of action are unknown, but when elucidated, might provide an important source of inhibitor molecules (e.g., a haloarchaeal transcription inhibitor). The halocin immunity genes, when isolated, may serve as selectable markers for halocin resistance. Finally, given that there are hundreds of halocin-producing strains that have been isolated (Meseguer et al., 1986; Torreblanca et al., 1994), and that three out of four halocins so far characterized can be desalted without affecting activity, the biotechnological implications of halocins are potentially vast, but remain unexplored.

ACKNOWLEDGMENTS

This work was supported by a Cowden Microbiology Fellowship to E.M.O., and by National Institute of Health grant GM52660 to R.F.S.

REFERENCES

Barefoot, S.F., K.M. Harmon, D.A. Grinstead, and C.G. Nettles. 1992. Bacteriocins, molecular biology, in *Encyclopedia of Microbiology*, Vol. I, J. Lederberg, Ed. Academic, San Diego. 191–202.

Betlach, M.C., J. Friedman, H.W. Boyer, and F. Pfeifer. 1984. Characterization of a halobacterial gene affecting bacterio-opsin gene expression. *Nucleic Acids Res.* 20:7949–7959.

Blanck, A. and D. Oesterhelt. 1987. The halo-opsin gene. II. Sequence, primary structure of halorhodopsin and comparison with bacteriorhodopsin. *EMBO J.* 6:265–273.

Borowitzka, L.J. 1981. The microflora. Adaptions to life in extremely saline lakes. *Hydrobiologia* 81:33–46.

Braun, V., H. Pilsl, and P. Groß. 1994. Colicins: structures, modes of action, transfer through membranes, and evolution. *Arch. Microbiol.* 161:199–206.

Cheung, J., K.J. Danna, E.M. O'Connor, L.B. Price, and R.F. Shand. 1997. Isolation, sequence, and expression of the gene encoding halocin H4, a bacteriocin from the halophilic archaeon *Haloferax mediterranei* R4. *J. Bacteriol.* 179:548–551.

DasSarma, S., U.L. RajBhandary, and H.G. Khorana. 1984. Bacterio-opsin mRNA in wild-type and bacterio-opsin-deficient *Halobacterium halobium* strains. *Proc. Natl. Acad. Sci. USA* 81:125–129.
Ebert, K. and W. Goebel. 1985. Conserved and variable regions in the chromosomal and extrachromosomal DNA of halobacteria. *Mol. Gen. Genet.* 200:96–102.
González-Pastor, J.E., J.L. San Millán, and F. Moreno. 1994. The smallest known gene. *Nature* 369:281.
Guijarro, J.I., J.E. González-Pastor, F. Baleux, J.L. San Millán, M.A. Castilla, M. Rico, F. Moreno, and M. Delepierre. 1995. Chemical structure and translation inhibition studies of the antibiotic microcin C7. *J. Biol. Chem.* 270:23520–23532.
Hall, M.J. and N.R. Hackett. 1989. DNA sequence of a small plasmid from *Halobacterium* strain GN101. *Nucleic Acids Res.* 17:10501.
Hoover, D.G. 1992. Bacteriocins: activities and applications, in *Encyclopedia of Microbiology*, Vol. I, J. Lederberg, Ed. Academic, San Diego. 181–190.
Izard, J.W., and D.A. Kendall. 1994. Signal peptides: exquisitely designed transport promoters. *Mol. Microbiol.* 13:765–773.
Jack, R.W., J.R. Tagg, and B. Ray. 1995. Bacteriocins of gram-positive bacteria. *Microbiol. Rev.* 59:171–200.
James, R., C. Lazdunski, and F. Pattus, Eds. 1992. *Bacteriocins, Microcins and Lantibiotics*. Springer-Verlag, New York.
Kleinkauf, H. and H. von Döhren. 1990. Nonribosomal biosynthesis of peptide antibiotics. *Eur. J. Biochem.* 192:1–15.
López-Garcia, P., J.P. Abad, C. Smith and R. Amils. 1992. Genomic organization of the halophilic archaeon *Haloferax mediterranei*: physical map of the chromosome. *Nucleic Acids Res.* 20:2459–2464.
Meseguer, I. and F. Rodriguez-Valera. 1985. Production and purification of halocin H4. *FEMS Microbiol. Lett.* 28:177–182.
Meseguer, I. and F. Rodriguez-Valera. 1986. Effect of halocin H4 on cells of *Halobacterium halobium*. *J. Gen. Microbiol.* 132:3061–3068.
Meseguer, I., F. Rodriguez-Valera, and A. Ventosa. 1986. Antagonistic interactions among halobacteria due to halocin production. *FEMS Microbiol. Lett.* 36:177–182.
Meseguer, I., M. Torreblanca, and F. Rodriguez-Valera. 1995. Specific inhibition of the halobacterial Na^+/H^+ antiporter by halocin H6. *J. Biol. Chem.* 270:6450–6455.
Moreno, F., J.L. San Millán, C. Hernández-Chico, and R. Kolter. 1995. Microcins, in *Genetics and Biochemistry of Antibiotic Production*, L.C. Vining and C. Stuttard, Eds. Butterworth-Heinemann, Newton, MA. 307–321.
Oren, A. 1993. Ecology of extremely halophilic microorganisms, in *The Biology of Halophilic Bacteria*, R.H. Vreeland and L.I. Hochstein, Eds. CRC, Boca Raton, FL. 26–53.
Rdest, U. and M. Sturm. 1987. Bacteriocins from halobacteria, in *Protein Purification: Micro to Macro*, R. Burgess, Ed. Alan R. Liss, New York. 271–278.
Rodriguez-Valera, F., G. Juez, and D.J. Kushner. 1982. Halocins: salt-dependent bacteriocins produced by extremely halophilic rods. *Can. J. Microbiol.* 28:151–154.
Ruepp, A. and J. Soppa. 1996. Fermentative arginine degradation in *Halobacterium salinarium* (formerly *Halobacterium halobium*): genes, gene products and transcripts of the *arcRABC* gene cluster. *J. Bacteriol.* 178:4942–4947.
Torreblanca, M., I. Meseguer, and F. Rodriguez-Valera. 1989. Halocin H6, a bacteriocin from *Haloferax gibbonsii*. *J. Gen. Microbiol.* 135:2655–2661.

Torreblanca, M., I. Meseguer, and F. Rodriguez-Valera. 1990. Effects of halocin H6 on the morphology of sensitive cells. *Biochem. Cell Biol.* 68:396–399.

Torreblanca, M., I. Meseguer, and A. Ventosa. 1994. Production of halocin is a practically universal feature of archaeal halophilic rods. *Lett. Appl. Microbiol.* 19:201–205.

Vining, L.C. and C. Stuttard, Eds. 1995. *Genetics and Biochemistry of Antibiotic Production.* Butterworth-Heinemann, Newton, MA.

von Heijne, G. 1984. How signal sequences maintain cleavage specificity. *J. Mol. Biol.* 173:243–251.

Section VI

Genetics and Genomics

25 Evolutionary Origins of the Haloarchaeal Genome

Robert L. Charlebois

CONTENTS

25.1 Introduction 309
25.2 Plasmids, Insertion Sequences, FI/FII, and Genetic Instability 309
25.3 Acquisition of Genes through Horizontal Transfer 311
25.4 Conclusion 314
Acknowledgments..314
References..314

25.1 INTRODUCTION

The field of genomics is in a state of rapid evolution. Maps have been overshadowed by sequences, and some of the most-characterized genomes belong to some of the least-characterized species, *Methanococcus jannaschii* being a case in point. Among the Archaea, it is the lineage of extreme halophiles whose genetics and genomics are best understood, despite the present lack of a complete sequence. In this chapter, I offer the current understanding of haloarchaeal genomes, and discuss how to apply the power of genomics technology to answer some outstanding issues.

25.2 PLASMIDS, INSERTION SEQUENCES, FI/FII, AND GENETIC INSTABILITY

It has long been known that haloarchaeal DNA is inhomogeneous in composition, fractionating in CsCl density gradients (Joshi et al., 1963; Moore and McCarthy, 1969) or on malachite-green bisacrylamide columns (Pfeifer et al., 1982) into a major higher-G+C fraction (FI), and a minor lower-G+C fraction (FII) comprising at least 10% of the total. Genomic profiling and restriction mapping later revealed that much of this FII DNA occurs in the form of plasmids (Charlebois et al., 1991; Ebert et al., 1984; Pfeifer et al., 1982; St. Jean et al., 1994) — though not all plasmids are FII — and some FII takes the form of islands embedded within otherwise-FI chromosomal DNA (Charlebois et al., 1991; Ebert and Goebel, 1985; Pfeifer and

Betlach, 1985). It has been proposed, though not proven, that these islands represent insertions (of plasmids?) into the chromosome (Pfeifer and Betlach, 1985; Charlebois et al., 1991), a plausible notion, given the difficulty in otherwise explaining why there should be compositionally discrete segments within chromosomal DNA. Plasmids have been known to integrate into chromosomal DNA in other species (e.g., Low, 1996), and the presence of many repeated sequences in haloarchaeal DNA (Cohen et al., 1992; DasSarma et al., 1983; Pfeifer et al., 1983; Sapienza and Doolittle, 1982; St. Jean and Charlebois, 1996) as well as an efficient system for homologous recombination (Cline et al., 1989) all support this idea. If FII islands are indeed integrated plasmids, we might also expect to find additional FI-like integrants whose composition might not so readily betray their location.

The function of haloarchaeal FII DNA has remained enigmatic, though the recent sequencing of pNRC100 (S. DasSarma, personal communication), a 200-kbp FII plasmid from *Halobacterium salinarum* (*halobium*) NRC-1 (Ng et al., 1991) will answer many questions. To date, we have known FII DNA primarily as a major repository for insertion sequences. Their activity, spilling over into nearby genes such as those encoding gas vesicles (DasSarma et al., 1988; Pfeifer et al., 1981); or transposing more broadly, as into sequences necessary for bacteriorhodopsin production (Betlach et al., 1983; DasSarma et al., 1983); is well-known (Charlebois and Doolittle, 1989; DasSarma, 1989; Pfeifer, 1986). Within FII DNA itself, transposition and insertion sequence-mediated rearrangement drive FII's dynamic nature (Pfeifer et al., 1988, 1989).

The extreme halophiles are not unique among the Archaea in possessing such a potentially unstable genome. We have discovered many families of insertion sequences within *Sulfolobus solfataricus* P2, at least some of which transpose at high frequency (Schleper et al., 1994; Sensen et al., 1996; unpublished data). Insertion sequences appear to be clustered in the *S. solfataricus* genome (Charlebois et al., 1996), as they are in the haloarchaeal genome, though there is no conveniently discernible compositional bias pointing to structurally inhomogeneous DNA. Plasmids, though present in some species of *Sulfolobus* (Zillig et al., 1996), are not as ubiquitous as in the haloarchaea.

What then is the relationship between plasmids, FII DNA, insertion sequences, and genetic instability? Copies of insertion sequences may move by transposition, or serve as targets for homologous recombination. Since haloarchaeal insertion sequences concentrate in FII DNA, one should expect a higher rate of rearrangement in FII relative to FI, as observed (Pfeifer et al., 1988, 1989). However, FI DNA is by no means free of insertion sequences (Cohen et al., 1992) nor of their disruptive influence (Betlach et al., 1983; DasSarma et al., 1983), yet FI (chromosomal) DNA seems evolutionarily stable and resistant to rearrangement (Hackett et al., 1994; López-García et al., 1995). The simplest explanation is that FI DNA is more "important" to the cell than is FII DNA, such that disruption of FI is more strongly counterselected than is disruption of FII.

Much FII DNA takes the form of plasmids, but some major haloarchaeal plasmids are compositionally FI, such as pHV3 from *Haloferax volcanii* (Charlebois et al., 1991). There has been dispute of early thoughts (Trieselmann and Charlebois, 1992) that FI plasmids might contain redundant or superfluous genes, different sorts of genes from those found in the chromosome, or nothing but plasmid-specific functions. The megaplasmid pHV3, for instance, contains genes whose functions are neither exotic nor extraneous; from 40 kbp of sequence, we identified seven oxidoreductases, three transporters, an aminotransferase, a homolog of hydantoinase, and a copy of ISH11 (accession numbers U95372-U95377 and unpublished data). Apart from the insertion sequence, these genes appear not to be repeated elsewhere in the *Hf. volcanii* genome (unpublished results). Furthermore, an intergeneric genomic comparison between *Hb. salinarum* GRB and *Hf. volcanii* DS2 revealed scattered homologies between chromosomes and plasmids not respecting any kind of functional distinction (St. Jean and Charlebois, 1996). These data suggest that a plasmid such as pHV3 is simply a portion of chromosome that replicates independently, perhaps as a strategy to speed replication of the genome and cell division (Stouthamer and Kooijman, 1993). Alternatively, a topological independence of replicons might facilitate the global switching of alternate sets of genes, depending on physiological need. The distinction between chromosome and megaplasmid seems equally blurred in other multi-replicon systems as well (Banfalvi et al., 1985; Suwanto and Kaplan, 1989), and their categorization remains rather subjective.

If, then, FI plasmids are simply extensions of the chromosome, what are FII plasmids and why are they so heavily infested with insertion sequences? Should they encode mostly selfish or otherwise unimportant functions, they may simply accumulate insertion sequences from an inability to counter their pressure. It could also be that the insertion sequences have "learned" to transpose preferentially into the lower-G+C DNA, a useful strategy given that FII is generally plasmid DNA, and that plasmids can in some cases be conjugative. Conjugative plasmids have been found in the Archaea (Schleper et al., 1995), though they have not yet been identified in the haloarchaea.

It has been proposed that genomic rearrangement accelerates adaptation in haloarchaea (Hackett et al., 1994), in keeping with the contextual model of gene expression (Charlebois and St. Jean, 1995). It has also been proposed that mutational tinkering with protein sequences contributes to adaptation — especially to variable hypersalinity (Dennis and Shimmin, 1997). Rather than being exclusive, these models are complementary. But does it make sense to assign positive roles to forces as disruptive as rearrangement and mutation? Perhaps these forces are ever-present, and in haloarchaea particularly potent, while in maladapted situations, forces that oppose disturbance are relaxed or otherwise overcome (Charlebois and St. Jean, 1995). Genomic rearrangement, mutation, partitioning into various replicons, mechanisms promoting genetic exchange, and an active endowment of "selfish DNA" might occur in haloarchaea not because of an advantage to instability, but because of the presence of strong stabilizing factors opposing disruption. In a zero-sum game, the polarity between these forces can be slight, or, as in haloarchaea, it can be great.

25.3 ACQUISITION OF GENES THROUGH HORIZONTAL TRANSFER

Similarities between insertion sequences in different species and even in different genera of haloarchaea argue that horizontal transfer of genetic materials does occur

(Ebert and Goebel, 1985; Pfeifer and Blaseio, 1990). Insertion sequences seem to be among the most "conserved" of haloarchaeal genes (as detected by hybridization), yet there is no reason to suppose that their sequences are slow to change and indeed there is evidence to the contrary, indicating rapid degeneration (Hofman et al., 1986).

Unless one invoked some advantage for chromosomally mediated interorganismal recombination — eukaryotic-style sex, in essence — it is difficult not to implicate the plasmids or phages, or even the insertion sequences themselves, in sponsoring horizontal transfer. The mere existence of such high numbers of active insertion sequences within the haloarchaeal genome strongly suggests that genetic exchange must be prevalent among at least some members of this lineage — without spread, insertion sequences have a frustrated purpose. Another line of evidence suggesting that genetic exchange does occur is the presence of restriction and modification systems in haloarchaea (Daniels and Wais, 1984; Patterson and Pauling, 1985), arguably for ensuring against phage infection, but more generally serving as a line of defense against all forms of intrusive DNA. Of course, the most compelling reason for suspecting genetic exchange among haloarchaea is the mating system described from *Haloferax volcanii* (Mevarech and Werczberger, 1985; Rosenshine and Mevarech, 1991; Rosenshine et al., 1989; Tchelet and Mevarech, 1994). Unfortunately, the genetics of this system are still unknown.

The implications of genetic exchange among haloarchaea, especially of the type that adds rather than replaces genes, are potentially profound. Genetic exchange not only results in the sharing of innovations, but also encourages innovation through the introduction of extra, heterologous functionality. DNA hybridization might fail to find a homolog within the same genome for a particular haloarchaeal gene, but this does not mean that the homolog is not present. If exchange occurs among distantly related species, then sequences that encode the same function can coexist but still be undetectable by DNA hybridization, and be resistant to gene conversion as well. With such duplication in a genome, whether from an intrinsic or an extrinsic source, the cell has freedom to explore new regions of sequence space and thereby to invent novel functionality or regulation. There are numerous examples of duplicate genes in haloarchaea (Antón et al., 1994; St. Jean and Charlebois, 1996), undoubtedly betraying a more extensive multiplication unseen by hybridization. Here, a genomic sequence would serve us immensely, since computer software finds homologs with much more sensitivity than does any wet-lab protocol.

The question arises about the breadth of sources contributing to the haloarchaeal genome. It may be safely assumed that haloarchaea exchange materials among themselves, but can they or did they receive DNA from other lineages? Transformation does not appear to be a viable means of gaining such DNA, since haloarchaea do not seem to be naturally transformable (Cline and Doolittle, 1987). Other important barriers include the need for gene products to be adapted to high concentrations of salt found intracellularly (Eisenberg and Wachtel, 1987), and the need for genes to be recognized by the (halo)archaeal-specific gene expression machinery. Thus, it seems unlikely that much non-haloarchaeal DNA could contribute to the genome. There was a time, however, when one of the barriers — salt adaptation of proteins — might not have been as important, and that is near the point in time when the haloarchaeal lineage was founded. Then, the constraints would have centered on the ability to mate with another type of cell or to be infected by one of its transducing phages, and on the ability to adapt gene-expression signals to the archaeal machinery (if the source of the DNA were non-archaeal).

There is a hint of distant gene immigration in the finding that some haloarchaeal gene products bear significant resemblance to those from cyanobacteria, including gas vesicles (Kinsman and Hayes, 1997; Walker et al., 1984; Walsby, 1994), ferredoxin (Hase et al., 1977; Pfeifer et al., 1993), and the DNA-binding antioxidant Dps (unpublished data). Another thing that needs explaining is how a predominantly aerobic heterotrophic extreme halophile can arise from a strictly anaerobic, methanogenic ancestor, as suggested by phylogenetic analysis (Woese, 1987). Perhaps the transition is not difficult, but it does require certain genes to be recruited or created. Genes must arise from other genes, but could any methanogen possess the genotype necessary to generate a haloarchaeal genome? Full genetic sequences from methanogens suggest not. Some functions, such as perhaps bacteriorhodopsin and its relatives, may have been created *de novo* (Shepherd, 1982), as some sort of "hopeful monster." Other genes, known to be homologous to genes in other lineages but which serve no obvious purpose in methanogens, either persisted unused throughout methanogenic history to be somehow resurrected within the haloarchaea and lost in modern methanogenic representatives, or they were imported from another lineage. Since the haloarchaeal energy metabolism appears phylogenetically to be recently derived, does this mean that, to import the genes necessary for chemoorganotrophy, it was necessary to search far afield? Chimeric origins for eukaryotes (Golding and Gupta, 1995; Zillig, 1991) and even for Archaea (Koonin et al., 1997) have already been proposed, to reconcile patterns of protein sequence conservation. Might the haloarchaeon represent such a chimera itself, perhaps a blend of some eukaryarchaeotic halophile with an aerobe such as a cyanobacterium? Should not only useful genes but also gene expression machinery be included in the donor's contribution, there would have been ample time to draw each new gene's expression and regulation into the purview of the dominant transcription mechanism, while the new gene's function continued to be enjoyed. Meanwhile, native and immigrant proteins alike would have been free to adapt to the hypersaline intracellular environment, as the haloarchaeon settled into its new niche. Perhaps relics of this ancient transition might still be present in the genome.

These, admittedly, are theoretical speculations. Still, hypotheses are useful, if not necessary, precursors to experimentation, and might serve to justify the effort and expense of such grand projects as genomic sequencing. Yet genomic technology need not produce costly finished sequences to be valuable to the research community. Comparative projects, especially, might benefit from the efficiency and throughput of sequencing laboratories to produce numerous "nucleotide-resolution genetic maps." My laboratory is currently engaged in such an approach of comparing ~ 50kbp from each of seven genomes, albeit from within the Sulfolobales rather than from within the *Halobacteriaceae* (unpublished). To understand the value of genomic organization, it must be useful to compare different genomes, just as it is useful to compare gene sequences to locate conserved and putatively important features. Genomes are more plastic than genes, however, and thus the gene-structure paradigm cannot be applied as such to genomic structure. At this early stage, our goal might focus on learning how to read a genome, as a prerequisite to understanding it. Then we should be better poised to evaluate the route by which an aerobic, chemoorganotrophic, extremely halophilic Archaeon could have come to be.

25.4 CONCLUSION

Haloarchaea have distinctive qualities that set them apart from other characterized Archaea. Focusing on the genome, there is inhomogeneity at the level of nucleotide/oligonucleotide composition, significant amounts of gene duplication, large families of active insertion sequences, and widespread use of megaplasmids. In addition, genes are present that confer specific adaptation to the hypersaline environment, genes that do not all necessarily occur in the methanogens from which the haloarchaeal lineage is believed to have arisen.

Plasmids, phages, and a mating system offer mechanisms by which new functions may be acquired rather than invented. Genetic exchange as an origin of new functionality is a hypothesis that can be tested using comparative genomic analysis. Alternatively, frequent gene duplication and genomic rearrangement may generate novelty intrinsically. As genomes continue to be sequenced, sufficient data should become available to enable us to understand the events surrounding the foundation of the haloarchaeal lineage, and by extension, the origin of other major biological innovations.

ACKNOWLEDGMENTS

I thank Dr. Andrew St. Jean for his critical reading of the manuscript. Work in my laboratory on haloarchaeal genomics has been funded by the Natural Sciences and Engineering Research Council of Canada.

REFERENCES

Antón, J., P. López-García, J.P. Abad, C.L. Smith, and R. Amils. 1994. Alignment of genes and *Swa*I restriction sites to the *Bam*HI genomic map of *Haloferax mediterranei*. *FEMS Microbiol. Lett.* 117:53–60.

Banfalvi, Z., E. Kondorosi, and A. Kondorosi. 1985. *Rhizobium meliloti* carries two megaplasmids. *Plasmid* 13:129–138.

Betlach, M., F. Pfeifer, J. Friedman, and H.W. Boyer. 1983. Bacterio-opsin mutants of *Halobacterium halobium*. *Proc. Natl. Acad. Sci. USA* 80:1416–1420.

Charlebois, R.L. and W.F. Doolittle. 1989. Transposable elements and genome structure in halobacteria, in *Mobile DNA*, M. Howe and D. Berg, Eds. American Society for Microbiology, Washington, DC. 297–307.

Charlebois, R.L. and A. St. Jean. 1995. Supercoiling and map stability in the bacterial chromosome. *J. Mol. Evol.* 41:15–23.

Charlebois, R.L., L.C. Schalkwyk, J.D. Hofman, and W.F. Doolittle. 1991. A detailed physical map and set of overlapping clones covering the genome of the archaebacterium *Haloferax volcanii* DS2. *J. Mol. Biol.* 222:509–524.

Charlebois, R.L., T. Gaasterland, M.A. Ragan, W.F. Doolittle, and C.W. Sensen. 1996. The *Sulfolobus solfataricus* P2 genome project. *FEBS Lett.* 389:88–91. (Erratum 398:343.)

Cline, S.W. and W.F. Doolittle. 1987. Efficient transfection of the archaebacterium *Halobacterium halobium*. *J. Bacteriol.* 169:1341–1344.

Cline, S.W., L.C. Schalkwyk, and W.F. Doolittle. 1989. Transformation of the archaebacterium *Halobacterium volcanii* with genomic DNA. *J. Bacteriol.* 171:4987–4991.

Cohen, A., W.L. Lam, R.L. Charlebois, W.F. Doolittle, and L.C. Schalkwyk. 1992. Localizing genes on the map of the genome of *Haloferax volcanii*, one of the Archaea. *Proc. Natl. Acad. Sci. USA* 89:1602–1606.

Daniels, L.L. and A.C. Wais. 1984. Restriction and modification of halophage S45 in *Halobacterium*. *Curr. Microbiol.* 10:133–136.

DasSarma, S. 1989. Mechanisms of genetic variability in *Halobacterium halobium*: the purple membrane and gas vesicle mutations. *Can. J. Microbiol.* 35:65–72.

DasSarma, S., U.L. RajBhandary, and H.G. Khorana. 1983. High-frequency spontaneous mutation in the bacterio-opsin gene in *Halobacterium halobium* is mediated by transposable elements. *Proc. Natl. Acad. Sci. USA* 80:2201–2205.

DasSarma, S., J.T. Halladay, J.G. Jones, J.W. Donovan, P.J. Giannasca, and N. Tandeau de Marsac. 1988. High-frequency mutations in a plasmid-encoded gas vesicle gene in *Halobacterium halobium*. *Proc. Natl. Acad. Sci. USA* 85:6861–6865.

Dennis, P.P. and L.C. Shimmin. 1997. Evolutionary divergence and salinity-mediated selection in halophilic Archaea. *Microbiol. Mol. Biol. Rev.* 61:90–104.

Ebert, K., W. Goebel, and F. Pfeifer. 1984. Homologies between heterogeneous extrachromosomal DNA populations of *Halobacterium halobium* and four new halobacterial isolates. *Mol. Gen. Genet.* 194:91–97.

Ebert, K. and W. Goebel. 1985. Conserved and variable regions in the chromosomal and extrachromosomal DNA of halobacteria. *Mol. Gen. Genet.* 200:96–102.

Eisenberg, H. and E.J. Wachtel. 1987. Structural studies of halophilic proteins, ribosomes, and organelles of bacteria adapted to extreme salt concentrations. *Ann. Rev. Biophys. Biophys. Chem.* 16:69–92.

Golding, G.B. and R.S. Gupta. 1995. Protein-based phylogenies support a chimeric origin for the eukaryotic genome. *Mol. Biol. Evol.* 12:1–6.

Hackett, N.R., Y. Bobovnikova, and N. Heyrovska. 1994. Conservation of chromosomal arrangement among three strains of the genetically unstable archaeon *Halobacterium salinarium*. *J. Bacteriol.* 176:7711–7718.

Hase, T., S. Wakabayashi, H. Matsubara, L. Kerscher, D. Oesterhelt, K.K. Rao, and D.O. Hall. 1977. *Halobacterium halobium* ferredoxin: a homologous protein to chloroplast-type ferredoxins. *FEBS Lett.* 77:308–310.

Hofman, J.D., L.C. Schalkwyk, and W.F. Doolittle. 1986. ISH51: a large, degenerate family of insertion sequence-like elements in the genome of the archaebacterium, *Halobacterium volcanii*. *Nucl. Acids Res.* 14:6983–7000.

Joshi, J.G., W.R. Guild, and P. Handler. 1963. The presence of two species of DNA in some halobacteria. *J. Mol. Biol.* 6:34–38.

Kinsman, R. and P.K. Hayes. 1997. Genes encoding proteins homologous to halobacterial Gvps N, J, K, F & L are located downstream of *gvpC* in the cyanobacterium *Anabaena flos-aquae*. *DNA Sequence* 7:97–106.

Koonin, E.V., A.R. Mushegia, M.Y. Galperin, and D.R. Walker. 1997. Comparison of archaeal and bacterial genomes: computer analysis of protein sequences predicts novel functions and suggest a chimeric origin for the archaea. *Mol. Microbiol.* 25:619–637.

López-García, P., A. St. Jean, R. Amils, and R.L. Charlebois. 1995. Genomic stability in the archaea *Haloferax volcanii* and *Haloferax mediterranei*. *J. Bacteriol.* 177:1405–1408.

Low, K.B. 1996. Hfr strains of *Escherichia coli* K-12, in *Escherichia coli and Salmonella: Cellular and Molecular Biology,* 2nd edition. F.C. Neidhardt, R. Curtiss III, J.L. Ingraham, E.C.C. Lin, K.B. Low, B. Magasanik, W.S. Reznikoff, M. Riley, M. Schaechter, and H.E. Umbarger, Eds. American Society for Microbiology, Washington, DC. 2402–2405.

Mevarech, M. and R. Werczberger. 1985. Genetic transfer in *Halobacterium volcanii*. *J. Bacteriol.* 162:461–462.

Moore, R.L. and B.J. McCarthy. 1969. Characterization of the deoxyribonucleic acid of various strains of halophilic bacteria. *J. Bacteriol.* 99:248–254.

Ng, W.-L., S. Kothakota, and S. DasSarma. 1991. Structure of the gas vesicle plasmid in *Halobacterium halobium*: inversion isomers, inverted repeats, and insertion sequences. *J. Bacteriol.* 173:1958–1964.

Patterson, N.H. and C. Pauling. 1985. Evidence for two restriction-modification systems in *Halobacterium cutirubrum*. *J. Bacteriol.* 163:783–784.

Pfeifer, F. 1986. Insertion elements and genome organization of *Halobacterium halobium*. *Syst. Appl. Microbiol.* 7:36–40.

Pfeifer, F. and M. Betlach. 1985. Genome organization in *Halobacterium halobium*: a 70 kb island of more (AT) rich DNA in the chromosome. *Mol. Gen. Genet.* 198:449–455.

Pfeifer, F. and U. Blaseio. 1990. Transposition burst of the ISH27 insertion element family in *Halobacterium halobium*. *Nucl. Acids Res.* 18:6921–6925.

Pfeifer, F., G. Weidinger, and W. Goebel. 1981. Genetic variability in *Halobacterium halobium*. *J. Bacteriol.* 163:783–784.

Pfeifer, F., K. Ebert, G. Weidinger, and W. Goebel. 1982. Structure and functions of chromosomal and extrachromosomal DNA in halobacteria. *Zbl. Bakt. Hyg., I. Abt. Orig. C* 3:110–119.

Pfeifer, F., M. Betlach, R. Martienssen, J. Friedman, and H.W. Boyer. 1983. Transposable elements of *Halobacterium halobium*. *Mol. Gen. Genet.* 191:182–188.

Pfeifer, F., U. Blaseio, and P. Ghahraman. 1988. Dynamic plasmid populations in *Halobacterium halobium*. *J. Bacteriol.* 170:3718–3724.

Pfeifer, F., U. Blaseio, and M. Horne. 1989. Genome structure of *Halobacterium halobium*: plasmid dynamics in gas vacuole deficient mutants. *Can. J. Microbiol.* 35:96–100.

Pfeifer, F., J. Griffig, and D. Oesterhelt. 1993. The *fdx* gene encoding the [2Fe–2S] ferredoxin of *Halobacterium salinarium* (*H. halobium*). *Mol. Gen. Genet.* 239:66–71.

Rosenshine, I. and M. Mevarech. 1991. The kinetic of the genetic exchange process in *Halobacterium volcanii* mating, in *General and Applied Aspects of Halophilic Microorganisms,* F. Rodriguez-Valera, Ed. Plenum, New York. 265–270.

Rosenshine, I., R. Tchelet, and M. Mevarech. 1989. The mechanism of DNA transfer in the mating system of an archaebacterium. *Science* 245:1387–1389.

Sapienza, C. and W.F. Doolittle. 1982. Unusual physical organization of the *Halobacterium* genome. *Nature* 295:384–389.

Schleper, C., R. Röder, T. Singer, and W. Zillig. 1994. An insertion element of the extremely thermophilic archaeon *Sulfolobus solfataricus* transposes into the endogenous β-galactosidase gene. *Mol. Gen. Genet.* 243:91–96.

Schleper, C., I. Holz, D. Janekovic, J. Murphy, and W. Zillig. 1995. A multicopy plasmid of the extremely thermophilic archaeon *Sulfolobus* effects its transfer to recipients by mating. *J. Bacteriol.* 177:4417–4426.

Sensen, C.W., H.-P. Klenk, R.K. Singh, G. Allard, C.C.-Y. Chan, Q.Y. Liu, F. Young, M. Schenk, T. Gaasterland, W.F. Doolittle, M.A. Ragan, and R.L. Charlebois. 1996. Organizational characteristics and information content of an archaeal genome: 156 kbp of sequence from *Sulfolobus solfataricus* P2. *Mol. Microbiol.* 22:175-191.

Shepherd, J.C.W. 1982. From primeval message to present-day gene. *Cold Spring Harbor Symp. Quant. Biol.* 47:1099-1108.

St. Jean, A. and R.L. Charlebois. 1996. Comparative genomic analysis of the *Haloferax volcanii* DS2 and *Halobacterium salinarium* GRB contig maps reveals extensive rearrangement. *J. Bacteriol.* 178:3860-3868.

St. Jean, A., B.A. Trieselmann, and R.L. Charlebois. 1994. Physical map and set of overlapping cosmid clones representing the genome of the archaeon *Halobacterium* sp. GRB. *Nucl. Acids Res.* 22:1476-1483.

Stouthamer, A.H. and S.A.L.M. Kooijman. 1993. Why it pays for bacteria to delete disused DNA and to maintain megaplasmids. *Antonie van Leeuwenhoek* 63:39-43.

Suwanto, A. and S. Kaplan. 1989. Physical and genetic mapping of the *Rhodobacter sphaeroides* 2.4.1. genome: presence of two unique circular chromosomes. *J. Bacteriol.* 171:5850–5859.

Tchelet, R. and M. Mevarech. 1994. Interspecies genetic transfer in halophilic archaebacteria. *Syst. Appl. Microbiol.* 16:578–581.

Trieselmann, B.A. and R.L. Charlebois. 1992. Transcriptionally active regions in the genome of the archaebacterium *Haloferax volcanii*. *J. Bacteriol.* 174:30–34.

Walker, J.E., P.K. Hayes, and A.E. Walsby. 1984. Homology of gas vesicle proteins in cyanobacteria and halobacteria. *J. Gen. Microbiol.* 130:2709–2715.

Walsby, A.E. 1994. Gas vesicles. *Microbiol. Rev.* 58:94–144.

Woese, C.R. 1987. Bacterial evolution. *Microbiol. Rev.* 51:221–271.

Zillig, W. 1991. Comparative biochemistry of *Archaea* and *Bacteria*. *Curr. Opin. Genet. Dev.* 1:544–551.

Zillig, W., D. Prangishvilli, C. Schleper, M. Elferink, I. Holz, S. Albers, D. Janekovic, and D. Götz. 1996. Viruses, plasmids and other genetic elements of thermophilic and hyperthermophilic Archaea. *FEMS Microbiol. Rev.* 18:225–236.

26 Expression of Ribosomal RNA Operons in Halophilic Archaea

Patrick P. Dennis

CONTENTS

26.1 Introduction 319
26.2 Operon Structure and Precursor rRNA Processing 320
26.3 Multiple rRNA Operon Promoters 320
26.4 The Folding of 16s rRNA 323
26.5 Multiple rRNA Operons in *Haloarcula marismortui* 325
26.6 Conclusion 328
References 329

26.1 INTRODUCTION

The three fundamental processes enshrined in the central dogma of molecular biology are replication, transcription and translation. Of these, the most complex is translation. Translation occurs on subcellular ribonucleoprotein particles called ribosomes, uses novel RNA-based chemistry, and involves the decoding of nucleotide sequence information carried on mRNA into amino acid sequence of protein. In halophilic Archaea, the process is further complicated by the hypersalinity of the intracellular milieu (between 2 and 5 molar in K^+ ions). Generally, such high salinity is devastating to biological systems because it disrupts ionic interactions through charge shielding, enhances hydrophobic interactions, and reduces the concentration of free water below the level normally required to sustain essential biological processes (Dennis and Shimmin, 1997 and references therein; Lanyi, 1974). In this chapter, I focus on some of the unusual and peculiar properties relating to biogenesis of ribosomal particles in halophilic Archaea — namely, the organization and transcription of genes encoding ribosomal RNAs, and the processing, folding, and assembly of the precursor rRNA transcript into functional ribosomes — and how these may in turn be related to halophilic adaptation. For a more general discussion of translation in the Archaea, the reader is referred to a recent review (Dennis, 1997).

26.2 OPERON STRUCTURE AND PRECURSOR rRNA PROCESSING

As expected from their slow growth characteristics, the genomes of halophilic Archaea contain only one or a small number of rRNA transcription units. The typical unit contains 16S, 23S and 5S rRNA genes and generally tRNA^{ala} and tRNA^{cys} genes in the spacer and distal positions (Hui and Dennis, 1985; Figure 26.1). The units are most commonly preceded by multiple, tandemly arranged promoters in the 5' flanking region and the 16S and 23S genes are surrounded by complementary sequences that provide the recognition features for the initial endonucleolytic processing of precursor (pre) rRNA. Through long-range interactions, the complementary sequences form long helical structures surrounding the 16S and 23S sequences. Each helix contains a highly conserved structural motif consisting of two three-base bulges on opposite strands and separated by four base pairs. The motif is cleaved by an unusual and interesting endonuclease releasing pre 16S and pre 23S rRNA from the primary transcript (Chant and Dennis, 1986). Little is known about the enzymes responsible for excision of 5S rRNA and the trimming of 16S and 23S rRNA. Excision of the tRNAs is inefficient and appears to utilize RNaseP and terminal nucleotidyl transferase activities to generate the mature tRNA 5' and CCA 3' ends.

The bulge-helix-bulge (BHB) endonuclease is a homodimer with a subunit size of 37 kDa (Kleman-Leyer et al., 1997). The enzyme is rigid in the structural specificity of its substrate, tolerating neither changes in the size of the bulges nor the number of base pairs between them; the two subunits bind and catalyze or facilitate the concerted cleavages between the second and third bases in each of the two bulges (Thompson and Daniels, 1990). The enzyme is also used to excise archaeal introns from the transcripts of intron containing rRNA and tRNA genes. In these instances, the exon-intron boundaries fold into the requisite structural motif. The cloning and characterization of the *Haloferax volcanii* gene and its encoded protein has led to the realization that the endonuclease protein is homologous to two of the subunits of the heterotetrameric and much more complex eukaryotic tRNA intron endonuclease (Kleman-Leyer et al., 1997). At this point, it is uncertain whether the primordial activity of this endonuclease family was to initiate the processing of pre rRNA or to excise introns from the transcripts of intron-containing genes.

By far the most profound and provocative aspect of these cleavage reactions is the possibility that the bulged helix motif may be an autocatalytic endonuclease — that is, a ribozyme — and that the function of the protein may be to simply facilitate the hydrolysis reaction, acting as a catalyst to accelerate the rate of an otherwise inefficient reaction (Belfort and Weiner, 1997). Efforts are currently under way to obtain an NMR structure for this motif (Peter Moore, Yale University, personal communication); the structure may provide valuable clues about the mechanism of endonucleolytic cleavage.

26.3 MULTIPLE rRNA OPERON PROMOTERS

Halophilic rRNA operons are generally preceded by multiple promoters (Dennis 1985; Mankin and Kagramanova, 1988; Mevarech et al., 1989). In *Halobacterium*

Expression of Ribosomal RNA Operons in Halophilic Archaea

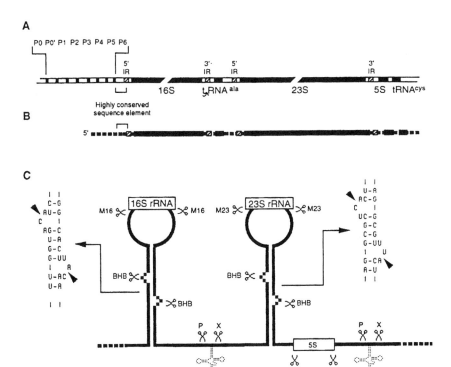

FIGURE 26.1 Genetic organization and processing of the transcript of the single copy ribosomal RNA operon of *Halobacterium cutirubrum*. A. The typical halophilic *rrn* operon from *Hb. cutirubrum* contains 16S, 23S and 5S genes (solid boxes) and an intergenic and distal tRNA gene (shaded boxes). The 16S and 23S genes are surrounded by inverted repeat sequences designated 5'IR and 3'IR (striped boxes). Transcription initiation occurs at potentially any one of eight tandem promoters in the 5' flanking region. A highly conserved 75 nucleotide long sequence located immediately 3' to the P6 promoter has been implicated in the folding of 16S rRNA. B. The primary transcript (pre rRNA) of the *rrn* operon is about 6,000 nt in length, contains the mature rRNA and tRNA sequences, and the signals required to initiate processing and folding. C. The inverted repeat sequences surrounding 16S and 23S rRNA, through long-range complementarity form helical structures in pre rRNA. These helices contain the recognition features for cleavage by the bulge-helix-bulge (BHB) endonuclease. The sequence and structure of the recognition motifs along with the sites of cleavage (▶) are indicated. Other sites of endonucleolytic cleavage within the pre rRNA are indicated: M16, M23, M5, uncharacterized maturases for 16S, 23S and 5S rRNAs; P, RNaseP, X, uncharacterized nuclease that cleaves 3' to tRNA sequences.

cutirubrum (renamed as *Halobacterium salinarum*), the single copy rRNA operon is preceded by eight promoters that are spread out over a 1 kb region of 5' flanking sequence. Why are multiple promoters maintained? Are all promoters used and required and is utilization dependent on extrinsic factors such as salt concentration, DNA topology or environmental stress? To address these questions, a nuclease protection assay is being developed to quantitate the *in vivo* utilization of each of these promoters (Figure 26.2). The assay uses a 3 kb RsrII fragment from the 5'

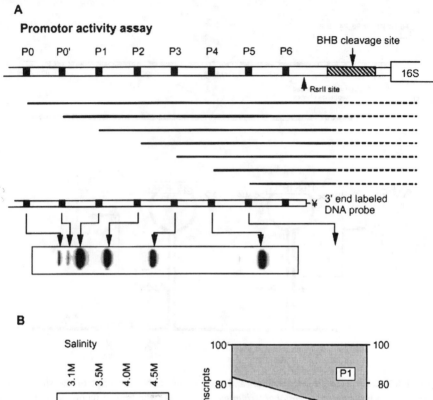

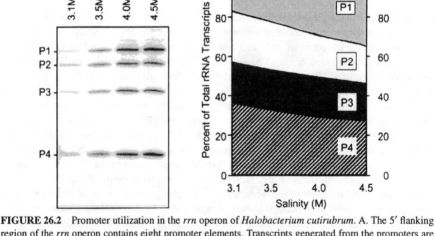

FIGURE 26.2 Promoter utilization in the *rrn* operon of *Halobacterium cutirubrum*. A. The 5′ flanking region of the *rrn* operon contains eight promoter elements. Transcripts generated from the promoters are complementary to a 3 kb RsrII restriction fragment. A nuclease protection assay using a 5′ end labeled RsrII fragment as probe detects six bands corresponding to P0, P0′, P1, P2, P3, P4 and P5. The P5 band is located lower down in the gel and the P6 promoter appears to be inactive. B. *Hb. cutirubrum* was grown exponentially in rich medium containing 3.1, 3.5, 4.0 or 4.5 M NaCl. The exponential mass doubling times of these cultures were about 30, 7, 6 and 6.5 hours, respectively. *Hb. cutirubrum* grows optimally in 4.0 M salt. Total RNA was isolated from each culture and used in an S1 nuclease protection assay with 5′ end labeled RsrII fragment as probe. The relative proportion of products resulting from protection by transcripts initiated at P1, P2, P3 and P4 was quantitated and illustrated as a function of salinity of the growth medium. Because of low abundancy, the P0 and P0′ products were not included in the quantitation.

flanking region that overlaps the eight promoters and ends just in front of the 16S processing helix. When 5' end labeled minus (−) DNA strand is used as probe, RNA transcripts derived from the eight promoters (PO, PO', P1, P2, P3, P4, P5 and P6) are expected to protect fragments of about 1131, 764, 605, 480, 359, 226, 93 and 27 nt in length. Under the *in vivo* conditions examined to date, the P5 and P6 promoters appear to be very weak and represent at most only a few percent of the total rRNA transcripts. For technical reasons, these minor protection products are not easily visualized. The other six products are clearly identifiable and easily quantitated (Figure 26.2A).

A unique feature of halophilic Archaea is that they grow in environments containing greater than 2 M salt and survive by maintaining a balance between the extracellular and intracellular ionic strength (Dennis and Shimmin, 1997; Kushner, 1985; Lanyi, 1974). During evolution, enzymes evolve to function optimally at a certain salt concentration. Deviations from the optimum result in diminution of activity. Does the activity of RNA polymerase and its associated factors or their collective ability to recognize and utilize a given promoter sequence therefore depend on the environmental salt concentration? To address this issue, *Hb. cutirubrum* was grown at 37°C in a nutrient-rich medium containing 3.1, 3.5, 4.0 or 4.5 M NaCl. Total RNA was extracted from mid-exponential phase cultures and analyzed for utilization of the rRNA operon P1, P2, P3 and P4 promoters; under these conditions, utilization of PO and PO' was negligible (Figure 26.2B). At low salt, P1 activity is moderate, representing only about 17% of the total transcripts. With increasing salt, the activity of P1 increases gradually at the expense of P2 and P4; in high salt, P1 represents about 35% of the transcripts. With additional refinements, the PO and PO' promoters can be added to the analysis and the absolute activity, in initiations per min, at each promoter can be obtained.

These experiments suggest that the multiplicity of rRNA promoters is required to ensure efficient rRNA transcription under a variety of physiological conditions. What is not so clear is whether this salinity affects changes in the initiation capacity of the transcriptional apparatus (i.e., RNA polymerase and its associated TBP and TFII factors), changes in DNA topology, or both. There is direct evidence to suggest that altering the superhelical density of the DNA by addition of novobiocin to an exponential culture dramatically activates P4 at the expense of P1 and P3. Finally, the presence of four promoters (PO, PO', P5 and P6) that exhibit little or no *in vivo* activity under conditions tested is intriguing. It could be that these promoters are functional and that the salinity or environmental stress conditions under which they are utilized has not yet been identified. Alternatively, possibly these are promoters that were previously active but have fallen into disuse through the accumulation of mutations. Having multiple promoter-like elements available to explore sequence spaces might have accelerated evolutionary processes and provided a selective advantage to the organism in adapting to the stresses caused by fluctuations in the salinity of the environment.

26.4 THE FOLDING OF 16S rRNA

Small subunit (16S or 18S) rRNAs contain a universally conserved pseudoknot structure that forms between the loop of the 5' terminal hairpin and the short

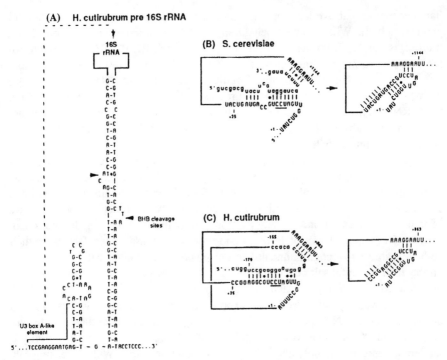

FIGURE 26.3 Involvement of the 5' external transcribed spacer sequence in 16S rRNA folding. A. Halophilic *rrn* operons contain a conserved 5' ETS sequence that is proposed to base pair with 16S rRNA sequences and aid in the folding of a 5' end pseudoknot structure. B. Pseudoknot formation in *Saccharomyces cerevisiae* is depicted. The 5' end of U3 sno RNA (lower case nucleotides) base pairs in an antiparallel direction simultaneously with the 5' end of 18S rRNA (nucleotides 9 to 23) and with the connector region (nucleotides 1139-1143) between the central domain and the major 3' terminal domain. Displacement of U3 RNA allows the pseudoknot (base pairing between the loop of the 5' terminal helix and the connector region) to form. C. Pseudoknot formation in *Hb. cutirubrum* is proposed to involve U3-like 5' ETS sequences (nucleotides –179 to –160) that base pair with the 5' end of 16S rRNA (nucleotides 9 to 23) and with the connector region (nucleotides 859 to 863). Rearrangement and release of the 5' ETS sequence allows the pseudoknot (base pairing between the loop of the 5' terminal helix and the connector region) to form.

connector region located between the central domain and major 3' terminal domains (Gutell et al., 1994). In eukaryotes, it has been suggested that formation of this pseudoknot is facilitated by base complementarity between a highly conserved region in U3 snoRNA and small subunit RNA pseudoknot sequences (Hughes, 1996; see Figure 26.3). In yeast and other eukaryotes, U3 is essential and U3 inactivation or depletion results in defects in 18S rRNA processing and 40S assembly. How do organisms without a functional U3-like sequence efficiently fold this pseudoknot structure in small subunit RNA?

A number of years ago it was noted that the 5' ETS regions of rRNA operons from three different genera of halophilic archaea (*Halobacterium*, *Haloferax* and *Haloarcula*) contain a 70 nucleotide-long sequence that is more highly conserved than are the 16S gene sequences themselves (Dennis, 1991). In reexamining the conserved halophile motif, a U3-like sequence element (Dennis et al., 1997) was recognized. In the single copy operon of *Hb. cutirubrum*, the element is located about 170 nt in front of the 16S rRNA gene, between the P6 promoter and the BHB processing site (Figure 26.3). In the pre rRNA transcript, this element has the potential to base pair simultaneously with the loop region of the 5' terminal helix and short internal connector sequence, thus bringing these two regions of the 16S rRNA into proximity (Figure 26.3). This structure, formed *in cis*, is analogous to the *trans* interaction proposed between eukaryotic U3 sno RNA and 18S rRNA (Hughes, 1996). In both cases, displacement of the U3-like element then allows formation of the conserved pseudoknot structure. Thus, halophilic Archaea appear to contain a cis-acting RNA chaperone sequence that assists in the correct folding of small subunit rRNA; in Eukarya the function is supplied *in trans* by U3 sno RNA. A similar U3-like sequence element has also been observed in the 5' ETS region of several bacterial rRNA operons (including those of *Escherichia coli*; Dennis et al., 1997). In the case of *E. coli*, mutations in the 5' ETS element result in the production of translation-defective 30S particles (Theissen et al., 1993). Attempts are currently being made to construct a second rRNA operon that can be transformed into *Hb. cutirubrum*. With this construct, the importance of this conserved sequence element in rRNA operon expression may be demonstrated.

26.5 MULTIPLE rRNA OPERONS IN *HALOARCULA MARISMORTUI*

Southern hybridization experiments with *HindIII*-digested genomic DNA from *Haloarcula marismortui* clearly revealed the presence of 10- and 20-kbp fragments containing rRNA genes (Mevarech et al., 1989). These two fragments were cloned; each contains a complete rRNA transcription unit. The two operons designated *rrn* A and *rrn* B have been completely sequenced, and *in vivo* patterns of processing of the transcripts from each operon have been analyzed by S1 nuclease protection and primer extension assays (Mylvaganam, 1996; Mylvaganam and Dennis, 1992). Remarkably, the two operons differ substantially. The *rrn* A operon is typical of other halophilic rRNA operons in that it contains tRNA[ala] and tRNA[cys] genes in the spacer and distal regions, it is preceded by a regiment of four active promoters in the 5' flanking region, and the complementary sequences surrounding the 16S gene contain the standard bulge-helix-bulge motif. In contrast, the *rrn* B operon lacks the two tRNAs, is preceded by only a single active promoter, and lacks the BHB recognition motif normally used to initiate processing of 16S rRNA (Figure 26.4).

For organisms that contain multiple genomic rRNA operons, the level of divergence between genes is generally less than one substitution per 1,000 nucleotides. This is because mechanisms such as recombination and gene conversion continually act to homogenize the gene sequences, thereby ensuring a homogenous population of ribosomal particles. It was therefore surprising to find that the 16S, 23S and 5S rRNA genes from the *rrn* A and *rrn* B operons exhibit a degree of divergence that is between 15- and 50-fold higher than normally observed (Table 26.1). The substitutions occur

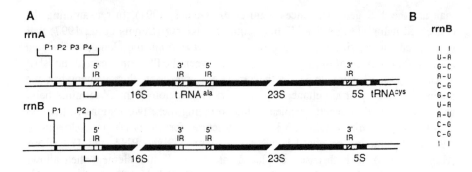

FIGURE 26.4 Comparison of the *rrn* A and *rrn* B operon structures from *Haloarcula marismortui*. A. The structure of the *rrn* A and *rrn* B operons is depicted; designations and abbreviations are as described in the legend to Figure 26.1. The 16S inverted repeats in the *rrn* B operon fail to form the expected bulge-helix-bulge motif and are indicated as open boxes. B. The sequence and structure of the *rrn* B 16S processing helix that would normally contain the BHB motif is illustrated.

TABLE 1
Nucleotide Sequence Heterogeneity Between 16S, 23S and 5S Genes of *Haloarcula marismortui*

Operons	Gene	Length (Nucleotides)	Nucleotide Substitutions	Sequence Divergence
rrn A:*rrn* B	16S	1472	74	5.0%
rrn A:*rrn* B	23S	2917	40	1.4%
rrn A:*rrn* C		2917	33	1.1%
rrn B:*rrn* C		2917	15	0.5%
rrn A:*rrn* B	5S	139	2	1.4%
rrn A:*rrn* C		139	—	0%
rrn B:*rrn* C		139	2	1.4%

only at nonessential or unimportant positions that are normally phylogenetically variable, and many occur as compensatory pairs (in helical regions of the RNA, i.e., both components of a base pair are changed but the base pair is maintained). The substitutions are not evenly distributed across a gene but rather tend to cluster within limited regions. For example, in the case of the 16S genes, 51 of the 74 substitutions occur within a portion of the central domain bounded by nucleotides 508 and 823. The significance of this sequence heterogeneity between the *rrn* A and *rrn* B operons is not known. Under standard laboratory conditions, both operons are expressed and both the *rrn* A and *rrn* B 16S sequences appear in active 70S ribosomal particles (Mylvaganam and Dennis, 1992).

There is some evidence to suggest that the genome of *Ha. marismortui* may contain a third rRNA operon, tentatively designated *rrn* C. In Southern hybridization with genomic DNA, a third substoichiometric band is sometimes observed. Moreover, a *Ha. marismortui* 23S rRNA gene sequence has been determined that is different from both the *rrn* A and *rrn* B 23S sequences we have determined (Brombach et al., 1989). Alignment of the three sequences revealed a total of 44 positions of nucleotide substitutions; 29 of the substitutions occur in *rrn* A, 11 of the substitutions occur in *rrn* B, and 4 of the substitutions occur in *rrn* C. This implies that the *rrn* C 23S gene is more of a composite of the *rrn* A and B 23S genes than an independently evolving entity. Indeed, the 16S-23S intergenic space and 5′ end of the 23S gene are very much like *rrn* B; the 3′ end of the 23S gene, the 5S gene, and the 3′ flanking region are very much like *rrn* A (Figure 26.5).

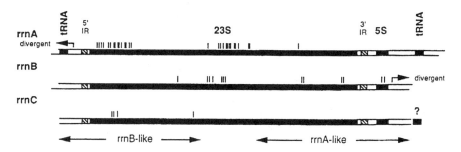

FIGURE 26.5 Nucleotide substitutions in the 23S and 5S genes of *rrn* A, *rrn* B and the putative *rrn* C of *Haloarcula marismortui*. The sequenced region of the *rrn* A, *rrn* B, and *rrn* C operons were aligned; the position of nucleotide substitutions unique to each sequence are indicated as vertical lines above the sequences. Other designations are as indicated in the legend of Figure 26.1. In the 16S-23S intergenic space, the last 178 nucleotides of all three operons are identical. Before this, *rrn* A has a tRNA^{ala} gene that is not present in *rrn* B and apparently not present in *rrn* C. In the 23S distal region, the three sequences are identical for the 139 nucleotides between 23S and 5S genes and for 61 nucleotides after the 5S gene. At this point, the *rrn* B sequence diverges from *rrn* A and *rrn* C. Within the 5S gene, *rrn* B has two unique substitutions. It is uncertain if *rrn* C contains the distal tRNA^{cys} gene, as this region was not sequenced.

Taken together, these data imply either that there is a third genomic operon or that occasional recombination between *rrn* A and *rrn* B generates a substoichiometric recombinant *rrn* C operon. To distinguish between these possibilities, 50 independent clones carrying rRNA genes were isolated from a lambda genomic library. The sequences of regions of frequent nucleotide substitution within the 16S and 23S genes are being determined to establish the identity of the operon on each clone. In addition, 50 independent PCR genomic clones of the 16S and the 23S variable regions are being generated and sequenced. If *rrn* C is a genomic operon, it should have unique sequence and structural features, and represent about one third of the total lambda and genomic PCR clones. If *rrn* C is a substoichiometric operon resulting from recombination between *rrn* A and *rrn* B, it will represent considerably less than one third of the total number of clones and will exhibit hybrid sequence and structural features derived from *rrn* A and *rrn* B.

Why are the *Ha. marismortui rrn* operons so different in their sequence, structure and processing features? Are these differences related to adaptation to salt? Do the different *rrn* operons provide additional flexibility to the organism in adapting to stresses resulting from fluctuating environments? These important questions beg to be addressed.

26.6 CONCLUSION

Halophilic Archaea are capable of exclusive exploitation of hypersaline environments that are between 2 and 5 molar in salt (Kushner, 1985). Under ideal conditions, doubling times between three and five hours are achieved (Chant et al., 1986). This is considerably slower than a typical bacterium like *Escherichia coli*, which can easily grow with a doubling time of 30 to 40 minutes (Bremer and Dennis, 1996). The slower rate of growth exhibited by halophiles is not a consequence of nutrient limitation, rather it reflects a lower efficiency of their metabolic apparatus, which has been adapted over the aeons to function in a high ionic strength intracellular milieu. Because these organisms are the exclusive inhabitants of hypersaline environments, the ability to cope effectively with high salinity, not rapid growth, is probably the major priority. Their evolutionary survival is not endangered by other non-archaeal halophilic organisms competing for the same ecological and nutritional niche.

Although hypersaline environments are often rich in radiant energy and organic nutrients, life is not easy — even for the halophilic Archaea. Water, the essential solvent for all life processes, is scarce because of salt ion hydration. Although the halophilic metabolic machinery has been adapted to function in high salt, it functions far less efficiently than the corresponding machinery of organisms growing under more standard physiological conditions. For example, in *E. coli*, ribosomes polymerize amino acids into proteins at a rate of about 20 per second. In *Hb. cutirubrum*, the rate is only about two amino acids per second (Bremer and Dennis, 1996; Chant et al., 1986). Thus, even with the same concentration of ribosomes in its cytoplasm, *Hb. cutirubrum* will have a doubling time 10 times longer than *E. coli*. Because the halophilic cell will have 10 times longer to reproduce its ribosome content of cell mass, it easily exists with a single copy *rrn* operon, whereas *E. coli* requires seven *rrn* operons to sustain rapid exponential growth.

Another important feature of hypersaline environments is their non-constancy (Dennis and Shimmin, 1997). Salinity is subject to constant fluctuations caused by physical processes such as solubilization-precipitation or dilution-evaporation. A given halophilic enzyme (or metabolic process such as protein synthesis) is highly adapted and magnificently engineered to function in high salt, but will undoubtedly exhibit a salt concentration activity profile; any deviation from the optimum is expected to lower activity. There are indications that suggest that many of the intriguing and unusual features of halophilic Archaea provide these organisms with additional flexibility in coping with fluctuations in environmental salinity. These include promoters to drive the transcription of rRNA operons, duplicated genes, and heterogenous *rrn* operons.

REFERENCES

Belfort, M. and A. Weiner. 1997. Another bridge between kingdoms: tRNA splicing in archaea and eucaryotes. *Cell* 89:1003–1007.

Bremer, P.P. and P.P. Dennis. 1996. Modulation of chemical composition and other parameters of the cell by growth rate, in *Escherichia coli and Salmonella: Cellular and Molecular Biology,* F.C. Neidhardt, R. Curtiss III, J.L. Ingraham, E.C.C. Lin, K.B. Low, B. Magasanik, W.S. Reznikoff, M. Riley, M. Schaechter and H. E. Umbarger, Eds. American Society for Microbiology, Washington, DC. 1553–1569.

Brombach, M., T. Specht, V.A. Erdmann, and N. Ulbrech. 1989. Complete nucleotide sequence of a 23S ribosomal RNA gene from *Halobacterium marismortui. Nucl. Acid. Res.* 17:3293.

Chant, J. and P.P. Dennis. 1986. Archaebacteria: transcription and processing of ribosomal RNA sequences in *Halobacterium cutirubrum. EMBO J.* 5:1091–1097.

Chant, J., I. Hui, D. de Jong-Wong, L. Shimmin, and P.P. Dennis. 1986. The protein synthesizing machinery of the archaebacterium *Halobacterium cutirubrum*: molecular characterization. *Syst. Appl. Microbiol.* 7:106–114.

Dennis, P.P. 1985. Multiple promoters for the transcription of the ribosomal RNA gene cluster in *Halobacterium cutirubrum. J. Mol. Biol.* 186:457–461.

Dennis, P.P. 1991. The ribosomal RNA operons of halophilic archaebacteria, in *General and Applied Aspects of Halophilic Microorganisms,* F. Rodriguez-Valera, Ed. Plenum, New York. 251–258.

Dennis, P.P. 1997. Ancient ciphers: translation in Archaea. *Cell* 89:1007–1010.

Dennis, P.P. and L.C. Shimmin. 1997. Evolutionary divergence and salinity-mediated selection in halophilic archaea. *Microbiol. Mol. Biol. Rev.* 61:90–104.

Dennis, P.P., A.G. Russell, and M. Moniz de Sá. 1997. Formation of the 5′ end pseudoknot in small subunit ribosomal RNA: involvement of U3-like sequences. *RNA* 3:337–343.

Gutell, R.R., N. Larsen, and C.R. Woese. 1994. Lessons from an evolving ribosomal RNA: 16S and 23S ribosomal RNA structures from a comparative perspective. *Microbiol. Rev.* 58:10–26.

Hughes, J.M.X. 1996. Functional base-pairing interaction between highly conserved elements of U3 small nucleolar RNA and the small ribosomal subunit RNA. *J. Mol. Biol.* 259:645–654.

Hui, I. and P.P. Dennis. 1985. Characterization of the ribosomal RNA gene cluster in *Halobacterium cutirubrum. J. Biol. Chem.* 260:899–906.

Kleman-Leyer, K., D.W. Armbruster, and C.J. Daniels. 1997 Characterization of the *Haloferax volcanii* tRNA intron endonuclease gene reveals a relationship between the archaeal and eucaryal tRNA intron processing system. *Cell* 89:839–847.

Kushner, D.J. 1985. The Halobacteriaceae, in *The Bacteria,* Vol. 8, C.R. Woese and R.S. Wolfe, Eds. Academic, New York. 171–214.

Lanyi, J.K. 1974. Salt dependent properties of proteins from extremely halophilic bacteria. *Bacteriol. Rev.* 38:272–290.

Mankin, A.S. and V.K. Kagramanova. 1988. Complex promoter pattern of a single ribosomal RNA operon of an archaebacterium, *Halobacterium halobium. Nucl. Acids Res.* 16:4679–4692.

Mevarech, M., S. Hirsch-Twizer, S. Goldman, E. Jacobson, H. Eisenberg, and P.P. Dennis. 1989. Isolation and characterization of the rRNA gene clusters of *Halobacterium marismortui*. *J. Bacteriol.* 171:3479–3485.

Mylvaganam, S. 1996. *Characterization of Ribosomal RNA Operons of Haloarcula marismortui*. Ph.D. thesis, University of British Columbia.

Mylvaganam, S. and P.P. Dennis. 1992. Sequence heterogeneity between the two genes encoding 16S rRNA from the halophilic archaebacterium *Haloarcula marismortui*. *Genetics* 130:399–410.

Theissen, G., L. Thelan, and R. Wagner. 1993. Some base substitutions in the leader of the *Escherichia coli* ribosomal RNA operon affect the structure and function of ribosomes. *J. Mol. Biol.* 233:203–218.

Thompson, L.D. and C.J. Daniels. 1990. Recognition of exon-intron boundaries by the *Halobacterium volcanii* tRNA intron endonuclease. *J. Biol. Chem.* 265:18104–18111.

27 A Model for the Genetic Exchange System of the Extremely Halophilic Archaeon *Haloferax volcanii*

Ron Ortenberg, Ronen Tchelet, and Moshe Mevarech

CONTENTS

27.1 Introduction 331
27.2 The Discovery of the Genetic Transfer System and
   its Basic Characteristics 332
27.3 Can Halobacterial Plasmids Be Used as Cytoplasmic
   Markers to Distinguish Between Bi-Directional
   Conjugation and Cellular Fusion? 333
27.4 Genetic Transfer of Selectable Plasmids 334
27.5 How Can We Reconcile Between the Apparently Conflicting
   Results Demonstrating the Immobility of the Indigenous
   Plasmids and the Movement of the Selectable Plasmids? 334
27.6 A Model for Halophilic Archaeal Mating 336
References 338

27.1 INTRODUCTION

The transfer of genetic information between organisms is one of the fundamental processes by which genetic variability is accomplished. In Eukarya, sexual transmission occurs when two haploid cells fuse to form a diploid cell. In the domain Bacteria, the horizontal transmission of genetic information involves either phage-mediated transduction, assimilation of naked DNA acquired from the environment, or cell contact-dependent DNA transfer, termed conjugation. In most cases, conjugation is mediated by plasmids that are mobilized in the process.

So far, genetic transfer has been demonstrated only in two Archaea. The first Archaeon for which genetic transfer has been reported is the extremely halophilic *Haloferax volcanii* (Mevarech and Werczberger, 1985). It was recently shown that a very similar system exists also in the thermophilic Archaeon *Sulfolobus acidocaldarius* (Grogan, 1996). Although much work was invested in the analysis of the genetic transfer system of *H. volcanii*, it is not absolutely clear whether the horizontal transfer of genes among archaeal cells resembles bacterial conjugation, or rather is the result of fusion between neighboring cells.

This chapter is dedicated to the presentation of the experimental work aimed at answering this intriguing question. Presented in the concluding section is a model for the genetic transfer process that adopts the view that the genetic transfer system in the halophilic Archaeon *H. volcanii* involves fusion between cells and is most compatible with the experimental results.

27.2 THE DISCOVERY OF THE GENETIC TRANSFER SYSTEM AND ITS BASIC CHARACTERISTICS

Haloferax volcanii is a prototrophic obligatory halophile that was originally isolated from the Dead Sea by Mullakhanbhai and Larsen in the mid 1970s (Mullakhanbhai and Larsen, 1975). This halophilic Archaeon grows in a simple defined medium containing, in addition to salts, a simple carbon source, such as glycerol, succinate or glucose, and ammonium or nitrate ions as nitrogen source. To be able to observe genetic transfer, random mutations were induced by treating *H. volcanii* cells with ethyl methanesulfonate, and auxotrophic mutants were isolated.

To test whether genetic transfer occurs, auxotrophic mutants were grown separately, then paired combinations of the mutants were mixed and the mixtures were filtered through nitrocellulose filters. The filters were placed on rich agar plates for up to 96 h and then the cells were washed, suspended in a salt solution, diluted, and plated on minimal agar plates. Prototrophic colonies were obtained in every cross at a frequency of about 10^{-6}, which is about 1,000 times higher than the frequency of the spontaneous mutation reversion rate. Since recombinants were obtained by every cross of two auxotrophic mutants, it is not possible to distinguish donor strains from recipient strains.

The efficiency of transfer of genetic information was dependent on the establishment of physical contact between the mutant cells. No transfer was observed when the two auxotrophic strains were grown in a shake culture, and only very low frequency of transfer was observed when the two strains were allowed only brief contact in a pellet. The transfer was insensitive to DNase, and no transfer was observed when the filtrate of conditioned medium of one mutant was supplied to the other. These last two observations eliminate the involvement of transformation or transduction processes in the genetic transfer.

Since the genetic transfer process in *H. volcanii* requires physical contact between cells, it resembles bacterial conjugation. This resemblance, however, is incomplete, since in bacterial conjugation a pore is formed between juxtaposed donor and recipient cells, and the donor cell that contains a conjugative plasmid transfers chromosomal genetic information into the recipient cell that lacks this plasmid. In the genetic-exchange system of *H. volcanii*, however, every cell can donate and receive genetic information.

27.3 CAN HALOBACTERIAL PLASMIDS BE USED AS CYTOPLASMIC MARKERS TO DISTINGUISH BETWEEN BIDIRECTIONAL CONJUGATION AND CELLULAR FUSION?

One explanation for the finding that in the genetic exchange system of *H. volcanii* there is no distinction between donor and recipient strains is that *H. volcanii* is able to carry out a novel form of bidirectional conjugation. Alternatively, the lack of classical donor and recipient strains can be explained by cellular fusion of neighboring cells. How can one distinguish experimentally between these two alternatives?

One way to distinguish between bidirectional conjugation and fusion is by using cytoplasmic markers. If in the process of genetic exchange cells fuse, all the recombinants will contain the cytoplasmic markers of both parents. If, however, only the chromosomes are transferred by conjugation, half of the recombinants will have the cytoplasmic markers of one parent and the other half will have the cytoplasmic markers of the other.

Because of the lack of proper cytoplasmic markers in prokaryotic microorganisms, an attempt was made to use plasmids as cytoplasmic markers (Rosenshine et al., 1989). The auxotrophic mutant strain WR335 containing two indigenous plasmids (pHV2 [Charlebois et al., 1987; Figure 27.1] and pHV11 [Rosenshine and Mevarech, 1989]) and requiring serine, was crossed with the auxotrophic mutant strain WR344 that lacks these plasmids and requires purine for growth. Prototrophic colonies were replicated on two sets of plates and were examined for the existence of the plasmids by colony hybridization. Fifty of the 96 colonies examined contained both plasmids and 46 colonies contained no plasmid. These results demonstrated

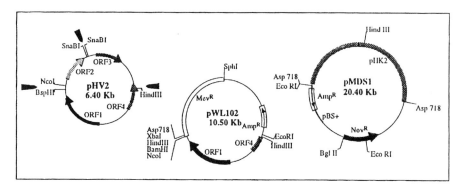

FIGURE 27.1 Partial restriction maps of the indigenous plasmids pHV2 (Charlebois et al., 1987), pWL102 (Lam and Doolittle, 1989) and pMDS1 (Holmes and Dyall-Smith, 1990). Arrows indicate the site of insertions of selectable marker into pHV2.

that the plasmids pHV2 and pHV11 are not mobilized by the genetic transfer system. The interpretation of these results was that the genetic exchange system resembles bacterial conjugation in which only the chromosomal DNA moves from one cell to the other, since if cell fusion had been involved, every prototrophic colony should have contained both plasmids. Another outcome of these results was the demonstration that both parental strains can serve either as donors or as recipients. In half the cases, the cells that contained the two plasmids were the recipients and in the other half, the cells that contained no plasmids were the recipients.

27.4 GENETIC TRANSFER OF SELECTABLE PLASMIDS

In the experiment described above, it was not possible to select for plasmid movement. What would have happened if a strain containing one selectable plasmid had been crossed with a strain containing another selectable plasmid and recombinants had been selected for the existence of both plasmids?

Recently, halobacterial selectable plasmids constructed by combining the origin of replication of indigenous halobacterial plasmids, halobacterial selectable markers and *Escherichia coli* plasmids became available. The auxotrophic strain WR1304 (requiring arginine and containing the plasmid pMDS1 [Holmes and Dyall-Smith, 1990; Figure 27.1], which confers resistance to novobiocin) was crossed with the auxotrophic strain WR1306 (requiring adenine and containing the plasmid pWL102 [Lam and Doolittle, 1989; Figure 27.1] which confers resistance to mevinolin), and colonies were selected either on plates containing both inhibitors or on minimal plates. Surprisingly, colonies were obtained at the same frequency on both selective plates. Moreover, it could be shown that cells that are resistant to both inhibitors contain both plasmids (Tchelet and Mevarech, 1994).

If plasmids do not move in the genetic transfer process (as was shown for the indigenous plasmids pHV2 and pHV11), how could cells containing both selectable plasmids be obtained? Moreover, if plasmids can move by a process resembling bacterial conjugation, it is expected that the frequency of recombinants selected for plasmid movement will be considerably higher than recombinants selected for chromosome movement, since in the latter case, the movement of the chromosomal marker should be followed by recombination of the chromosomal marker into the recipient chromosome.

27.5 HOW CAN WE RECONCILE BETWEEN THE APPARENTLY CONFLICTING RESULTS DEMONSTRATING THE IMMOBILITY OF THE INDIGENOUS PLASMIDS AND THE MOVEMENT OF THE SELECTABLE PLASMIDS?

There is an apparent discrepancy between the experiments that show that the indigenous plasmids are not transferred in the genetic exchange process and the experiments that show that selectable plasmids are transferred. If the mode of genetic transfer in halobacteria is by "conjugation" and the indigenous plasmids cannot move, how is it possible that the selectable plasmids that had been originally constructed by subcloning of the origin of replication of the indigenous plasmids acquired the ability of movement by conjugation?

This discrepancy can be solved if we assume that the genetic exchange in *H. volcanii* is, after all, the result of cellular fusion between neighboring cells. According to this model, it is obvious why the selectable plasmids are "mobile." The inability of the indigenous plasmids to move from one cell to the other, however, can be explained by assuming the existence in the original indigenous plasmids of specific genes whose functions interfere with the ability of the plasmid to be exchanged between the mating cells. These putative genes have probably been deleted in the process of construction of the selectable plasmids.

To substantiate this model, several plasmids were constructed in which insertions were introduced at different loci in pHV2 to try to inactivate the putative genes responsible for its "untransferability" phenotype. The indigenous plasmid pHV2 has four long Open Reading Frames (ORFs); three of them are coded from one strand and the fourth is coded on the other strand. Of these ORFs only ORF1, which codes for a putative 809 amino acid protein, shares substantial sequence identity with a known gene — *repH*. The gene *repH* is implicated as responsible for replication of the halobacterial plasmid pNRC100 (Ng and DasSarma, 1993) and is absolutely required for replication of this plasmid in halobacteria.

Insertions were introduced separately into three places in pHV2, as shown in Figure 27.1. Auxotrophic mutants containing each one of these plasmids were crossed with other auxotrophic mutants lacking any plasmid, and recombinants were selected on minimal plates. More than 95% of the prototrophic recombinants contained the plasmid, indicating a tight linkage between the transfer of the plasmid and the transfer of the chromosomal marker. It seems, therefore, that every interruption in the integrity of pHV2 allows its transfer in the genetic exchange system.

What might be the physiological meaning of this bizarre untransferability phenotype? Although pHV2 has only about six copies per cell and contains no beneficial role for the cell, it is an extremely stable plasmid (Charlebois et al., 1987). It is therefore probable that pHV2 has a special maintenance mechanism and the various ORFs are involved in the maintenance and stability of the plasmid. If so, it is expected that each of the plasmids in which insertions were introduced will be less stable than pHV2.

The stability of these plasmids was determined by growing cells containing the plasmids for 28 generations in rich medium without selection, and then plating the cells on rich plates and isolating colonies on rich plates containing the selectable marker. It was found that the stability of each of the plasmids is significantly lower than that of the original indigenous plasmid pHV2 from which they were derived.

These results establish an inverse correlation between plasmid stability and transferability. The indigenous plasmids are extremely stable and are transferred during mating at a very low frequency. On the other hand, the selectable derivative plasmids are less stable and are transferred during mating at a much higher frequency.

27.6 A MODEL FOR HALOPHILIC ARCHAEAL MATING

Any model to describe the events involved in mating in *H. volcanii* must account for two important features: (1) that the transfer of genetic material is bidirectional (Mevarech and Werczberger, 1985), thus it is impossible to distinguish donor cells from recipient cells; (2) that plasmids and chromosomal markers are transferred in tight linkage and at similar frequencies. These two properties of the genetic exchange are compatible with a model in which the *H. volcanii* mating system resembles the mating system of Eukarya, where cells involved in mating fuse to form diploid cells.

Since the mating system requires physical contact between cells, it seems that the first step in the process is adherence of the cells (Figure 27.2, stage I). Kinetic studies of the genetic exchange (Rosenshine and Mevarech, 1991) have shown that even a brief contact between *H. volcanii* cells is sufficient to enable the process to occur. The frequency of genetic exchange was found, however, to increase with time, and reaches a plateau after about eight hours (Rosenshine and Mevarech, 1991). Recently, a recessive mutation in mating was isolated. *H. volcanii* auxotrophic mutants also carrying this mutation were able to exchange chromosomal markers only with normal auxotrophic mutants (unpublished data). It is therefore likely that the ability to adhere is determined by a specific surface component of the cell. This adherence interaction might also be responsible for the genus specificity of the

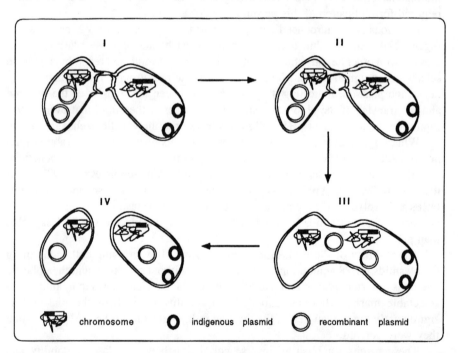

FIGURE 27.2 Schematic presentation of the various stages in the model for genetic exchange in *H. volcanii*: I - Establishment of physical contact between cells; II - The creation of cytoplasmic bridges; III - Expansion of the bridge creating a fused cell; IV - Division of the fused cell.

genetic exchange system. It was found (Tchelet and Mevarech, 1994) that *H. volcanii* transfers selectable plasmids to *Haloferax mediterranei* but not to *Haloarcula marismortui* or *Halobacterium salinarum* (*halobium*).

Electron micrographs of *H. volcanii* have shown (Mullakhanbhai and Larsen, 1975; Rosenshine et al., 1989) that cells that grow in colonies are connected by cytoplasmic bridges. It was argued earlier (Mullakhanbhai and Larsen, 1975) that these bridges are the result of a failure in the completion of cell division. It was later demonstrated (Rosenshine et al., 1989) that cytoplasmic continuity is established *de novo* between cells that grow together on solid surfaces. It seems, therefore, that the second step in the mating process is the establishment of the cytoplasmic bridges (Figure 27.2, stage II).

However, not every two cells that have cytoplasmic continuity can exchange genetic markers. It was shown that artificial removal of the cellular envelopes in the middle of mating by reducing the Mg^{2+} ion concentration increased the frequency of recombinants by a factor of about 100 (Rosenshine et al., 1989). Thus, it was concluded that under natural conditions only about 1% of the connected cells complete the fusion process (Figure 27.2, stage III).

Following fusion, the cellular cytoplasmic components, including chromosomes and some plasmids, can move from one cell to the other, establishing a transient diploid stage at which recombination between chromosomes can occur (Rosenshine and Mevarech, 1991). If no selection pressure is applied, when the fused cell divides (Figure 27.2, stage IV) the two chromosomes will segregate due to the lack of a mechanism that ensures the cosegregation of the two chromosomes to daughter cells (unlike the situation in diploid Eukarya cells). The plasmids, on the other hand, will be inherited by the daughter cells because of their high copy number.

Perhaps the least substantiated aspect of the cell fusion model for genetic transfer in halophilic Archaea is the fact that indigenous plasmids are transferred from one cell to its partner only at low frequency. If cells fuse in the mating process, what restricts plasmid movement? Possibly the indigenous plasmids are bound to the cell membrane as a part of their segregation mechanism and are not free to move in the cytoplasm. It is noteworthy, therefore, that when selectable markers were inserted either into ORF4, between ORF1 and ORF2, or between ORF2 and ORF3, the plasmids developed instability and at the same time were transferred in tight linkage with the chromosome. The functional importance of ORF2, ORF3, and ORF4 in preventing the efficient transfer of pHV2 in mating was also demonstrated by complementing this function *in trans* (our unpublished results). The exact role of these three ORFs is still unclear and is under investigation. According to this hypothesis, the indigenous plasmids are restricted to the parental compartment of the fused cells and remain that way after septation of the cells.

In conclusion, this communication presents further evidence that genetic transfer in *H. volcanii* involves cell fusion between cells and resembles mating in Eukarya. However, whereas eukaryal cells can be distinguished according to their mating types, in *H. volcanii* no mating type has been detected. Every *H. volcanii* cell appears competent to fuse with any other cell of the same species. The nature of the cellular interactions and the basis for their specificity is unclear, but seems to involve contacts between surface proteins. Moreover, conditions that promote fusion between cells exist within developing colonies. Thus, advantageous mutations in one cell within a colony might therefore be distributed at high frequency to other cells. Also, the fact that interspecies genetic transfer occurs at high frequency provides an effective mechanism for horizontal gene transfer.

REFERENCES

Charlebois, R.L., W.L. Lam, S.W. Cline, and W.F. Doolittle. 1987. Characterization of pHV2 from *Halobacterium volcanii* and its use in demonstrating transformation of an archaebacterium. *Proc. Natl. Acad. Sci. USA.* 84:8530–8534.

Grogan, D.W. 1996. Exchange of genetic markers at extremely high temperatures in the archaeon *Sulfolobus acidocaldarius. J. Bacteriol.* 178:3207–3211.

Holmes, M.L. and M.L. Dyall-Smith. 1990. A plasmid vector with selectable marker for halophilic archaebacteria. *J. Bacteriol.* 172:756–761.

Lam, W.L. and W.F. Doolittle. 1989. Shuttle vector for the archaebacterium *Halobacterium volcanii. Proc. Natl. Acad. Sci. USA.* 86:5478–5482.

Mevarech, M. and R. Werczberger. 1985. Genetic transfer system in *Halobacterium volcanii. J. Bacteriol.* 162:461–462.

Mullakhanbhai, M.F. and H. Larsen. 1975. *Halobacterium volcanii* spec. nov., a Dead Sea halobacterium with moderate salt requirement. *Arch. Microbiol.* 104:207–214.

Ng, W.-L. and S. DasSarma. 1993. Minimal replication origin of the 200-kilobase *Halobacterium* plasmid pNRC100. *J. Bacteriol.* 175:4584–4596.

Rosenshine, I. and M. Mevarech. 1989. Isolation and partial characterization of plasmids found in three *Halobacterium volcanii* isolates. *Can. J. Microbiol.* 35:92–95.

Rosenshine, I. and M. Mevarech. 1991. The kinetic of the genetic exchange process in *Halobacterium volcanii* mating, in *General and Applied Aspects of Halophilic Microorganisms,* F. Rodriguez-Valera, Ed. Plenum, New York. 265–270.

Rosenshine, I., R. Tchelet, and M. Mevarech. 1989. The mechanism of DNA transfer in the mating system of an archaebacterium. *Science* 245:1387–1389.

Tchelet, R. and M. Mevarech. 1994. Interspecies genetic transfer in halophilic archaebacteria. *Syst. Appl. Microbiol.* 16:578–581.

INDEX

^{13}C depletion
  from atmospheric CO_2 invasion, 109, 114
  Baertschi effect, 109, 111, 114, 117
  degree of evaporation, 109, 111, 112—114
  DIC depletion correlation, 114—117
  dissolved methane in Orca Basin, 150
  gypsum precipitate, 113, 114
  and microbial mats, 111, 114—117
  study methodology, 111, 112
^{14}C-methanol
  metabolized in Dead Sea sediment experiment
    methanogenic activity in some sediment, 150—151, 153—154, 157
    oxidation of methanol sediment, 152—153
    radioisotope studies, 151—154
  methylotrophic precursors, 150, 156—157
  in Solar Lake, 154—157
16S rRNA sequencing, 50
  alkaliphilic Archaea, 16
  analysis of preserved Dead Sea isolates, 144
  biodiversity recognition, 3
  cloning and sequencing studies for analysis, 28
  environments for study, 27—28
  folding of operons, 228, 323—325
  in *Haloarcula*, 22
  phylogenetic position determination, 144, 145
  phylotype analysis in crystallizer ponds, 31—33
  polar lipid studies, 133
  pseudoknot formation, 324
  salinity domain diversity, 31
  saltern crystallizer pond studies, 3, 28—29
2-dehydro-3-deoxyphosphogluconate aldolase, 268
2-keto-3-deoxygluconate kinase, 268
2-oxoacid dehydrogenase
  DHLipDH enzyme component, 241
  discovered in Archaea, 246, 247
  and pyruvate dehydrogenase, 239—240

A

A-ATPases, *see also* ATPases
  azide effects, 276—277
  characteristics, 273—274, 278
  inhibitor effects, 275—276
Acetate, 8
  and ATP synthesis, 276
  proposed use as culture growth media, 69
Acetyl-CoA via PDHC, 239—241
Acridine orange, 277, 297
Acrodictys, 80
Actinomycetes, 78, 82
Actinopolyspora halophila, 218, 219
ADI pathway, 260—261
  arc gene cluster transcription analysis, 253—255
  in Archaea and Bacteria, 252, 256
  fermentation
    Archaea and Bacteria similarities, 250—251
    degradation via ADI pathway, 252
    gene cluster from *Hb. salinarum*, 252—253, 254
    ornithine antiporter, 252, 253, 256
Age dating of microbial samples, 56—57, 67
Algae bloom in the Dead Sea, 129—130
  chlorophyll index, 131—132
  inoculum sources, 130—132
Algae in ESSA, 43
Algal index, 131
Alginite, 103
Alicante saltern (Spain)
  halobacterial phylotypes found, 23, 33
  inorganic nutrient concentrations, 42
  microbial ecological studies, 41—43
Alkaliphilic Archaea, 13, 36
  16S rRNA sequencing genes, 16
  genera, 15—16

Natronobacterium taxonomy change
    proposal, 16—17, 20—21
  spacer regions, 16
Amino acids
  ATPases ancestor, 273—274
  catalytic activity and spatial
    arrangement, 245
  organic nutrient, 49
  ratio of acidic to basic residues in
    Dca sequences, 209—210,
    211
  sequence for transferrin proteins,
    204
  sequence of GvpA, 282—283
  signal peptide in Hal H4, 298—299
Ammonia, in Bonaire saltern
  collections, 44
Amplified Ribosomal DNA Restriction
  Analysis (ARDRA)
  correlation between detected bands
    and diversity, 31
  phylotype detection, 28
  of saltern ponds, 29—31
Anabaena sp.
  gas vesicles in, 281
  GvpA and GvpC genes isolated, 283
  gvp gene arrangement, 285—286
  sucrose-phosphate synthase activity,
    179
Anaerobic hypersaline environments, 4
Ancient rock formations, 59—61
Ancient salt formations
  culture growth media, 68—69
  evolutionary studies, 70
  isolation difficulties, 65—66
  microbial sampling, *see* Micro-
    biological sampling
  presence of bacteria in, 53—54
Anhydrite, 91
Anoxic basinal bottoms, 92—93,
    94, 95
Anoxygenic phototrophic bacteria, 190
Antarctica, 3, 281
Antiporters
  arginine/ornithine, 252, 253, 256
  Na^+/H^+
    genes, 164—165
    H-NS involvement, 171—172
    *nha*A expression control,
      169—171
    *nha*B gene activity, 164
    NhaR, 169—171
    pH regulation, 164, 166—169
    proteins NhaA and B, 165—166
Arab Potash Co., 123

Archaea (*Halobacteriaceae*), *see also*
    individual genera and
    species
  2-oxoacid dehydrogenase discovery,
    246, 247
  A-ATPases
    azide effects, 276—277
    characteristics, 273—274, 278
    inhibitor effects, 275—276
  acidic residue preponderance, 209
  ADI pathway detected, 252, 256
  alkaliphilic members, 13, 36
    16S rRNA sequencing genes, 16
    genera, 15—16
    Natronobacterium taxonomy
      change proposal, 16—17,
      20—21
    spacer regions, 16
  bacteriophages, 133—134,
    222—223
  bacteriorhodopsin, 221
  biodiversity of, 2—3, 23, 50
  bloom in the Dead Sea, 4—5, 15,
    124, 129—130
    bacteriophages in, 133—135
    characteristics of, 132—133
    chlorophyll index, 131—132
  carotenoid pigments, 2
  chemotaxonomic criteria for
    classification, 13—14
  chimeric origins proposed, 313
  color display, 1—2, 42, 124,
    129—130
  community density regulation,
    133—134
  DHLipDH function, 241—242
  environments for study, 27—28,
    328—329
  enzymes compared to Dca and TTf,
    209
  evolutionary changes, 220—221
  in forest soils, 23
  genera, 13—15
  gene-specific regulation systems,
    251—252, 256
  genomes, 7—8
  glycine betaine production,
    188—189
  glycolipids, 132—133
  halocins produced by, 296
  halorhodopsin, 6, 221
  history of, 217—219
  interaction between salt and
    proteins, 7
  internal chloride concentrations, 221

Index 341

isolated in subterranean brines, 54
lysogenic phages, 134—135
methanogenic, 5, 222
non-alkaliphilic classification, 36
osmotic equilibrium strategies, 5—6
phylogenetic tree, 17, 250
phylotypes, novel, in natural
  environments, 23
in poikilotrophic microorganism
  sites, 78
polar lipid composition in Dead Sea,
  132—133
properties and structure, 219—222
pyruvate oxidoreductase, 240, 241
rate of growth, 328
red, in Dead Sea, 129—130
retinal proteins, 6
ribosomal RNA operons, 319
  16S folding, 323—325
  chaperone sequence, 325
  in *Ha. marismortui*, 325—328
  multiple promotors, 320—323
  structure and precursor
    processing, 320
salinity domain diversity, 29—31
signature bases, 16, 18—20
species in Dead Sea, 4—5, 132
structural protein GvpA, 282—283
transcription
  Eukarya initiation similarities,
    250
  Eukaryal TBP differences, 258
  gene expression levels, 250—251
  TATA box initiation, 256—257
  TBP conservation in Eukarya,
    257—258
  transformation systems, 251
  virus-like particles as regulators, 134
ARDRA, *see* Amplified Ribosomal
  DNA Restriction Analysis
Arginine catabolism (*arc*) genes
  cloned and sequenced from *Hb.
    salinarum*, 252—253, 254
  transcription analysis, 253—255
Arginine deiminase, *see* ADI pathway
Artemia salina (brine shrimp), 29, 41,
  43
Arthrobacter sp., 183, 268
Aseptic sampling techniques, 53
Asphalts, 95, 98—99
Athalassohaline system, 40, 41
Athens (Greece), 82
Atmospheric CO_2
  ^{13}C depletion from, 109, 114
  reequilibration and E_{Br}, 113

ATPases, 7
  common amino acid ancestor back-
    ground, 273—274
  synthesis of ATP in vesicles
    azide effect, 276—277, 281—282
    inhibitor effect, 274—276, 278
  types, 273—274
Auxotrophic mutants
  in genetic transfer studies, 332
  in *Hf. volcanii* mating, 336
  used to demonstrate bacterial
    conjugation, 333
Azide, ATP synthesis inhibitor,
  276—277, 281—282

B

Ba_1 strain, 143
Bacillus sp.
  β-galactosidase sequence similarity,
    267—268
  in preserved Dead Sea isolates,
    144—145
  transcription of general stress
    proteins, 197
Bacteria (halophilic)
  ADI pathway found in, 252, 256
  bacteriophages, 222—223
  biodiversity of, 2—3, 23, 29—31
  characteristics, 222
  chemoheterotrophic, 189, 190—191
  color display, 1—2, 42
  in the Dead Sea, 5
  environments used for study, 27—28
  evolutionary changes, 220—221
  glycine betaine production,
    188—189
  history of, 217—219
  methanogenic, 8
  osmotic equilibrium strategies, 5—6
  and PDHC, 239—241
  properties and structure, 220—222
  salinity domain diversity, 29—31
  salt adaptation strategy, 187
  in saltern system, 50
  transfer of genetic information, 331
Bacterial milking in ectoine production,
  192
Bactericidal proteins, *see* Bacteriocins
Bacteriocins, 7, 295
  description of, 295—296, 298
  production benefits, 303
Bacterio-opsin gene expression, 251
Bacteriophages, 50
  of Archaea halobacteria, 222—223

role in Dead Sea bacterial densities, 133—134
Bacteriorhodopsins, 6, 7
 in Dead Sea, 132, 133
 FII DNA sequences, 310
 in halophilic Archaea, 221
 and horizontal genetic transfer of genomes, 313
Bacterioruberins
 in Dead Sea water samples, 135—136
 in *Natrialba*, 15
Baertschi effect, 109, 111, 114, 117
Baja California (Mexico), 92, 114, 300
β-carotene, 2, 8
Beggiatoaceae, 90
Benthic microbial mats, *see* Microbial mats
Bergey's Manual of Determinative Bacteriology, 2—3, 21
Betaine-aldehyde dehydrogenase, 183
β-galactosidase, 7, 251
 gene cloning
 E. coli studies, 266
 ORFs surrounding *bgaH*, 268
 sequence of *bgaH*, 267—268
 Hf. alicantei isolate studied, 266—267, 268, 269
 transformation of *Hf. volcanii*, 269
BHB (bulge-helix-bulge) endonuclease, 320, 321, 325
Bidirectional conjugation, 332, 333—334
Bolivinids, 97
BIOLOG GN, 48—49
Biomass accumulation, 91—92
β-ketoacly synthase, 211, 212
Black fungi, 82
Black Sea, The, 90
Black yeast (*Phaeococcus* sp.), 80
Bonaire solar salt plant (Netherlands Antilles), 40, 42, 50
 cultivation results, 44, 45
 data analysis, 45
 ecological zones, 44—46
 model food web, 46
 nutrient chemistries, 44
 similarity matrix, 47
Brine shrimp (*Artemia salina*), 29, 41, 43
Bromides in ^{13}C depletion study, 112
Bromine in the Dead Sea, 123, 140
Bulge-helix-bulge (BHB) endonuclease, 320, 321, 325
Buoyancy from gas vesicles, 281

C

Calothrix spp., 183, 281, 283
Cambrian age, 70
Canada solar salt plants
 Horn River, 95
 Saanich Inlet, 100, 101
Capillary methods for microbial sampling, 67—68, 69
Carbamate kinase (CK), 252
Carbonate sediments, 90, 91
Carbonic anhydrase, *see* Dca
Carboniferous reservoir rocks, 99
Carbon isotopic depletion, *see* Baertschi effect
Carbon sources of culture medium, 35, 36
Carbonylcyanide m-chlorophenylhydrazone (CCCP), 274
Cargill solar salt plant (USA), 47—49, 50
Carnallite, 123—125
Carotenoid pigments
 in Archaea, 2
 in *Dunaliella*, 8
Castile Formation (USA), 57
Catabolic ornithine transcarbamylase (cOTCases), 252, 256
Cave paintings, 77, 82
CCCP (carbonylcyanide m-chlorophenylhydrazone), 274
Cellular fusion in genetic transfer, 332, 333—334, 337—338
Chaperone sequence, RNA, 325
Chemiosmotic energy, 164
China, 15, 22
Chlamydomonas
 in Dead Sea, 140—141
 reinhardtii, 207, 208, 209
Chloride pumps, 6
Chlorophyll index, 131
Choline
 dehydrogenase, 183
 oxidation, 183, 189
Chromobacterium marismortui
 reclassification, 143
Chromohalobacter-Deleya-Halomonas group, 188
Chromohalobacter marismortui
 reclassification, 143
Chroococcidiopsis, 80, 81
Ciliate in Dead Sea isolates, 142
Citrulline, 225
CK (carbamate kinase), 252

Clostridium discovered in Dead Sea, 139
Co$_2$, atmospheric
    ^{13}C depletion from, 109, 114
    reequilibration and E$_{Br}$, 113
CO$_2$ limitation and Dca accumulation, 207, 209
Colicins produced by *E. coli*, 296
Color display of halophiles, 1—2
    Archaea bloom in the Dead Sea, 129—130
    indicator of Bacteria vs. Archaea, 42
    red Archaea, 124
Compatible solutes
    accumulated during salt stress adaptation, 177—178
    biosynthetic pathways for ectoine, 194, 195
    cyanobacteria studies, 188
    ectoine production
        bacterial milking, 192—193
        genetic studies with transposon mutagenesis system, 188, 194—195
        from *Halomonas elongata*, 189, 191—193
        hydroxyectoine from *Marinococcus*, 193—194
        genomic organization of ectoine genes, 195—196
    glycine betaine, 188—189
    osmoregulated expression in *E. coli*, 196—197
    osmoregulated expression of ectoines, 196—197
    osmotic equilibrium function, 5, 188
    stabilizing effect on biomolecules, 188
Conjugation, 331, 332—333
Copiotroph, 78
cOTCases (catabolic ornithine transcarbamylase), 252, 256
Crenarchaeota, 23
Crimea Sea, 150
Crystallizer ponds, *see* Saltern crystallizer ponds
Crystals
    in ancient salt formations
        culture growth media, 68—69
        evolutionary studies, 70
        isolation difficulties, 65—66
        microbial sampling, *see* Microbiological sampling
        presence of bacteria in, 53—54
        deposition cycle, 58—61

halite
    fluid inclusions, 58—59, 62
    gaseous phase formed, 62
    proof of primary sediments, 59—61
    primary
        diagnostic features of, 55
        fluid-inclusion bands, 58, 61
        isolating viable cultures, 66, 67
Culturing methods
    effect of medium on saltern studies, 48, 49
    bias in microorganism recovery, 32—33
    mediums, 35, 36, 68—69, 332
    PCR amplification comparison, 33, 35
    population bias, 34, 35—36
Cuprophane membrane, 193
C-vac region promotors, 284—285, 287, 289
Cyanobacteria
    behavior in biofilm cultures, 79
    compatible solute studies, 188
    in the Dead Sea, 142
    ectoine synthesis, 189
    gas vesicles in, 281
    gvp gene resemblance to nv-*gvp* gene, 285—286
    mats, 2, 90
        anoxia or basin formation, 92—93, 94, 95
        biomass accumulation, 91—92
        isotope-distribution difference, 97
        primary producers, 29
    osmoprotective compounds
        gene cloning in GG synthesis, 181—183
        GG biosynthesis, 179—181, 188
        glycine betaine biosynthesis, 183—184
        salt resistance limits, 178—179
        sucrose biosynthesis, 179
        trehalose biosynthesis, 179
        types, 178
    poikilotrophic microorganisms, 82
    resemblance of haloarchaeal gene products to, 313
    salt adaptation events, 184
    solar salt plants, 43
        proteins GvpA and C, 282—283
        water potential achieved, 177—178
Cys scanning, 166—169
Cytoplasmic markers in genetic transfer systems, 333, 337

D

Darwin, Charles, 40
Dca (duplicated carbonic anhydrase)
　accumulation regulation, 207, 209
　alignment with animal carbonic
　　anhydrases, 207, 208
　Archaea enzyme comparison, 209
　new halotolerant protein, 209
　ratio of acidic to basic amino acid
　　residues, 209—210, 211
　TTf common features, 209
DCCD (Dicyclohexylcarbodiimide), 274
Dead Sea, The
　16S rRNA sequencing analysis, 144
　algae bloom, 129—130
　　chlorophyll index, 131—132
　　inoculum sources, 130—132
　annual cycles, 123
　Archaea bloom, 15, 129—130
　　bacteriophages in, 133—135
　　characteristics of, 132—133
　　chlorophyll index, 131—132
　bacterial species, new, 143
　bacteriorhodopsins, 133
　bacterioruberins, 135—136
　bromine from, 123, 140
　color display, 2, 42, 124, 129—130
　composition and size, 121—122, 140, 141
　cyanobacteria in, 142
　depositional environments, 93, 98—99
　diatoms, 142
　double-diffusive mixing, 136
　Ein Gedi beach, 135
　evolution
　　future chemical, 124—125
　　recent, 122—124
　halite precipitation changes, 123, 124, 125
　holomictic conditions, 123, 130, 136
　Jordan Valley Rift, 98—99
　meromictic conditions, 123, 124
　meromixis, 130
　methanogenesis
　　methanogenic activity in some
　　　sediment, 150—151,
　　　153—154, 157—158
　　oxidation of methanol sediment, 152—153
　　radioisotope studies, 151—154
　microbiological studies, 140—143
　microbiota characterization
　　categories of, 141—142
　　distribution, 131—132
　　extreme halophiles, 142, 143—144
　　halotolerant, 142
　　moderate halophiles, 142, 144—145
　oil generation, 93, 98—99
　overturn of stratification, 123, 124, 130, 135—136
　phylogenetic tree of isolates, 144, 145
　polar lipid analysis of Archaea, 132—133
　products from, 122—123, 140
　proposed uses of, 125—126
　Senonian Formation deposits, 93, 98—99
　sterility belief, 139—140
　stratification termination evidence, 135—136
　temperature and rainfall impact on, 50
　types of halophilic microorganisms, 4—5
　virus-like particles occurrence, 133—135
　watercolumn structure, 135
Dead Sea Works Ltd., 123
Degree of evaporation (E_{Br}), 109, 111, 112—114
Deposition cycle of crystals, 58—61
Desalinization of sea water, 126
Deserts and poikilotrophic
　　microorganism, 78—79, 82
Devonian period reservoir rocks, 99
DHLipDH (Dihydrolipoamide
　　dehydrogenase), 240
　expression in Hf. volcanii, 241—242
　function in halophilic Archaea, 241—242
　PDHC operon component, 246, 247
　protein molecular model, 242—243
　protein solvation, 245
　site-directed mutagenesis, 243—245
　structural basis, 245
Dialysis bioreactor, 193
Diamino acids, 189
Diatoms, 29, 93, 142
DIC (Dissolved inorganic carbon)
　　depletions
　　^{13}C depletion correlation, 114—117

degree of evaporation, 109
  behavior during evaporation, 111
  function of, 112—114
  invasion of atmospheric CO_2, 109
  isotopic depletion by oxidation, 114
Dicyclohexylcarbodiimide (DCCD), 274
Dihydrofolate dehydrogenase, 245
Dihydrolipoamide dehydrogenase, see DHLipDH
Dimastigamoeba, 142
Dimethyl sulfate methylates, 171
Dimethylsulfide (DMS), 4, 151, 171
Dimethylsulfoniopropionate, 4
Dissimilatory sulfate reduction, 4, 90, 92
Dissolved inorganic carbon, see DIC
Distal promotor, see TATA box
DNA-DNA hybridization
  FI and FII composition, 309—311, 312
  or halobacteria species identification, 144
  Marinococcus classification, 193
Domanic solar salt plant (Volgo-Ural area), 93
Double-diffusive mixing, 136
Drilling and coring
  crystal sampling techniques, 66—68
  microbiological sample retrieval, 63—64
Dunaliella (green alga), 2, 6
  carotenoid pigments, 8
  in the Dead Sea, 5
    annual cycles, 123—124, 136
    bloom origination from resting cells, 132
    chlorophyll index, 131—132
    decline in community size, 134
    microbial studies, 140—142
  parva, 129—130
  discovery in salterns, 41
  fatty acid elongase, 210—211, 212
  in high brine waters, 43
  historical documentation, 218
  osmotic balance in glycerol, 6, 203
  in saltern salinity spectrum, 29
  salt-induced proteins
    Dca, 207—210, 211
    TTf, 204—207, 209
  salt tolerance mechanisms, 203, 204
  viridis found in water samples, 142
  VLCFAs, 210—211, 212
Duplicated carbonic anhydrase, see Dca

E

Ectoines (compatible solutes), 6, 197
  bacterial milking, 192—193
  bacterial species known to synthesize, 189, 190—191
  biosynthetic pathways, 194, 195
  direct synthesis, 194—195
  gene organization, 195—196
  Hm. elongata as ideal producer strain, 192
  osmoregulated expression in E. coli, 196—197
  production strategies
    hydroxyectoine from Marinococcus, 193—194
    synthesis of Hm. elongata, 189, 190, 191—193
  products from, 8
  salt tolerance restored with synthesis, 194—195
Ectothiorhodospira spp., 143, 189
Eilat saltern (Israel)
  clones and isolates in crystallizer ponds, 31, 32
  evaporation pan map, 110
  isotopic depletion in brines, 109, 114
Ein Gedi beach (Dead Sea), 135
Enterococcus hirae, 169
Entner-Doudoroff pathway, 268—269
Environment, extremophile, 75—78
Enzymes, halophilic, 241
  ADI pathway, 260—261
    arc gene cluster transcription analysis, 253—255
    in Archaea and Bacteria, 252, 256
    fermentation, 250—251, 252
    gene cluster from Hb. salinarum, 252—253, 254
    ornithine antiporter, 252, 253, 256
  Archaeal compared to Dca and TTf, 209
  ATPases
    common amino acid ancestor background, 273—274
    synthesis of ATP in vesicles, 274—277, 278
    types, 273—274
  β-galactosidase, 7, 251
    gene cloning, 266, 268, 267—268
    Hf. alicantei isolate studied, 266—267, 268, 269
    transformation of Hf. volcanii, 269
  BHB endonuclease, 320, 321, 325
  carbamate kinase, 252

cOTCase, 252, 256
DHLipDH, 240
 expression in *Hf. volcanii*, 241—242
 function in halophilic Archaea, 241—242
 PDHC operon component, 246
 protein molecular model, 242—243
 protein solvation, 245
 site-directed mutagenesis, 243—245
 structural basis, 245
  malate dehydrogenase, 7, 229
 amino acid arrangement and interaction, 245
 crystallographic studies, 233—234
 obtained from *E. coli*, 229
 solvation-stabilization model, 231—233
 stability and solvation characterization, 229—231
 structure-stability relationships, 233, 234
 thermodynamic activation parameters, 229—231
RNase E, 235, 256
RNase P, 320
thermotolerance, 63
Escherichia coli, 5
 *arc*R gene expressed, 253
 cloned genes in antiporter study, 164—165
 coding for β-galactosidase, 266
 colicin produced by, 296
 ectoines, 196—197
 genomic organization of ectoine genes, 195
 glycine betaine synthesis, 183
 halobacteria selectable plasmids movement, 334
 Hf. volcanii shuttle vector and rRNA operon promotor, 242
 malate dehydrogenase obtained from, 229
 mechanosensitive channels, 193
 osmoregulated expression, 196—197
 protected by osmolytes, 219
 rate of growth, 328
 rRNA operon folding, 325
 trehalose analogies to sucrose, 179
ESSA (Exportadora de Sal) (Mexico), 42, 43
Ethanol as a sterilizing agent, 64—65
Eukarya, 23
 16s rRNA pseudoknot formation, 324
 algae in the Dead Sea, 5
 and Archaea TBP differences, 258
 chimeric origins, 313
 Dunaliella, see *Dunaliella*
 PDHC characteristics, 239—241
 similarity to Archaea transcription initiation, 250
 TBP conservation in Archaea, 257—258
 transfer of genetic information, 331, 336—338
Euryhaline, 29
Eutrophic microbial communities, 77, 79
Evaporation, degree of (E_{Br}), 109, 111, 112—114
Evaporites
 origins of Salado rocks, 57—59
 permeability, 57
 positive features in isolation studies, 63
 trapping mechanisms, 58, 61
Exportadora de Sal (ESSA) (Mexico), 42, 43
Extremophile environment, 75—78

F

F-ATPases
 azide effects, 276—277
 characteristics, 273—274, 278
 inhibitor effects, 274—276
Fatty acids elongase, 210—211, 212
Ferredoxin, 7, 240, 241, 245
Fibronectin type III motif, 269
FI/FII DNA haloarchaeal genomes, 309—311
Finland, 23
Firmicutes, 189
Fish meal as culture medium, 35
Fluid inclusions
 chemistry changes with time, 62—63
 in evaporites, 58—59
 hydrocarbons, 61
 microdrill extraction, 66—68
 specimen selection criteria, 61—62
Forest soils and haloarchaea, 23
Frankia-like actinomycetes, 78
Free energy in protein unfolding, 228
Fumarate respiration, 254—255
Fungi in extreme environments, 77

G

Galactosylglycerol, 181
Galilee (Lake Kinneret), 121
Gas vesicles, 7
  ATP synthesis
    azide effect, 276—277, 281—282
    inhibitor effects, 274—276, 278
  buoyancy from, 281
  in cyanobacteria, 281
  FII DNA activity, 310
  gene clusters, 251—252
  Gvp proteins
    activator function of E, 290
    regulatory proteins D and E, 288—289
    structural proteins A and C, 282—283
  lack of formation in low salt concentrations, 287
  p-vac transformation in *Hf. volcanii*, 285
  resemblance of haloarchaeal gene products to, 313
  shape and arrangement, 282
  synthesis of
    mc-vac region of *Hf. mediterranei*, 283—284, 287
    nv-vac region of *N. vacuolatum*, 285—286
    p-vac and c-vac region of *Hb. salinarum*, 284—285, 287, 289
    vac region expression, 286—287, 290
General stress proteins, 197
Gene-specific regulation
  ADI pathway, 260—261
  arc gene cluster transcription analysis, 253—255
  in Archaea and Bacteria, 252, 256
  gene cluster from *Hb. salinarum*, 252—253
  gene expression, 250—251
  general transcription apparatus
    TATA box, 256—257
    TBP, 257—260
  model systems for halophilic Archaea, 251—252
Genetic exchange, *see* Genetic transfer
Genetic instability of haloarchaeal genomes, 309—311
Genetic transfer
  bidirectional conjugation, 332, 333—334
  cellular fusion, 332, 333—334, 337—338, 337—338
  cytoplasmic activity, 333, 337
  horizontal, in haloarchaeal genomes, 311—314, 331—332
  immobile indigenous vs. mobile selectable plasmids, 334—335
  systems, 336—338
  physical contact requirement, 332—333
Genomes, haloarchaeal, 7—8
  FI/FII DNA, 309—311
  genetic instability, 309—311
  horizontal gene transfer, 331—332
    evidence of occurrence, 311—312
    implications of genetic exchange, 312
    sources, 312
    theoretical speculations, 313—314
  insertion sequences, 309—311, 312
  plasmids, 309—311, 314
Geochemical tracers, 100—102, 103, 104
Geodermatophilus group, 77, 80, 82
Geological history
  age determination techniques, 56—57, 67
  importance of knowing for studies, 55
  isolation verification, 63
  rock formation conditions, 62
  stratigraphy considerations of underground crystal samples, 56
Geological time, *see* Geological history
GFOR (glucose-fructose oxidoreductase), 268
GG, *see* Glucosylglycerol
GGP (glucosylglycerol-phosphate), 179—181
Glucose-fructose oxidoreductase (GFOR), 268
Glucosylglycerol (GG), 6
  biosynthesis in cyanobacteria, 188
  enzyme system experiments, 179—181
  regulatory role of NaCl concentration, 180—181
  gene cloning
    salinity influence on regulatory factors, 182
    sucrose in mutants, 182
    transporter activity, 182, 183

osmoprotectant, 178
synthesis and regulation in
cyanobacteria, 188
Glucosylglycerol-phosphate (GGP),
179—181
Glutamate betaine
osmoprotectant in *Calothrix*, 183
role of, 181
salt tolerance, 178
Glycerol
Hf. volcanii growth medium, 332
lack of in Cargill solar salt plant, 49
no methanogenic substrate,
153—154
osmotic balance in *Dunaliella*, 6,
203
proposed use as culture growth
media, 69
Glycine betaine, 4, 6
biosynthesis in cyanobacteria, 183
compatible solute for halophiles,
188—189
osomoprotectants, 219
salt tolerance, 178
transporter activity with GG, 178
Glycine cleavage system, 241
Glycolipids
assignment of halobacteria isolates,
13—14
differences among Archaea,
132—133
in *Halobaculum* polar lipids, 15
in *Haloborubrum*, 14
Hb. trapanicum morphology,
21—22
in proposed natronobacteria
taxonomy change, 20—21
Great Salt Lake (USA), 3
color display, 1—2
halocin S8 origin, 302
probability of methanogenesis, 150
temperature and rainfall, 50
Green algae, see *Dunaliella*
Green River Formation (USA), 91
Growth media in microbiological
samples, 68—69
Guerrero Negro (Baja California), 92,
114, 300
Gvp proteins
A, gas vesicle structural protein,
282—283
C, gas vesicle structural protein,
282—283
D, regulation of gas vesicle
formation, 288—289

E, activator function, 290
E, regulation of gas vesicle
formation, 289
Gypsum
in fluid composition analysis, 61
new precipitate in ^{13}C depletion
study, 113, 114
precipitate in salterns, 50
precipitate in salt pans, 39—40
in Salado rocks, 58
in saltern salinity spectrum, 29

H

Halite
beds in Salado basin, 58
crystals
fluid inclusions, 58—59, 62
gaseous phase formed, 62
proof of primary sediments,
59—61
in depositional environments, 91
precipitation in Eilat, 109
precipitation in the Dead Sea, 123,
124, 125
Haloanaerobiaceae, 5
Haloarchaeal genomes, *see* Genomes,
haloarchaeal
Haloarcula, 13
hispanica, 144, 278
marismortui, 7, 15, 143
BHB endonuclease, 325
importance of amino acid
arrangement and
interaction, 245
malate dehydrogenase,
229—234, 235
multiple rRNA operons,
325—328
RNase E studies, 235
in Dead Sea, 132
mukohataei discovery from 16S
rRNA analysis, 22
phylogenetic relationship in
crystallizer ponds, 32—33
in preserved Dead Sea isolates, 144
RNA chaperone sequence,
324—325
Halobacteriaceae, *see* Archaea
Halobacterium, 13
cutirubrum, see *Halobacterium
salinarum*
halobium, see *Halobacterium
salinarum*
halocin Hal R1 production, 300

Index

phylogenetic relationships in
crystallizer ponds, 32—33
in preserved Dead Sea isolates,
143—144
salinarum, see *Halobacterium salinarum*
similarity matrix, 47
trapanicum
comparison to *Natrialba*, 15
differences in reporting
morphology and
glycolipids, 21—22
Halobacterium salinarum, 2—3
ADI pathway, 260—261
arc genes cloned and sequenced,
252—253, 254
and ATP level, 252
citrulline as *arc* gene transcription
inducer, 255
comparison to bacterial, 256
tbp gene cloned and sequenced,
257—258
transcript induction in *arc* gene
cluster, 254—255
bacteriophages isolated, 134
function of haloarchaeal FII DNA,
310
gas vesicle formation
c-vac region, 289
gas vesicle shape, 282
p-vac and c-vac gas vesicles
region, 284—285
p-vac region expression and
transcription, 287
genomic comparison to *Hf. volcanii*,
311
GvpA and C genes isolated, 283
halocin-sensitive indicator strain,
297
in Huelva saltern, 43
Natrialba comparison, 15
polymerization rate, 328
retinal proteins, 6
RNA
chaperone sequence, 324—325
polymerase activity, 323
ribosomal operon, 242
ribosomal operon multiple
promotor, 320—323
sequencing clones, 31
spacer regions, 16
Halobacteroides halobius, 143
Halobaculum, 13
determination of, 15
glycolipid in polar lipids, 15

gomorrense, 15, 133
Halocins, 7, 50
activity assays, 297
gene expression, 252
H4
analysis results, 296—297
comparison to halocin H6,
299—300
production in *Hf. mediterranei*,
297—299
signal peptide, 298—299
structural gene cloning and
sequencing, 298—299
H6, 299—300
Hal R1
characterization, 300—301, 302
size and composition, 301
produced by Archaea, 296
production benefits, 303—304
S8, 301, 302—303
Halocline in the Dead Sea, 123
Halococcus, 13
in Alicante sea water, 43
glycolipid in polar lipids, 15
morrhuae, 2—3, 21—22
similarity matrix, 47
Haloferax, 13
alicantei, 266
β-galactosidase isolate study,
266—267, 268, 269
novobiocin resistance marker,
283
gibbonsii and halocin H6
production, 299—300
glycolipid in polar lipids, 15
mediterranei, 7, 278
gas vesicle formation in,
282—284
genetic map of gas vesicle gene
clusters, 284
GvpA and C genes isolated, 283
halocin H4 production and
activity, 297—299
mc-gvpD transcripts found, 287
salt-dependent synthesis of gas
vesicles, 287
phylogenetic relationship in
crystallizer ponds, 32—33
in preserved Dead Sea isolates, 144
RNA chaperone sequence,
324—325
volcanii, see *Haloferax volcanii*
Haloferax volcanii, 15, 143, 144
2-oxoacid dehydrogenase discovery,
246

arcC gene expressed, 253
ATP synthesis, 278
bgaH introduction, 269
BHB endonuclease, 321
 in the Dead Sea, 132
DHLipDH
 cloned and sequenced, 241—242
 molecular model, 242
 and E. coli shuttle vector, 242
 functional TATA boxes selected, 257
 genetic studies, 8
 genetic transfer system
  bidirectional conjugation, 332, 333—334
  cellular fusion, 332, 333—334, 337—338
  cytoplasmic activity, 337
  discovery of basic characterizations, 332
  horizontal, 312
  immobile indigenous vs. mobile selectable plasmids, 334—335
  mating system, 336—338
  physical contact requirement, 332—333
  properties of, 336
  selectable plasmid movement, 334
 genomic comparison to Hb. salinarum, 311
 glycerol as growth medium, 332
 minimal salt concentration for growth, 219
 PDHC operons, 246
 plasmid composition, 310—311
 p-vac transformations, 285
 recipient strain for transformation experiments, 283
 site-directed mutagenesis, 243—245
 tbp genes in genomes, 258
 transformation using mc-gvpA fragment, 284
 vac region expression, 286—287, 290
Halomonas, 3
 canadensis, 219
 elongata
  ectoine synthesis, 189, 190, 191—193, 194—195
  ideal ectoine producer strain, 192
  osomoprotectants, 219
  properties of, 189, 191—192
  transposon mutagenesis, 188, 194—195
 halmophila, 143, 145
 israelensis, 219
 similarity matrix, 47
Halomonas-Deleya complex, 145
Halophilic microorganisms, see also Archaea; Bacteria; Eukarya
 in acidic residues, 209
 biochemistry, 6—7
 biodiversity, 2—3, 23
 biogeochemistry, 4
 biotechnology applications, 2, 8
 color display, 1—2
 in the Dead Sea, 4—5
 ecology of, 3—4
 genetics and genomics, 7—8
 molecular biology, 3, 6—7
 osmotic equilibrium strategies, 5—6
 osmotic stress adaptation, 2
 phototrophic adaptation strategy, 187
 retinal proteins, 6
 salt concentration adaptation, 5—6
 solvation-stabilization process
  malate dehydrogenase, 229—231, 233—234
  protein stability, 227—229
  RNase E, 235
 structure of halophilic communities, 3
Halorhodopsin, 6, 221
Halorubrobacterium, see Halorubrum
Halorubrum, 13—14
 determination of, 14
 gomorrense, 143
 mobility disagreements, 14
 proposed natronobacteria taxonomy change, 20—21
 saccharovorum, 273—274
  acridine orange in, 277
  ATP synthesis in vesicles, 274—278
  similarity to cultured halobacteria, 35
 sodomense, 15, 143
  bacteriorhodopsin produced by, 133
  in the Dead Sea, 132
  spacer regions, 16
 trapanicum/Hb. trapanicum strains, 21—22
High-salt casein medium (HSC), 44, 48
Histidine
 Cys scanning, 166—167
 H225 replacement studies, 166—169

Index

membrane topology, 168—169
NEM alkylation, 167
role in NhaA pH response, 166
TMS, 167, 169
H-NS involvement in *nha*A regulation, 171—172
Hofmeister series, 233, 245
Holomictic conditions in the Dead Sea, 123, 130, 136
Hopanes, 101
Horizontal gene transfer, 311—314, 331—332
Horn River (Canada), 95
HSC (High-salt casein medium), 44, 48
Huelva saltern (Spain), 43
Hydrocarbons in fluid inclusions, 61
Hydrocoleum sp., 81
Hydroelectric energy from Dead Sea hydrostatic potential, 125
Hydrophobic interactions, 232—233, 319
Hydroxyectoine
  in ectoine production, 189
  evidence of direct ectoine synthesis, 194—195
  production from *Marinococcus*, 193—194
Hypersaline depositional environments
  benthic microbial communities, 92—93, 94, 95
  biomass accumulation, 91—92
  geochemical parameters in case studies, 95—96
  geochemical tracers
    chemical biomarkers, 101, 102, 103, 104
    sulfur enrichment, 100—102
  Jordan Valley Rift, 98—99
  link between halophilic microorganisms and kerogen source rocks, 93
  Monterey Formation
    anoxic environments and kerogens, 96—97
    lack of sulfate-containing minerals, 98
    microbial mats and bituminous rocks, 96—97
    secondary sulfur enrichment, 97—98
    sedimentary organic matter, 96—97
    sulfur isotopic distribution, 98
  oil generation source rocks, 89—91
  precambrian stromatolites, 91, 92—93
  published research, 90—91
  Senonian Formation, 98—99
  Tataria and Perm basins
    chemical characterization, 99—100
    origin, 99

I

Inhibitors in ATP synthesis, 274—276, 278
Inorganic nutrient concentrations in salterns, 42, 43
Insertion sequences
  haloarchaeal genomes, 309—311
  sequence change speed, 312
Internally triplicated transferrin, *see* TTf
Intracellular ionic environments, 5
In vitro protein synthesis, 222
Iron uptake and TTf, 206—207
Isofloridoside, 181
Isolation of crystals
  determination from underground sampling, 55
  difficulty in obtaining a viable organism, 65—66
  factors affecting samples, 63
  geologic age determination techniques, 56—57
  growth media, 68—69
  microdrill extraction, 66—68
  origins of Salado rocks, 57—58
  permeability of evaporites, 57
Isotopes
  depletion by oxidation of organic matter, 109, 111, 114, 117
  discrimination, 92
  fractionation of, 97, 100
Israel Salt Company, 109, 110, 111
Israel solar salt plants, 122
  Eilat, 31, 32, 109, 110, 114
  Jordan Valley Rift, 98—99
  Lake Kinneret, 121
  Negev desert, 77, 84

J

Japan, 14
Jordan, 122
Jordan River, 141
Jordan Valley Rift (Israel), 98—99
Jurassic period, 70, 91

K

KDG kinase, 268
KDPG aldolase, 268
Kerogens, 91
   benthic and planktonic
      contributions, 92
   from biomass mats, 93
   carbon isotope enrichment, 100
   formation source, 90
   in Monterey Formation, 96—97
   sulfur isotopic distribution, 98
Klebsiella sp., 296

L

L-2,4-diaminobutyric acid, 194—195
Lactic acid bacteria, 252
Lactoferrin, 204
Lagoons, *see* Sabkhas
Lake Kinneret (Israel), 121
La Malá (Spain), 41—43
LANDSAT and biota information, 131
L-ectoine in ectoine production, 189
Leeuwenhoek, Antoni van, 40
Lichens, 83
Light energy, 6, 8
Lipids, 93
   in high salt concentrations, 221
   indication of methogens in Orca Basin, 150
Lipoate acetyl-transferase, 240
Lipoic acid, 241
Lisan Lake (Dead Sea area), 121
Louann saltern (USA), 57
Lysogenic phages in the Dead Sea, 134—135

M

Magnesium, 2, 123
Malate dehydrogenase (MalDH), 7
   amino acid arrangement and interaction, 245
   crystallographic studies, 233—234
   obtained from *E. coli*, 229
   solvation-stabilization model, 231—233, 235—236
   stability and solvation characterization, 229—231
   structure-stability relationships, 233, 234
   thermodynamic activation parameters, 229—231

Marine bacteria in salt tolerance research, 218
Marine phanerogames, 29
Marinococcus
   classification by DNA-DNA hybridization, 193
   in ectoine production, 189
   halophilus, 6
      biosynthesis of ectoine, 195—196
      change in salt requirement, 218—219
      minimal salt concentration for growth, 219
   hydroxyectoine production
      dialysis cultivation technique, 193
      downstream processing steps, 194
MCA (Modified Casamino Acid), 48
Mcc (Microcins), 296
   halocin Hal R1, 300—301
   halocin S8, 302—303
Mc-*gvp*A gene, 283—284, 288—289
Mc-vac region
   DNA region in *Hf. mediterranei*, 283—284
   promotors, 287
Mechanosensitive channels in *E. coli*, 193
Media, growth, in microbiological samples, 68—69
Mediterranean–Dead Sea Company Ltd., 125
Mediterranean Sea, 50, 95
Megaplasmid, 311, 314
Melanotransferrin iron uptake role, 206—207
Meromictic conditions, 123, 124
Meromixis in the Dead Sea, 130
Methane
   in Dead Sea sediment experiment, 149, 151—153
   formation in anaerobic hypersaline environments, 4
   in Great Salt Lake, 150
   in Solar Lake, 156
Methanobacterium thermo-autotrophicum, 278
Methanococcus, 256
Methanococcus jannaschii, 250, 251, 258
Methanogenesis
   bacterial, 8
   in the Dead Sea, 5
      methanogenic activity in some sediment, 150—151, 153—154, 157—158

Index

oxidation of methanol sediment, 152—153
radioisotope studies, 151—154
glycine betaine production in Archaea, 189
and horizontal genetic transfer of genomes, 313
in hypersaline environments, 149—150
radioisotope studies
Dead Sea, 151—154
Solar Lake, 154—157
Methanogens, 23
archaeal ribosome research, 222
methylotrophic, 150, 156—157
presence indicated by lipids, 150
salinity domain diversity, 31
Methanol in Dead Sea sediment, 152, 153
Methanosarchina barkeri, 278
Methionine, 150, 153, 157
Methylated amines, 4
Methylotrophic bacteria, 8
Mexico solar salt plants
Baja California, 92, 114, 300
ESSA, 42, 43
Guerrero Negro, 92, 114, 300
Orca Basin, 150
Microbial ecology
Bonaire solar salt plant, 40, 50
cultivation results, 44, 45
data analysis, 45
ecological zones, 44—46
model food web, 46
nutrient chemistries, 44
similarity matrix, 47
Cargill solar salt plant, 47—48, 50
solar salterns vs. hypersaline lake systems, 50
Microbial mats
^{13}C depletion, 111, 114—117
atmospheric CO_2 invasion from photosynthesis, 113
biogeochemistry, 4
community redistribution in depletion study, 112
cyanobacteria, 2, 90
anoxia or basin formation, 92—93, 94, 95
biomass accumulation, 91—92
isotope-distribution difference, 97
primary producers, 29
DIC correlation, 112—117
in saltern salinity spectrum, 29
in salt lakes, 2

similar to laminations, 96
Microbiological sampling, 54
crystal sample value and integrity, 55—56
fluid inclusions
chemistry changes with time, 62—63
in evaporites, 58
hydrocarbons, 61
microdrill extraction, 66—68
specimen selection criteria, 61—62
geological isolation, 63
geological study parameters
age dating, 56—57, 67
origin of rocks, 57—58
permeability of evaporites, 57
retrieval techniques
capillary methods, 67—68, 69
crystal drilling, 66—68
culture growth conditions, 70
isolating viable cultures, 65—66
isolation media, 68—69
mine samples, 64
remote drilling, 63—64
sterility control, 64
rock formation changes, 62
trapping mechanisms in evaporites, 58, 61
Microcins (Mcc), 296
halocin Hal R1, 300—301
halocin S8, 302—303
Microcoleus sp., 81
Microdrill techniques, 66—68
Miocene period, 96
Modified casamino acid (MCA), 48
Molecular interactions in extreme halophiles, *see* Solvation-stabilization hypothesis
Molecular techniques
16S rRNA cloning and sequencing, 28
application in solar salterns, 50
ARDRA, 28, 29—31
culture bias, 34, 35—36
multi-pond saltern studies, 28—29
PCR amplification and culturing comparison, 27, 28, 31, 33, 35
phylotypes in crystallizer ponds, 31—33
salinity domain similarity level, 29—31
Monodictys, 80
Monterey Formation (USA)

anoxic environments and kerogens, 96—97
lack of sulfate-containing minerals, 98
microbial mats and bituminous rocks, 96—97
Miocene period, 93
secondary sulfur enrichment, 97—98
sedimentary organic matter, 96—97
sulfur isotopic distribution, 98
Murals and poikilotrophic microorganisms, 77, 82
Mycobacterium ADI pathway found, 252

N

Na^+-ATPase, 169
Na^+/H^+ antiporters
    activity inhibited by halocin H6, 300
    genes, 164—165
    H-NS involvement, 171—172
    *nha*A gene
        expression control, 169, 172
        interaction with Na^+ and NhaR, 170
        Na^+/H^+ antiporter activity, 164
        NhaR footprint, 170—171
    *nha*B gene activity, 164
    NhaR
        footprint, 170—171
        /*nha*A interaction, 170, 172
        regulatory gene, 169—170
    pH regulation, 164, 172
        H225 replacement studies, 166—169
        histidine residue, 166
    proteins NhaA and B, 165—166
National Science Foundation, 77
Natrialba spp., 13
    determination of, 14—15
    in proposed natronobacteria taxonomy change, 20—21
Natronobacterium, 13
    16S rRNA analysis, 16
    phylogenetic tree, 17
    proposed taxonomy change, 16—17, 20—21
    signature bases, 16, 19
    species of, 15
    vacuolatum
        expression of nv-*gvp* gene cluster, 286—287
        gas vesicle formation, 282

        nv-vac region gas vesicle synthesis, 285—286, 290
        salt-dependent synthesis of gas vesicles, 287
Natronococcus, 13
Negev desert (Israel), 77, 84
NEM (N-ethyl maleimide)
    alkylation of Cys replacements, 167
    ATPases inhibitor, 273—274
New Mexico (USA), 55—58
Nickel, 100
Nitrate/nitrite in Bonaire saltern collections, 44
Nitrogen correlation to saltern bacterial population, 46
Novobiocin, 283, 323
Nucleic acid-protein interactions, 235
Nutrient chemistries and microbial populations, 44, 50
Nv-vac region of *Natronobacterium*, 285—286, 290

O

Oil deposits, *see* Oil generation
Oil generation
    halophilic microorganisms and kerogen source rocks, 93
    hypersaline depositional environments
        benthic microbial communities, 92—93, 94, 95
        biomass accumulation, 91—92
        published research, 90—91
    Jordan Valley Rift, 98—99
    microbial degradation, 90
    Monterey Formation, 96—97
    open sea models, 90
    origination theories, 89—90
    Senonian Formation, 98—99
    source rocks, 89—91, 93
Oil shale formation, 90, 98—99
Oligotroph microorganisms, 76, 77
    classification by survival strategy, 79
    growth features, 79
Orca Basin (Gulf of Mexico), 150
Orenia marismortui, 143
ORFs (Open Reading Frames), 267, 268—269, 335, 337
Organic osmolytes, 6
Ornithine antiporter, 252, 253, 256
Osmoprotective compounds, *see also* Compatible solutes
    cyanobacteria, 177—178

Index 355

gene cloning in GG synthesis, 181—183
GG biosynthesis, 178—181
glycine betaine in biosynthesis, 183—184
salt resistance limits, 178—179
sucrose biosynthesis, 179
trehalose biosynthesis, 179
in *Hm. elongata* glycine betaine, 219
Osmoregulated expression of cloned genes, 196—197
Osmotic equilibrium, 2, 5—6; *see also* Compatible solutes
Osmotic solutes, *see* Compatible solutes
Overturn in the Dead Sea, 123, 124, 130, 135—136
Ovotransferrin, 204

P

Paleozoic period, 99
Paradox basin (USA), 91
PCR amplification
 culture method comparison, 33, 35
 molecular technique, 27, 28, 31
PDHC, *see* Pyruvate, PDHC
Peptide antibiotics, 295
Peptide, signal, 298—299
Perm basin (Volgo-Ural, Russia), 99—100
Permian-age salt formations, 54, 56
 considerations for selecting culture growth media, 68
 primary sediments, 59
 reservoir rocks in Tatarian and Perm basins, 99
Petroleum, *see* Oil generation
PGS (Phosphatidylglycerosulfate), 13—14
Phaeococcus sp. (black yeast), 80
Phaeosclera, 80
Phosphate, 44, 50
Phosphatidylglycerosulfate (PGS), 13—14
Photosynthesis and atmospheric CO_2 invasion, 113
pH regulation
 antiporter sensitivity, 166, 172
 azide effects on ATPases, 276—277
 histidine H225 response, 166—169
 inhibitor effects on ATPases, 274—276
 and lack of antiporters, 164
 sodium pumps, 5

Phylogenetic tree
 constructed for TBP sequence, 257
 Dead Sea isolate positions, 144, 145
 of halophilic Archaea, 17, 250
Pits and pipes in trapping mechanisms, 59—61
Plankton
 in benthic microbial communities, 92
 source of organic matter, 90
Planococcus halophilus, 218—219
Plasma membrane proteins, *see* Salt-induced proteins
Plasmids
 conjugation mediation, 331
 as cytoplasmic markers, 333
 genetic transfer of selectable, 334—335
 haloarchaeal genomes, 309—311, 314
 horizontal gene transfer, 312
 immobile indigenous vs. mobile selectable, 334—335
 stability vs. transferability, 333
 transfer from *Hf. volcanii*, 337
Poikilotrophic microorganisms
 adaptation, 4
 biofilm cultures, 79
 on cave paintings, 77, 82
 characterization of, 80—82, 84
 chemoorganotrophic representatives, 80—81, 82
 cyanobacteria, 92
 definition, 76—77, 84
 in deserts, 78—79, 82
 detected genera, 82
 environment, 75—78
 on murals, 77, 82
 photosynthetic representatives, 80—81, 82
 proto-lichenic, 83
 rock inhabiting, 79—83
 survival strategy, 80—84
 trehalose production, 81—82
Polar lipids
 16S rRNA analysis of halophilic Archaea in Dead Sea, 132—133
 assignment of halobacteria isolates, 13—14
 in the Dead Sea, 15, 132—133
 glycolipids, 15
 in *Halobateriaceae*, 2—3
 modified during cultivation, 50
Polysulfides, 92, 93

Porphyrins, 100
Potash
  in depositional environments, 91
  industrial activities in the Dead Sea, 124
  product from the Dead Sea, 123, 140
Poterioochromonas malhanensis, 181
Precursor rRNA, 320
Primary crystals
  fluid-inclusion bands, 58, 61
  isolating viable cultures, 66, 67
Pristane/phytane ratio, 100, 101, 102, 104
Prokaryotes, *see also* Archaea; Bacteria
  biochemistry and molecular biology, 6—7
  biodiversity information by PCR, 27
  compatible osmotic solutes, 5
  diversity of, 28
Proline, 189
Promotors in rRNA operons, 320—323
Proteins
  antibiotics, *see* Bacteriocins; Halocins
  folding of 16s rRNA, 228
  halophilic structural basis, 245
  salt-induced
    Dca, 207—210, 211
    TTf, 204—207, 209
  solvent interactions, 228
  stabilization, 227—229
Proteobacteria, 189
Protonophore, 274, 277
Proton pumps, 6
Proton-translocation, *see* ATPases
Pseudomonal spp.
  ADI pathway studies, 252, 256
  resemblance to *Hf. volcanii* DHLipDH enzyme, 242
Psychrophiles, 76
Purple sulfur bacteria, 2
P-vac gene expression, 284—285, 287, 289
Pycnocline in the Dead Sea, 124
Pyrite formation, 92, 96, 98
Pyruvate metabolism, 7
  converted to acetyl-CoA, 239—241
  decarboxylase, 240
  dehydrogenase (PDHC), 239—241
  and 2-oxoacid dehydrogenase, 239—240
    characteristics, 243—245
    operon in halophilic Archaea, 246
    DHLipDH

expression in *Hf. volcanii*, 241—242
  in halophilic Archaea, 241—242
  PDHC operon component, 246, 247
  protein molecular model, 242—243
  site-directed mutagenesis, 243—245
  structural basis, 245
oxidoreductase in Archaea, 240, 241

Q

Quaternary ammonium compounds, 178

R

Radioisotope studies of methanogenesis
  in Dead Sea, 151—154
  in Solar Lake, 154—157
Radiometric ages, 56—57, 63
Radiotracer experiments, 150
Recrystallized salt, 54, 55
Retinal proteins, 6
Ribosomal RNA operons, 319
  16S rRNA folding, 323—325
  E. coli-Hf. volcanii shuttle vector, 242
  genome studies, 8
  in *Haloarcula marismortui*
    *rrn*A and B operons, 325—326
    sequence heterogeneity, 326
    third genomic operon, 327—328
  multiple promotors
    purpose in *Hb. salinarum*, 320—323
    requirement for, 323
    RNA polymerase activity, 323
  structure and precursor processing, 320
RNA
  analysis of transcript appearance, 286
  chaperone sequence, 325
  polymerase activity, 323
  recognition and halophilic RNase E, 235
RNase E
  enzyme in all domains, 235
  transcription processing of bacterium *P. aeruginosa*, 256

RNase P, 320
Rock paintings, poikilotrophic microorganisms site, 82
Rock salt, 39, 54
rRNA, see Ribosomal rRNA operons
Russia solar salt plants (Volgo-Ural region)
  Domanic, 93
  Perm basin, 99—100
  Tataria basin, 99—100

S

Saanich Inlet (Canada), 100, 101
Sabkhas, 92, 93, 95
Saccharomyces cerevisiae, 323—325
Salado Formation (USA)
  culture growth media, 68—69
  fluid inclusions, 61—62
  geological study parameters
    age determination techniques, 56—57
    origin of rocks, 57—58
    permeability of evaporites, 57
  halite crystals
    fluid inclusions, 58—59, 61
    proof of primary sediments, 59—61
  isolating viable cultures, 66, 67
  maximum burial depth
    thermal history, 62—63
  underground sampling studies, 55
Salinity domain diversity, 29—31
Salinivibrio costicola, 219
Saltern crystallizer ponds, 8
  16S rRNA sequencing studies, 3, 28—29
  clones and isolates, 31, 32
  color display, 1—2, 1—2
  molecular technique studies, 28—29
  novel Archaea phylotypes, 23
  phylotype analysis, 31—33, 31—33
  salinity spectrum, 29
Salt-induced proteins
  Dca, 207—210, 211
  TTf, 204—207, 209
Salting-in/out, 228—229
Salt mines
  ecology studies, 4
  extreme halophiles in, 54
San Diego (USA), 43
Scytomema, 81
Secondary sulfur enrichment, 100—102

Senonian Formation bituminous rocks (Israel), 93, 98—99
Sensory rhodopsins, 6, 282
Sequence element of transcriptional start site, see TATA box
Serum transferrin, 204, 206
Shuttle vector of *E. coli/Hf. volcanii*, 242
Signal peptides, 298—299
Signature bases of halophilic Archaea, 18—20
Single-hit kinetics, 298
Site-directed mutagenesis, 242, 243—245
Smackover Formation (USA), 91
Smectites, 91, 92
Sodium/proton antiporters, 5
Sodium pumps, 5
Solar Lake (Sinai), 4
  active methanogenesis, 150
  cyanobacteria mats, 92—93, 94, 95
  radioisotope studies, 154—157
  sulfur enrichment, 100, 101
Solar saltern, 28, 50
Solar Salt Lake, see Solar Lake
Solar salt plants
  discovery of microorganisms, 39—41
  early collection methods, 39
  inorganic nutrient concentrations, 43
  microbial ecological studies
    Alicante, 41—43
    Bonaire, 40, 42, 44—47, 50
    Cargill, 47—49, 50
    community use of saltern compounds, 48—49
    Western hemisphere, 43
Solvation–stabilization hypothesis
  evaporation, 41
  malate dehydrogenase
    crystallographic studies, 233—234
    obtained from *E. coli*, 229
    solvation–stabilization model, 231—233
    stability and solvation characterization, 229—231
    structure–stability relationships, 233, 234
    thermodynamic activation parameters, 229—231
  protein stability
    salt effects, 228—229
    solvent interactions, 227—228
  RNase E, 235

Source rocks for petroleum, 89—91, 93
Spacer regions in alkaliphilic Archaea, 16
Spain solar salt plants
  Alicante
    halobacterial phylotypes found, 23, 33
    inorganic nutrient concentrations, 42
    microbial ecological studies, 41—43
  Huelva, 43
  La Malá, 41—43
Spirulina subsalsa, 183
Steranes, 101
Sterility control in microbial sampling, 64, 65
Stigonema, 80, 81
Stratification overturn in the Dead Sea, 135—136
Stromatolites, 91, 92—93
Sucrose
  biosynthesis in cyanobacteria, 179
  in GG mutants, 182
  osomoprotective compound in cyanobacteria, 178
  trehalose analogy in *E. coli*, 170
Sulfate-reducing bacteria, 2, 93, 97—98
  competition for hydrogen and acetate, 150
  isotopic discrimination, 92
Sulfolobus sp.
  archaeal transcription initiation, 256—257
  genetic transfer in, 332
  genome studies, 313
  insertion in genome sequences, 310
  limited similarity to BgaH, 268
Sulfur, 95, 98—99, 100—102
Sulfur-dependent thermophiles, 23
Surface energy/tension, 245
Synechococcus sp., 178, 179, 183—184
Synechocystis sp., 6, 179, 181, 182
Syrian–African valley rift (Dead Sea area), 140

T

Taiwan, 14
Tasmanales algae, 96
Tasmanite shales, 96, 103
TATA box, 250
  binding protein (TBP), 250
    conserved sequence in *arc*B and *arc*C, 255
    present upstream in *Hb. salinarum*, 256
    protein DNA interactions, 259—260
    sequence element study, 256—257
    TBP conservation in Eukarya and Archaea, 257—258
Tataria basin (Volvo-Ulga, Russia), 99—100
Temperature
  affect on cultural population, 36
  free energy of unfolding as a function of, 228
Temporal variation, 29—31
Terminal lakes, 121
Texas (USA)
  geological study parameters, 56—58
  underground sampling studies, 55
Thalassohaline brines, 40, 141
Thermodynamics of unfolding, 227—228
Thermophiles, 76
Thermotolerance of enzymes, 63
Thiophenes, 101
TMS (trans membrane segments)
  in antiporter genes, 164, 165
  length, 169
  quantifying, 167
Transcription
  and archaeal gene expression, 250—251
  general apparatus elements
    TATA box, 256—257
    TBP, 257—260
  induction of *arc* gene cluster, 252—254
  mc-gvpD found in *Hf. mediterranei*, 287
  signals homology to eukaryotic system, 266
Transferrin, *see* TTf
Transformation systems for Archaea, 251
Translation, 319
Trans membrane segments (TMS)
  in antiporter genes, 164, 165
  length, 169
  quantifying, 167
Transposon mutagenesis, 188, 194—195
Trapping mechanisms for evaporites, 58—61
Trehalose
  6-phosphate synthase, 181

biosynthesis in cyanobacteria, 179
consequence sequences, 197
osmoprotective compounds in cyanobacteria, 178
produced by poikilotroph microorganisms, 81—82
sucrose analogy in *E. coli*, 179
Tricyclic compounds, 101, 103
Trimmatostroma, 80
Triterpanes, 96
Trulline in *arc* gene transcription, 255
TTf (triplicated transferrin)
  animal transferrin structural characteristics, 205, 206
  Archaea enzyme comparison, 209
  comparison of algal proteins to animal transferrins, 204—206
  Dca common features, 209
  iron uptake mechanism, 206—207
  new halotolerant protein, 209
  ratio of acidic to basic residues, 209—210
Turgor pressure, 222—223, 282
Type-IIs kerogens, *see* Kerogens
Tyro Basin (Mediterranean Sea), 95

U

Underground sampling, 55
Unicellular green algae, see *Dunaliella*
Urea, 44
USA solar salt plants
  Cargill, 47—49, 50
  Castile Formation, 57
  Great Salt Lake, 1—2, 3, 50, 150, 302
  Green River Formation, 91
  Louann saltern, 57
  Monterey Formation, 93, 96—98
  New Mexico, 55—58
  Paradox basin, 91
  Salado Formation
    culture growth media, 68—69
    fluid inclusions, 61—62
    halite crystals, 58—59, 61, 59—61
    isolating viable cultures, 66, 67

maximum burial depth, 62—63
parameters for geological studies, 56—58, 57
underground sampling studies, 55
Smackover Formation, 91
Texas, 55—58
Western Salt Company, 42, 43
Yellowstone National Park, 23

V

Vac-region expression in *Hf. volcanii*, 286—287, 290
Vanadium, 100
V-ATPases
  azide effects, 276—277
  characteristics, 273—274, 278
Very long-chain fatty acids (VLCFAs), 210—211, 212
Vesicles, see Gas vesicles
Vibrio alginolyticus, 169
Virus-like particles in the Dead Sea, 133—135
VLCFAs (very long-chain fatty acids), 210—211, 212
Volgo–Ural region (Russia), 93, 99—100
VO/Ni ratio, 95, 96, 98—99, 100

W

Water column, 123
  dissolved methane in Solar Lake, 156
  and halophilic blooms, 130
  structure indicated by distribution of biota, 135
Western Salt Company (WSC) (USA), 42, 43

Y

Yeast extract as a culture medium, 35
Yellowstone National Park (USA), 23

Z

Zymomonas, 268